Lecture Notes in Control and Information Sciences

Edited by A. V. Balakrishnan and M. Thoma

For further listing of published volumes please turn over to inside of back cover.

Lecture Notes in Control and Information Sciences

Edited by A.V. Balakrishnan and M. Thoma

IFIP

42

Advances in Filtering and Optimal Stochastic Control

Proceedings of the IFIP-WG 7/1 Working Conference
Cocoyoc, Mexico, February 1–6, 1982

Edited by
W.H. Fleming and L.G. Gorostiza

Springer-Verlag Berlin Heidelberg GmbH 1982

Editors
Wendell H. Fleming
Division of Applied Mathematics
Brown University
Providence, Rhode Island 02912
USA

Luis G. Gorostiza
Departamento de Matemáticas
Centro de Investigación y de Estudios Avanzados del IPN
Apartado Postal 14–740
México 07000, D.F.
México

ISBN 978-3-662-13531-0 ISBN 978-3-540-39517-1 (eBook)
DOI 10.1007/978-3-540-39517-1

Originally published by Springer-Verlag Berlin Heidelberg New York in 1982.

2061/3020-543210

PREFACE

This volume contains contributions to a conference on filtering, optimal stochastic control, and related topics, held in Cocoyoc, Mexico in February 1982. The conference was held under the auspices of IFIP WG 7.1. The host institution was the Centro de Investigacion y de Estudios Avanzados, whose assistance was appreciated by conference participants. In addition, we wish to thank the following sponsoring organizations in Mexico, whose generous support was essential to the success of the conference:

Centro de Investigación y de Estudios Avanzados del IPN
Campañía Nestlé
Consejo Nacional de Ciencia y Tecnología
Fondo de Fomento Educativo BCH
Instituto de Investigaciones Eléctricas
Instituto Politénico Nacional
Secretaría de Asentamientos Humanos y Obras Públicas
Secretaría de Comercio
Secretaría de Educación Pública
Secretaría de Turismo
Universidad Autónoma Metropolitana-Azcapotzalco
Universidad Autónoma Metropolitana-Iztapalapa

Our thanks also go to other members of the International Program Committee and of the Local Organizing Committee for their valuable assistance in arranging the conference, and to Y-C Liao, R. McGwier, and S-J Sheu for their help in editing this volume.

In addition to specialists in nonlinear filtering and stochastic control, several outstanding probabilists working in related fields came to the conference. Their presence was an important element in its success, both through the formal lectures presented and through informal discussion with other participants. Conference speakers included senior scientists of long experience and energetic younger people. As put by one participant, we had both "old coyotes and young lions"*.

The intent of the conference organizers was to focus not only on the mathematical aspects of the theory, but also on some newer areas of application and on numerical techniques of approximate solution to problems in filtering and stochastic control. We think that this objective was fairly well met.

Wendell H. Fleming
Luis G. Gorostiza

---

* This simile was in part suggested by the fact that Cocoyoc means "place of the coyotes" in the Náhuatl language, and in part by the name of one young speaker.

ADDRESSES OF CONTRIBUTORS

BARAS, J. S.
Deparment of Electrical Engineering
University of Maryland
College Park, MD 20742
U. S. A.

BENES, V. E.
Bell Laboratories
Murray Hill, NJ 07974
U. S. A.

BENSOUSSAN, A.
INRIA
Domaine de Voluceau - Rocquencourt
B. P. 105
78150 Le Chesnay
FRANCE

BLANKENSHIP, G. L.
Department of Electrical Engineering
University of Maryland
College Park, MD 20742
U. S. A.

CLARK, J. M. C.
Department of Electrical Engineering
Imperial College
London, SW7 2BT
ENGLAND

DAVIS, M. H. A.
Department of Electrical Engineering
Imperial College
London, SW7 2BT
ENGLAND

DAWSON, D. A.
Department of Mathematics and Statistics
Carleton University
Ottawa K1S 5B6
CANADA

EL KAROUI, N.
Ecole Normale Supérieure
3, rue Boucicaut
92260 Fontenay-aux Roses
FRANCE

ELLIOTT, R. J.
Department of Pure Mathematics
The University of Hull
Hull HU5 2DW
ENGLAND

FLEISCHMANN, K.
Akademie der Wissenschaften der DDR
Zentralinstitüt für Mathematik
und Mechanik
DDR - 1080 Berlin, Mohrenstrasse 39
GERMAN DEMOCRATIC REPUBLIC

FLEMING, W. H.
Division of Applied Mathematics
Brown University
Providence, RI 02912
U. S. A.

GOROSTIZA, L. G.
Departamento de Matemáticas
Centro de Investigación y de Estudios
Avanzados, IPN
Apartado Postal 14-740
México 14, D. F.
MÉXICO

HAUSSMANN, U. G.
Department of Mathematics
University of British Columbia
Vancouver, B. C., V6T 1W5
CANADA

HELMES, K.
Institüt für Angewandte
Mathematik
Universität Bonn
5300 Bonn, Wegelerstr. 6-10
FEDERAL REPUBLIC OF GERMANY

HIJAB, O.
Department of Mathematics and Statistics
Case Western Reserve University
Cleveland, OH 44106
U. S. A.

KURTZ, T. G.
Department of Mathematics
University of Wisconsin
Madison, WI 53706
U. S. A.

KUSHNER, H. J.
Division of Applied Mathematics
Brown University
Providence, RI 02912
U. S. A.

LIONS, P-L.
Ceremade, Paris IX University
Place de Lattre de Tassigny
75775 Paris Cedex 16
FRANCE

MANDL, P.
Department of Probability and Mathematical Statistics
Charles University
Sokolovská 83
186 Prague 8
CZECHOSLOVAKIA

MARCUS, S. I.
Department of Electrical Engineering
University of Texas at Austin
Austin, TX 78712
U. S. A.

MAZZIOTTO, G.
Centre National d'Etudes des Télécommunications
92 131 - ISSY LES MOULINEAUX
FRANCE

MENALDI, J-L.
Department of Mathematics
Wayne State University
Detroit, MI 48202
U. S. A.

MITTER, S. K.
Department of Electrical Engineering and Computer Science and
Laboratory for Information and Decision Systems
Massachusetts Institute of Technology
Cambridge, MA 02139
U. S. A.

NISIO, M.
Department of Mathematics
Faculty of Sciences
Kobe University
Rokkodai-machi, Nada-Ku
Kobe 657
JAPAN

PARDOUX, E.
U.E.R. de Mathematiques
Université de Provence
3 Place Victor-Hugo
13331-Maiselle Cedex 3
FRANCE

PLISKA, S. R.
Department of Industrial Engineering and Management Science
Northwestern University
Evanston, IL 60201
U. S. A.

PRAGARAUSKAS, H.
Institute of Mathematics and Cybernetics
Academy of Sciences of the Lithuanian SSR
232 600 Vilnius 54, K. Pozelos Str.
U.S.S.R.

QUADRAT, J-P.
Domaine de Voluceau-Rocquencourt
B. P. 105
78150 Le Chesnay
FRANCE

RISHEL, R. W.
Department of Mathematics
University of Kentucky
Lexington, KY 40506
U. S. A.

SAZONOV, V. V.
Steklov Mathematical Institute
Academy of Sciences of the USSR
42 Vavilova Street
Moscow - B 333
U.S.S.R.

SHENG, D. D.
Bell Laboratories
Holmdel, NJ 07733
U. S. A.

STROOCK, D. W.
Mathematics Department
University of Colorado
Boulder, CO 80309
U. S. A.

VARADHAN, S. R. S.
Courant Institute of Mathematical Sciences
New York University
251 Mercer Street
New York, NY 10012
U. S. A.

CONTENTS

EXISTENCE, UNIQUENESS AND TAIL BEHAVIOR OF SOLUTIONS TO ZAKAI EQUATIONS WITH UNBOUNDED COEFFICIENTS*

W.E. Hopkins, Jr., J.S. Baras and G.L. Blankenship
Electrical Engineering Department
University of Maryland
College Park, Maryland 20742

ABSTRACT

Conditions are given to guarantee the existence and uniqueness of solutions to the Duncan-Mortensen-Zakai equation for nonlinear filtering of multivariable diffusions with unbounded coefficients. Sharp upper and lower bounds on the tail of conditional densities are also obtained. A methodology is described to treat these problems using classical p.d.e. methods applied to the "robust" version of the DMZ equation. Several examples are included.

* Supported in part by ONR Contract N00014-79-C-0808.

1. INTRODUCTION AND STATEMENT OF THE PROBLEM

Recently the problem of filtering a diffusion process x(t) from nonlinear observations y(t) in additive Gaussian noise has been studied by analyzing an unnormalized version of the conditional distribution of x(t) given the past of y(.). If this conditional distribution is absolutely continuous with respect to Lebesgue measure, then it has a density which satisfies a linear stochastic partial differential equation known as the Duncan-Mortensen-Zakai (DMZ) equation [1]. Background information on this equation and other aspects of the nonlinear filtering problem may be found in [2]. In the present paper we focus on existence-uniqueness results for the DMZ equation and on tail estimates of the resulting solutions. Our motivation for these problems stems primarily from the following areas: (a) numerical algorithms for the solution of DMZ equation and their subsequent implementation by special purpose array processors; (b) numerical evaluation of the Kallianpur-Striebel path integral representation of the solution; (c) accuracy and convergence analysis in asymptotic expansions of the solution.

In cases where the process x(t) evolves in a bounded domain in $\mathbb{R}^n$, or when the state space is unbounded but the coefficients of the DMZ equation are bounded, a satisfactory existence-uniqueness theory is available [3]-[5]. More recently, existence-uniqueness of solutions has been established for filtering problems having "strongly" unbounded coefficients. In [6] polynomial observations are studied via a related optimal control problem. In [7] classical results for fundamental solutions of parabolic equations with unbounded coefficients are applied to the "robust" version of the DMZ equation [1], [2], for scalar diffusions. In [8], [9], the method of [7] is extended to multidimensional problems. Furthermore in [7]-[9] tight estimates of the tail behavior of solutions are obtained. In the present paper we review and summarize the method and main results of [7]-[9].

To set the problem, consider the pair of Itô stochastic differential equations

$$\begin{aligned} dx(t) &= f(x,t)dt + g(x,t)dw(t), \quad x(0) = x_0 \\ dy(t) &= h(x,t)dt + dv(t) \qquad\quad , \quad y(0) = 0 \end{aligned} \tag{1}$$

where $x(t), w(t) \in \mathbb{R}^n$, $y(t), v(t) \in \mathbb{R}^m$, $w(\cdot)$ and $v(\cdot)$ are mutually independent Wiener processes independent of x_0, and x_0 has a density $p_0(\cdot) \in L^1(\mathbb{R}^n) \cap C^0(\mathbb{R}^n)$. The coefficients f,g,h are assumed to satisfy $f^i \in HC^{1,0}_{\ell oc}(\mathcal{D})$, $h^i \in HC^{2,1}_{\ell oc}(\mathcal{D})$, where $\mathcal{D} = \mathbb{R}^n \times (0,T]$ and $HC^{i,j}_{\ell oc}$ denotes the space of functions having locally Hölder continuous derivatives of order i in x and j in t. Furthermore the generator L of the diffusion process $x(\cdot)$ is assumed to be uniformly elliptic; that is, there exist continuous functions $\theta_i(x,t)$ $i=1,2$, and a constant $\theta_0>0$ such that for all $\xi\in\mathbb{R}^n$ and $(x,t)\in\mathcal{D}$ $\theta_0|\xi|^2\leq\theta_1(x,t)|\xi|^2\leq\sum_{i,j=1}^{n}\alpha^{ij}(x,t)\xi_i\xi_j\leq\theta_2(x,t)|\xi|^2$, where $(\alpha^{ij})\triangleq\frac{1}{2}gg^T$. The filtering problem for (1) is to estimate statistics of $x(t)$ given the σ-algebra $\mathcal{F}_t^y=\sigma\{y(s)|0\leq s\leq t\}$. Equivalently we can compute the conditional distribution of $x(t)$ given $\mathcal{F}_t^y$.

Formally, the conditional density of $x(t)$ given $\mathcal{F}_t^y$ is given by

$$p(x,t) = U(x,t)/\int_{\mathbb{R}^n} U(x,t)dx$$

where $U(x,t)$ is a solution of the DMZ equation

$$\begin{aligned} dU &= (L^* - \tfrac{1}{2}|h|^2)Udt + \langle h,dy(t)\rangle U, \quad (x,t)\in\mathcal{D} \\ U(x,0) &= p_0(x) \quad , \quad x\in\mathbb{R}^n \end{aligned} \tag{2}$$

Here L^* is the adjoint of L and (2) is written using Stratonovich calculus.

It is well known, particularly in recent studies, that if one introduces the transformation

$$V(x,t) = U(x,t)\exp(-\langle h(x,t), y(t)\rangle) \tag{3}$$

then V satisfies a classical, linear, parabolic partial differential equation, parametrized by the observation paths $y(\cdot)$, which is called the "robust" version of the DMZ equation [1], [2]:

$$\frac{\partial V}{\partial t}(x,t) = \sum_{i,j=1}^{u} A^{ij}(x,t)V_{x_ix_j}(x,t) + \sum_{i=1}^{n} B^i(x,t)V_{x_i}(x,t) +$$

$$C(x,t)V(x,t), \quad (x,t)\in\mathcal{D} \tag{4}$$

$$V(x,0) = p_0(x) \quad , \quad x\in\mathbb{R}^n.$$

The functions $B^i(x,t)$ are pointwise linear functions of $y(t)$, while $C(x,t)$ is pointwise a quadratic function of $y(t)$. Since the paths of $y(\cdot)$ are Hölder continuous, (4) is a classical p.d.e. and classical results on existence-uniqueness of fundamental solutions for linear parabolic equations due to Besala [10] can be fruitfully applied to the robust DMZ equation. Direct analysis of the DMZ equation is rather complicated. However, since the transformation (3) is invertible, positivity preserving, and (2), (4) are linear, existence, uniqueness, and tail behaviour of solutions of the DMZ equation may be obtained by analyzing (4) instead.

2. OUTLINE OF THE METHOD

Our method can be applied to a variety of problems with unbounded coefficients. We outline briefly here the main steps. We shall refer to the zeroth order coefficient of a parabolic equation like (4) as the potential (i.e. $C(x,t)$ in (4)). We use a result of Besala [10] which asserts that if we can find a "weight" of the form $\Phi(x,t) = \exp(\psi(x,t))$ such that the function $U=V\Phi$, satisfies a parabolic p.d.e. where the potential term and the potential term of the adjoint are nonpositive for $(x,t)\in\mathcal{D}$, then there exist a classical fundamental solution for (4). Furthermore integral growth estimates for the fundamental solution are given in [10].

To apply this idea to (4) we must consider stopping time partitions of the interval $[0,T]$; there are special cases where stopping times are not needed. These partitions are defined as follows. Given a small positive number $\varepsilon>0$, we define $0 = t_0<t_1<\ldots<t_N = T$ via

$$t_0 = 0$$
$$t_{k+1} = \begin{cases} \inf\limits_{t_k<t<T} \{t:|y(t)-y(t_k)| =\varepsilon\}, \text{ whenever inf exists} \\ T, \text{ otherwise} \end{cases} \tag{5}$$
$$N = \min \{k:t_k=T\}.$$

Here ε will be fixed by other considerations. Then on each set $\mathcal{D}_k \triangleq \mathbb{R}^n\times(t_k,t_{k-1}]$, $0\le k\le N-1$ we introduce the transformations

$$u^k(x,t) = U(x,t) \exp(\phi^k(x,t) - \langle h(x,t), y(t)\rangle). \tag{6}$$

Then u^k satisfies

$$u^k_t = \mathcal{L}^k u^k, \quad (x,t)\in\mathcal{D}_k, \quad 0\le k\le N-1$$

$$u^k(x,t_k) = \begin{cases} p_0(x)\exp(\phi^0(x,0)) & , k = 0 \\ u^{k-1}(x,t_k)\exp(\phi^k(x,t_k) - \phi^{k-1}(x,t_k)), & 1\le k\le N-1 \end{cases} \tag{7}$$

where

$$\mathcal{L}^k u^k = \sum_{i,j=1}^n a^{ij} u_{x_i x_j} + \sum_{i=1}^n b^{ik} u_{x_i} + c^k u$$

$$b^{ik} = - f^i + 2\sum_{j=1}^n a^{ij}_{x_j} + 2\sum_{j=1}^n a^{ij}(\langle h_{x_j}, y\rangle - \phi^k_{x_j}) \tag{8}$$

$$c^k = c^k_{ess} - \sum_{i,j=1}^n a^{ij}(\phi^k_{x_i x_j} - \langle h_{x_i x_j}, y\rangle) -$$

$$- 2\sum_{i,j=1}^{n} a^{ij}_{x_i}(\phi^k_{x_j} - \langle h_{x_j},y\rangle) + \sum_{i,j=1}^{n} a^{ij}_{x_i x_j} - \sum_{i=1}^{n} f^i_{x_i} - \langle h_t,y\rangle$$

$$c^k_{ess} = \frac{1}{2}|g\nabla\phi^k - g(\langle h_{x_i},y\rangle)_{i=1}^n +$$

$$+ (g^{-1})^T f|^2 - \frac{1}{2}|h|^2 - \frac{1}{2}|(g^{-1})^T f|^2 + \phi^k_t .$$

Thus u^k satisfy linear parabolic equations of the same type as (4). We then choose ϕ^k on each subinterval so as to render the potential term c^k of $\mathcal{L}^k$ as well as the potential term of $(\mathcal{L}^k)^*$ nonpositive, and apply Besala's results.

The guidelines for this construction are relatively simple to understand. The actual computations and resulting conditions depend on the case at hand. First one poses appropriate assumptions on f,g,h so that c^k and the adjoint potential are dominated by c^k_{ess} as $|x|\to\infty$. Then one fixes the parameters in ϕ^k (in particular ϕ^k_t) so that c^k_{ess} is ≤ 0. To construct ϕ^k one first constructs the piece of ϕ^k corresponding to the Fokker-Planck equation for $x(\cdot)$. This is natural since the diffusion itself must be well-behaved: no explosions, nice solutions of the Fokker-Planck equation, etc. Indeed if we set h=0 in (4) we get the Fokker-Planck equation. The construction of this first part of ϕ (here we mean that typically $\phi=\phi_1 + \phi_2$) brings us naturally into contact with Khas'minskii's test for nonexplosions and generalizations thereof. To construct ϕ_2, after obtaining ϕ_1, one returns to the full expressions for c_k and the adjoint potential and appropriately balances growth conditions on h with those of f,g, so as to make both potentials ≤ 0. For uniqueness we use classical weak maximum principles. The procedure gives us upper bounds on the tails by means of the uniqueness class identified. To obtain lower bounds on the tails we choose ϕ^k such that the potential c^k of $\mathcal{L}^k$ is positive; i.e. using classical comparison theorems for linear parabolic equations. We refer to the references cited above [7]-[9] for the details of these constructions. In the remainder of the paper we summarize the results in particular cases.

3. THE SCALAR CASE n=1, m=1

In this section we describe our results for the scalar case including polynomial nonlinearities. Our assumptions on the coefficients in (1.6) are stated in terms of the original functions f,g,h. To state these succinctly, we will use the relative order notation:

<u>Definition</u>. Let $F,G,:R\to R$ and

$$L = \limsup_{|x|\to\infty} |F(x)/G(x)| \varepsilon [0,\infty]$$

Then $F = O(G)$ if $L<\infty$ and $F = o(G)$ if $L \to 0$.

The coefficients of the diffusion x are assumed to satisfy

(A1) $f\in C^1(R)$, $g\in C^2(R)$, f_x, g_{xx} are locally Hölder continuous;

(A2) $g(x)\geq\lambda>0, \forall x\varepsilon R$ and some λ;

(A3) $-\int_0^x (f/g^2)(\xi)d\xi \geq M, \forall x\varepsilon R$ and some M;

(A4) $(f/g^2)_x = o(f^2/g^4)$, $f_x = o(f^2/g^2)$; and

(A5) the martingale problem for (f,g) is well-posed.

The last condition implies that the stochastic differential equation for x has a unique weak solution for all $t\geq 0$. A sufficient condition for this is the existence of a Lyapunov function for the backwards Kolmogorov equation associated with the process x[11]. If the integral in (A3) diverges to $+\infty$ as $|x|\to\infty$, then it could serve as the Lyapunov function. If the martingale problem is not well posed, then the process x may have "explosions" (escape times which are finite with probability one). In this case the conditional distribution of x(t) given $\mathcal{F}_t^y$ may have singular components which are not computed by the DMZ-equation.

The observation function h is assumed to satisfy:

(B1) $h\in C^2(R)$, h_{xx} is locally Hölder continuous;

(B2) either $g^2 h_{xx}$, $(g^2 h_x)_x = o(h^2)$, or
$g^2 h_{xx}$, $(g^2 h_x)_x = o(g^2 h_x^2)$

(B3) either $gh_x = O(h)$ or $gh_x = o(f/g)$;

(B4) either $(g^2)_{xx} = o(h^2)$ or $(g^2)_{xx} = o(f^2/g^2)$;

(B5) one of the two mutually exclusive cases holds:

(i) either $h = O(f/g)$ or $h = O(gh_x)$; or

(ii) both $f/g = o(h)$ and $gh_x = o(h)$; in addition,
gh_x, $g_x h = o(h^2)$

(B6) in case (B5)(i),

$$\lim_{|x|\to\infty} \max\{|h(x)|, -\int_0^x (f/g^2)(\xi)d\xi\} = +\infty;$$

and in case (B5)(ii),

$$\lim_{|x|\to+\infty} \left|\int_0^x (h/g)(\xi)d\xi\right| = +\infty$$

Remarks. (1) the growth conditions are relatively easy to understand in the case when f,g,h are polynomials, especially $f(x) = f_0 x^j$, $g(x) =$

$g_0(1+x^2)^k$, $h(x) = h_0 x^\ell$.

(2) The conditions (A1)-(B6) are not necessary; different choices of the weight functions used in the proofs would lead to different growth restrictions. In fact, one could consider optimizing the choice of the weight functions. Here the weights are chosen as

$$\phi^k(x,t) = \psi^k(x) - \gamma^k t \tag{9}$$

where

$$\psi^k(x) = \alpha\phi_1(x) + \beta_1^k\phi_2(x) + \beta_2[1 + \phi_2^2(x)]^{\frac{1}{2}}$$

$$\phi_1(x) = \begin{cases} -\int_0^x (f/g^2)(\xi)d\xi & \text{in case (B5)(i)} \\ 0 & \text{in case (B5)(ii)} \end{cases} \tag{9a}$$

$$\phi_2(x) = \begin{cases} h(x) & \text{in case (B5)(i)} \\ \int_0^x (h/g)(\xi)d\xi & \text{in case (B5)(ii)} \end{cases}$$

The parameters $\alpha, \beta_2, \{\beta_1^k, \gamma^k, t_k\}_{k=0}^{\infty}$ will be functionals of the path $y(t)$, $t \geq 0$; for their explicit definition see [7].

Assumptions (A3) and (B6) together with the constraints $\alpha>0$, $\beta_2>0$, $\beta_2>|\beta_1^k|$, imply that the weight functions $\psi^k(x)$ diverge to $+\infty$ as $|x|\to\infty$. The remaining growth conditions serve to identify the dominant terms (as $|x|\to\infty$) in the potential $c^k(t,x)$ in (7) and in the potential of the adjoint of (7). Assumption (B3) permits us to select the functions ψ^k and the constants γ^k so that these potentials are nonpositive. This in turn permits the use of a maximum principle.

Under these assumptions we show that the robust equation (4) has a fundamental solution which may be used to construct a unique solution to the DMZ-equation within a certain class of functions. To describe this class, we define the constants

$$\eta_i = \limsup_{|x|\to\infty} |g\cdot(\phi_i)_x|/[h^2+f^2/g^2]^{\frac{1}{2}} \tag{10}$$

$$\nu_i = \liminf_{|x|\to\infty} |g\cdot(\phi_i)_x|/[h^2+f^2/g^2]^{\frac{1}{2}},\quad i = 1,2$$

The assumptions imply η_1, $\nu_1 \in [0,1]$ and η_2, $\nu_2 \in [0,\infty)$ when (B5)(i) holds, while $\eta_1 = 0 = \nu_1$, $\eta_2 = 1 = \nu_2$ when (B5)(ii) holds. The assumption that either (B5)(i) or (ii) holds implies $(\nu_1 + \nu_2)>0$.

Theorem 1. Suppose (A1) - (A5), (B1) - (B6) hold. Let $p_o(x)$ be continuous, $p_o(x)\geq 0$, and assume that there exist constants $\theta_i>0$, i=1,2, such that $0<\theta_1\eta_1 + \theta_2\eta_2<1$, and

$$p_o(x)\exp[\theta_1\phi_1(x) + \theta_2|\phi_2(x)|]\leq M, \ \forall x\varepsilon R \tag{11}$$

and some $M<\infty$. Then for any constants $\tilde{\theta}_i, 0<\tilde{\theta}_i<\theta_i$, i=1,2, there exists a unique solution to the DMZ-equation (2) within the class of functions satisfying

$$\lim_{|x|\to\infty} \sup U(x,t) \exp[\tilde{\theta}_1\phi_1(x) + \tilde{\theta}_2|\phi_2(x)|] = 0, \ \forall t\geq 0 \tag{12}$$

This solution satisfies $U(t,x) = U^k(t,x)$, $t\varepsilon(t_k,t_{k+1})$

$$U^k(x,t) = e^{h(x)y(t)}\int_{-\infty}^{\infty}\Gamma_k(t,x;z,t_k)U^{k-1}(t_k,z)dz \tag{13}$$

$$U^0(x,0) = p_o(x), \quad k = 1,2,\ldots$$

where Γ_k is the fundamental solution of (7).

Theorem 2. Suppose (A1) - (A5), (B1) - (B6) hold, and assume when case (B5)(i) holds with $\nu_1>0$, $\nu_2>0$, that $-f(x)$ sgn(x) and $h_x(x)$ sgn(xh(x)) are non-negative for $|x|$ sufficiently large. Let $p_o(x)$ satisfy the conditions in Theorem 1, and suppose further that there exist $M_0>0, K_0>0$ such that

$$M_0 \exp[-K_0\phi(x)]\leq p_o(x), \forall x\varepsilon R \tag{14}$$

where

$$\phi(x) = \phi_1(x) + |\phi_2(x)| \tag{15}$$

Then for any $T<\infty$, there exist positive constants M_1, M_2, K_1, K_2, which may depend on the path $\{y(t), 0\leq t\leq T\}$ such that the solution of the DMZ-equation given by (13) satisfies

$$M_1 \exp[-K_1\phi(x)]\leq U(x,t)\leq M_2 \exp[-K_2\phi(x)] \tag{16}$$

$$\forall (x,t)\varepsilon R\times[0,T]$$

To illustrate our results and make contact with other recent work on nonlinear filtering (e.g., [6],[12]), we consider a class of systems with polynomial f,h.

So let f,h, be polynomials with f odd and stable, i.e.,

$$f(x) = \sum_{i=0}^{2q-1} f_i x^i, \quad -f_{2q-1}>0 \tag{17}$$

$$h(x) = \sum_{j=0}^{s} h_j x^j,\ h_s \neq 0$$

where q,s are positive integers. Suppose

$$g(x) = g_o(1+x^2)^{r/2},\ g_o > 0 \tag{18}$$

where $r \in [0,\infty)$. Our conditions for existence and uniqueness and estimates of the asymptotic behavior of the density depend on whether or not g(x) is globally Lipschitz and on the degree of h(x) relative to the degree (or stability) of f(x). There are two cases covered by Theorems 1-2.

Case 1: $r \in [0,1]$, $q>r$, $s \geq 1$, $q \geq 1$.

The restrictions (A1)-(A5),(B1)-(B6), applied here require g(x) to satisfy a linear growth constraint $r \in [0,1]$, that f be at least a cubic polynomial, $q \geq 2$, where g(x) is of linear growth, r=1, and that h(x) be non-constant.

Then from Theorem 2 for any $0 \leq t_1 \leq t_2$, there exist constants M_i, K_i depending on the observation path such that

$$M_1 \exp[-K_1|x|^\rho] \leq U(x,t) \leq M_2 \exp[-K_2|x|^\rho] \tag{19}$$

where

$$\rho = \begin{cases} s-r+1 & , r<1 \text{ and } r+s>2q-1 \\ \max[s,2(q-r)], & \text{otherwise} \end{cases} \tag{20}$$

We refer to [7] for details. Although it is not covered in the present case, the situation r=0, f and its first two derivatives are bounded, and h is asymptotic to a non-constant polynomial, can be easily treated by adapting the arguments in Theorems 1-2. In particular, the inequalities (19) hold with $\rho = s+1$, and this result overlaps [6] [12]. For example, if f=0, g=1, $h(x) = h_s x^s$, $h_s \neq 0$, then

$$0 < K_2 < |h_s|/(s+1),\ \rho = s+1. \tag{21}$$

This was obtained by Sussmann for s=3 in [12].

Case 2: $r>1$, $q>r+\frac{1}{2}s$, $s \geq 1$, $q \geq 2$.

Here g(x) is of super linear growth, f(x) is at least a cubic polynomial, and h(x) is dominated, as indicated, by the dynamics of the state process. In this case the asymptotic behavior of the conditional density is the same as that of the *a priori* density (of x(t)).

4. THE SCALAR BILINEAR FILTERING PROBLEM

The example presented here illustrates a different grouping of terms and different growth conditions, necessary for this class of problems. It is presented here as another application of the method described earlier.

Consider the system

$$\begin{aligned} dz(t) &= f(z(t))dt + z(t)dw(t) \\ dy(t) &= h(z(t))dt + dv(t) \\ z(0) &= z_o,\ y(0) = 0,\ 0\leq t\leq T<\infty \end{aligned} \tag{22}$$

with z_o having density $p_o(z)$, and z_o, w,v mutually independent as before. Since z(t) will eventually be trapped in either the positive or negative half space, we shall arrange that $z(t)\in[0,\infty)$ by taking $f\in C^1(0,\infty)$ satisfying

$$(C1)\quad \begin{cases} f(z) \leq K(1+z) \text{ for some } K>0 \\ \\ f(0)\geq 0 \end{cases} \tag{23}$$

and by taking $p_o(z)$ defined on $(0,\infty)$ and continuous and integrable there. We also assume that $h\in C^2(0,\infty)$ with h_z, h_{zz} locally Hölder continuous (and so, bounded at zero).

We impose the following growth conditions on f,h.

(C1') $f_z(z)$ is bounded and locally Hölder continuous

(C2) $\lim\limits_{z\downarrow 0} [f(z)/z] = >0$

(C3) $\lim\limits_{z\to\infty} |h(z)|/\log z = +\infty$

(C4) $\lim\limits_{z\to\infty} [h_{zz}(z)/h_z^2(z)] = 0$

(C5) for some constants K_i, M_i, i=1,2

$$M_1 + K_1|zh_z(z)|\leq|h(z)|\leq M_2 + K_2|zh_z(z)|$$

Note that these conditions are satisfied when f(z) is affine and h(z) is a non-constant polynomial; we do not consider the case h(z) constant. The assumptions that f,h and their derivatives are bounded at the origin are made for convenience only. They can be relaxed by introducing more complex growth restrictions. The other assumptions (C2)-(C5) are essential (to our method).

The DMZ-equation associated with (22) is

$$dU(z,t) = [\tfrac{1}{2}(z^2U)_{zz} - (fU)_z - \tfrac{1}{2}h^2U]dt + h\,U\,dy(t)$$
$$U(z,0) = p_o(z)\ ,\quad (z,t)\in[0,\infty)\times[0,T] \tag{24}$$

Because the generator for the diffusion x in (22) is not uniformly elliptic our theorems are not directly applicable. If we make the logarithmic change of coordinates $x=\log z$ and let $W(x,t) = U(e^x,t)$, $x\in R$, (24) becomes

$$dW(x,t) = \{\tfrac{1}{2}W_{xx} + [3/2 - e^{-x}f(e^x)]W_x$$
$$+ [1-f_z(e^x) - \tfrac{1}{2}h^2(e^x)]W\}dt + h(e^x)Wdy(t) \tag{25}$$
$$W(x,0) = p_o(e^x)$$

The methods of Besala [10] as used in the proofs of Theorems 1 and 2 can be directly applied to the robust version of this equation. The proofs of Theorems 1-2 go through when the weight functions $\psi^k(x)$ are defined by (9) with $x=\log z$ and

$$\phi_1(z) = -\int_1^z f(\xi)/\xi^2 d\xi$$
$$\phi_2(z) = h(z) \tag{26}$$

Similarly as before define

$$\nu_1 = \eta_1 = \lim_{z\downarrow 0}|f/z|/[h^2 + f^2/g^2]^{\frac{1}{2}}$$
$$= 1 \text{ by (C2)} \tag{27}$$

$$\nu_2 = \liminf_{z\to+\infty} z|h_z(z)|/[h^2 + f^2/g^2]^{\frac{1}{2}}\in(0,\infty)$$
$$\eta_2 = \limsup_{z\to+\infty} z|h_z(z)|/[h^2 + f^2/g^2]^{\frac{1}{2}}\in(0,\infty) \tag{28}$$

Then we have obtained [7] the following result.

Theorem 3. Suppose (C1)-(C5) hold. Let $\theta_i>0$, $0<\theta_1 + \eta_2\theta_2<1$, and suppose $p_o(z)$ satisfies

$$p_o(z)\le M_1 \exp[\theta_1\int_1^z (f(\xi)/\xi^2)d\xi - \theta_2|h(z)|] \tag{29}$$

for all $z\in(0,\infty)$ and some $M_1>0$. Then for any $\hat{\theta}_i<\theta_i$ the DMZ-equation (24) has a unique solution in the class of functions satisfying, for all $t\ge 0$

$$\limsup U(z,t)\exp[-\hat{\theta}_1\int_1^z(f(\xi)/\xi^2)d\xi + \hat{\theta}_2|h(z)|] = 0 \tag{30}$$

Moreover, if there exist constants, M_2, $\hat{\theta}_i$, $0<\hat{\theta}_i<\theta_i$, such that for all

$z \in (0,\infty)$

$$M_2 \exp\left[\hat{\theta}_1 \int_1^z (f(\xi)/\xi^2)d\xi - \hat{\theta}_2 |h(z)|\right] \le p_o(z) \tag{31}$$

then for all $t \geq 0$ the solution $U(z,t)$ is asymptotic to

$$\exp\left[\int_1^z (f(\xi)/\xi^2)d\xi - |h(z)|\right] \tag{32}$$

in the sense of (16) in Theorem 2.

For example, when $f(z) = az + b$, assumption (C2) implies either $b>0$ or $a>0$ and $b = 0$. Then

$$\phi_1(z) = bz^{-1} - b - a \log z$$

and, whenever (29) (30) are satisfied, $U(z,t)$ is asymptotic to

$$\begin{array}{ll} \exp[-|h(z)| - b/z] , & \text{if } b>0 \\ z^a \exp[-|h(z)|] , & \text{if } b=0 \text{ and } a>0 , \end{array} \tag{33}$$

5. THE MULTIVARIABLE CASE

In [8] [9] the previous results have been generalized to multivariable diffusions. We shall briefly describe one of the cases here and refer the reader to [8], [9] for other cases and further details. The assumptions on the functions f,g,h are of two types basically: relative order relations implying the potentials of $\mathcal{L}^k$ and its adjoint are dominated by certain terms in c^k_{ess} as $|x|\to\infty$, and inequalities providing for control of these dominant terms. These assumptions may be stated succintly using the following definitions.

Definition. Let $f,g, \in C(\mathbb{R}^n \times [0,T])$ with $g \geq 0$. Then $f = o_b(g)$ if for every $\varepsilon>0$ there exists a constant $K(\varepsilon)$ such that for all $(x,t)\in\mathcal{D}$

$$|f(x,t)| \le \varepsilon g(x,t) + K(\varepsilon)$$

Definition. A nonnegative function $r \in H^2_{loc}(\mathbb{R}^n)$ is said to be a scale function if

(i) there exist positive constants θ^1, R such that $|\nabla r(x)|^2 \geq \theta^1$ for all $|x| \geq R$

(ii) $\lim\limits_{R\to\infty} \min\limits_{|x|=R} r(x) = +\infty$

We shall use on occas ion the notation

$$A_r(x,t) \triangleq 2 \sum_{i,j=1}^n a^{ij}(x,t) r_{x_i}(x) r_{x_j}(x) = |g(x,t)\nabla r(x)|^2 \geq 0.$$

Definition. Let $\mathcal{P}$ denote the collection of pairs of functions $(F(z,t), r(x))$ satisfying

$F \in HC^{2,1}_{loc}(\mathbb{R}\times[0,T])$, $r \in H^2_{loc}(\mathbb{R}^n)$ and $\mathscr{P}$(i) $F_z(z,t) = o_b(F^2(z,t))$

$$\int_0^{r(x)} F(z,t)dz, \quad \int_0^{r(x)} F_t(z,t)dz = o_b(F^2 A_r)$$

$$\mathscr{P}(\text{ii})\ F \sum_{i,j=1}^n a^{ij} r_{x_i x_j}, \ F \sum_{i,j=1}^n a^{ij}_{x_i} r_{x_j} = o_b(F^2 A_r)\ .$$

Definition. Two time-varying vector fields $f_1(x,t)$, $f_2(x,t)$ are said to be compatible if there exists a constant $R > 0$ such that $\sum_{i,j=1}^n a^{ij}(x,t) f_1^i(x,t) f_2^i(x,t) \geq 0$, for all $|x| \geq R$, $t \in [0,T]$. If we now let $(\bar{a}^{ij})$ denote the inverse of (a^{ij}) we can state the assumptions on the coefficients f,g of the diffusion as follows. We let θ_1, θ_2 be such that

$$\theta_1(x,t)|\xi|^2 \leq \sum_{i,j=1}^n a^{ij}\xi_i\xi_j \leq \theta_2(x,t)|\xi|^2.$$

Hypothesis F. There exist a scale function $r(x)$, nonnegative functions $\bar{F}$, $\underline{F} \in HC^{1,1}_{loc}(\mathbb{R}\times[0,T])$ satisfying $0 \leq \bar{F} \leq \underline{F}$, and constants $F_0 \geq 0$, $F_1 > 0$, $R > 0$ such that $(\bar{F},r)$, $(\underline{F},r) \in \mathscr{P}$, and for all $|x| \geq R$, $t \in [0,T]$,

(i) $\bar{F}(r(x),t) \leq (F_0 - \sum_{i=1}^n f^i(x,t) r_{x_i}(x))/A_r(x,t)$

(ii) $\underline{F}(t(x),t) \geq (-F_0 - \sum_{i=1}^n f^i(x,t) r_{x_i}(x))/A_r(x,t)$

(iii) $A_r(x,t)\bar{F}^2(r(x),t) \geq 2F_1[-F_0 + (\frac{\theta_2}{\theta_1} \sum_{ij=1}^n \bar{a}^{ij} f^i f^j)(x,t)]$

(iv) $\mathrm{div}(f)$, $\sum_{i,j=1}^n a^{ij}_{x_i x_j}(x,t) = o_b(\bar{F}^2 A_r)$

Hypothesis G. There exist constants $0 \leq \mu \leq \nu \leq 2$ and positive constants R, K_0, K_1 such that for all $|x| \geq R$, $t \in [0,T]$,

$$K_1|x|^{2-\nu} \leq \theta_1(x,t)$$

$$|a^{ij}(x,t)| \leq K_0|x|^{2-\mu}$$

$$|a^{ij}_{x_i}(x,t)| \leq K_0|x|^{1-\nu}$$

$$|a^{ij}_{x_i x_j}(x,t)| \leq K_0$$

If hypothesis G is not satisfied, it will be assumed that

Hypothesis L. $\lim_{|x|\to\infty} \int_0^{r(x)} \bar{F}(x,t)dx = +\infty$, uniformly in $[0,T]$.

One sees easily that F(i) is a generalization of Khas'minskii's test for explosions. In fact we have the following result [8] [9].

Theorem 4. If either F(i), (iii) and G hold, or F(i) and L hold, then the martingale problem for the pair (f,g) is well-posed.

Next the observation nonlinearity is assumed to satisfy the following conditions. Let

$$h^+(x,t) = ([\vec{1} + h^k(x,t)]^{\frac{1}{2}})_{k=1}^m$$

where $\vec{1} = (1,1,\ldots,1)^T \in \mathbb{R}^n$.

Hypothesis H. There exists a scale function $s(x)$, nonnegative functions $\overline{H}$, $\underline{H} \in HC^{2,1}_{loc}(\mathbb{R}\times[0,T])$ satisfying $0 \le \overline{H} \le \underline{H}$, and constants $R>0$, $H_0 \ge 0$ such that $(\overline{H},s)$, $(\underline{H},s) \in \mathscr{P}$ and for all $|x| \ge R$, $t \in [0,T]$,

(i) $\overline{H}^2(s(x),t) \le (|h(x,t)|^2 + H_0)/A_s(x,t)$

(ii) $\underline{H}^2(s(x),t) \ge (|h(x,t)|^2 - H_0)/A_s(x,t)$

(iii) $|h_t| = o_b(|h|^2)$, $|h_t| = o_b(\overline{H}^2 A_s)$

$$\sum_{i,j=1}^n a^{ij}|h_{x_i x_j}|, \quad \sum_{i,j=1}^n a^{ij}_{x_i}|h_{x_j}|, \quad \sum_{i,j=1}^n a^{ij}\langle h^+,\vec{1}\rangle_{x_i x_j},$$

$$\sum_{i,j=1}^n a^{ij}_{x_i}\langle h^+,\vec{1}\rangle_{x_j} = o_b\Big(\sum_{i,j=1}^n a^{ij}|h_{x_i}||h_{x_j}|\Big).$$

(iv) $\sum_{i,j=1}^n |a^{ij}||h_{x_i}||h_{x_j}| = o_b(\overline{H}^2 A_s + \overline{F}^2 A_r)$

(v) $|h(x,t)| = o_b\Big(\int_0^{s(x)} \overline{H}(z,t)\,dz\Big)$

Finally we need the following hypothesis which help control the growth of the potentials.

Hypothesis I.

(i) $x_i|h_{x_i}| = o_b(\overline{H}^2 A_s)$

(ii) $\sum_{i=1}^n f^i|h_{x_i}| = o_b(\overline{H}^2 A_s + \overline{F}^2 A_r)$

(iii) $\overline{F}\sum_{i,j=1}^n a^{ij} r_{x_i}|h_{x_j}|$, $\overline{H}\sum_{i,j=1}^n a^{ij} s_{x_i}|h_{x_j}| = o_b(\overline{H}^2 A_s + \overline{F}^2 A_r)$

$\underline{F}\sum_{i,j=1}^n a^{ij} r_{x_i}|h_{x_j}|$, $\underline{H}\sum_{i,j=1}^n a^{ij} s_{x_i}|h_{x_j}| = o_b(\underline{H}^2 A_s + \underline{F}^2 A_r)$

To compute lower bounds on the density we shall use

Hypothesis K. (i) ∇r and ∇s are compatible

(ii) $\vec{x}$ is compatible with both ∇r and ∇s.

We can now state the following existence result. For a proof see [9].

Theorem 5. (Existence of fundamental solutions). Suppose hypotheses F, H, I (ii) and I (iii) hold. Then for each Hölder continuous path $\{y(t),\ 0\leq t\leq T\}$ of the observation process, there exists a classical fundamental solution of the robust DMZ equation (4).

To describe our results for uniqueness and tail behavior we need the definition of certain function classes, given below.

Definition. Let $f\in C(\mathbb{R}^n \times [0,t])$ and $\phi\in C(\mathcal{D})$. Then $f\in\mathcal{E}(\phi)$ if $\exists$ constant K

s.t. $|f|\leq K \exp(\phi)$ for all $(x,t)\in\mathcal{D}$, and

$f\in\mathcal{E}_0(\phi)$ if $\lim\limits_{|x|\to\infty} |f| \exp(-\phi) = 0$, uniformly in $t\in[0,T]$.

Let

$$\bar{\psi}^0(x,t) = \begin{cases} \log|x| \ , & \text{if } \mu=0 \\ & , \text{ for } |x|\geq R \\ |x|^{\mu}/\mu, & \text{if } \mu\in(0,2] \\ \text{any } C^{\infty}\text{-time invariant extension for } |x|\leq R \end{cases}$$

$$\bar{\psi}^1(x,t) = \int_0^{r(x)} \bar{F}(z,t)dz$$

$$\bar{\psi}^2(x,t) = \int_0^{s(x)} \bar{H}(z,t)dz$$

Theorem 6. [9] (Existence and uniqueness of solutions). Suppose F, G,H,I hold and that p_o is continuous, nonnegative and integrable. Then there exist positive constants a_1, $\underline{\alpha}_i$, $\bar{\alpha}_i$, $i = 1,2$ such that whenever

$$p_o(x)\in \mathcal{E}\ (-\alpha_0\bar{\psi}^0(x) - \alpha_1\bar{\psi}^1(x,0) - \alpha_2\bar{\psi}^2(x,0))$$

for some constants $\alpha_0\in(0,\bar{\alpha}_0)$, $\alpha_1\in(\underline{\alpha}_1,\bar{\alpha}_1)$, $\alpha_2\in(0,\bar{\alpha}_2)$, the DMZ equation has a unique solution in

$$\mathcal{E}_0\left(\frac{\bar{\alpha}_0}{1-a_1t}\bar{\psi}^0(x) - \underline{\alpha}_1\bar{\psi}^1(x,t) + (1-\varepsilon)\underline{\alpha}_2\bar{\psi}^2(x,t)\right)$$

for all $\varepsilon\in(0,1)$. If instead of hypothesis G, hypothesis L holds the result remains valid with $\alpha_0 = \bar{\alpha}_0 = 0$ and $\underline{\alpha}_1$ chosen arbitrarily small.

Theorem 7. [9] (Lower bounds). Suppose the same hypotheses hold as in theorem 6, including the growth assumption on p_o. Suppose in addition that $\mu>0$ and hypothesis K holds.

(a) If $\mathrm{supp}(p_o(x))\supseteq(|x|\leq R)$, where R is the maximum of the radii R in F,G,H,K, then there exist constants M, α_i, $i=0,1,2$, depending on the path of the observation process $y(\cdot)$, such that

$$U(x,t) \geq M \exp[-\frac{1}{t}(1 + \sum_{i=0}^{2} \alpha_i \psi^i(x,t))].$$

(b) If there exist constants $M^0, \alpha_i^0 > 0$, $i = 0,1,2$, such that for all $x \in \mathbb{R}^n$

$$p_0(x) \geq M^0 \exp[-\sum_{i=0}^{2} \alpha_i^0 \psi^i(x,0)]$$

then there exist constants M, α_i, $i=0,1,2$, with M depending on the observation path such that

$$U(x,t) \geq M \exp[-\sum_{i=0}^{2} \alpha_i^0 \psi^i(x,t)].$$

There are several other cases where the method described here has been successfully applied. We refer to [9] for the details.

We close this section with a class of relatively simple but interesting examples. Consider the problem of a scalar observation of a two dimensional Wiener process,

$$dx(t) = dw(t)$$

$$dy(t) = h(x_1,x_2)dt + dv(t), \quad y(t) \in \mathbb{R}^1.$$

Then with $\overline{F} = \underline{F} = 0$, the natural upper and lower bounds of Theorems 6, 7 are of the form

$$\exp(-A(x_1^2 + x_2^2) - Bs(x)).$$

Examples of pairs of functions (h,s) for which Theorems 5-7 hold include the following: $h = (x_1^{2p} + x_2^{2q})^{\frac{1}{2}}$, $s = c(\sqrt{1 + x_1^{2p+2}} + \sqrt{1 + x_2^{2q+2}})$; $h = ax_1^{2p} + bx_2^{2p}$, $s = c(h^2 \rho^{1-2p} + \rho)$ where $\rho = (x_1^2 + x_2^2)^{\frac{1}{2}}$. Here p,q are positive integers, a,b,c positive numbers.

REFERENCES

[1] M.H.A. Davis, S.I. Marcus, "An introduction to nonlinear filtering", in Stochastic Systems: The Mathematics of Filtering and Identification and Applications, M. Hazewinkel, J.C. Willems, eds., NATO Adv. Study Inst. Ser., Reidel, Dordrecht, 1981, pp. 53-76.

[2] M. Hazewinkel, and J.C. Willems, eds., The Mathematics of Filtering and Identification and Applications, NATO Adv. Study Inst. Ser., Reidel, Dordrecht, 1981.

[3] N.V. Krylov, B.L. Rozovskii, "On the conditional distribution of diffusion processes", Izvestia Akad. Nauk, SSSR, Math. Ser., 42, 1978, pp. 356-378.

[4] E. Pardoux, "Stochastic partial diff. eqs. and filtering of diff.

processes", Stoch., 3, 1979, p. 127.

[5] D. Michel, "Regularite des lois conditionnelles en theorie du filtrage non-lineaire et calcul des var. stoch." J. Funct. Anal., 41, 1981, pp. 8-36.

[6] W.H. Fleming, S.K. Mitter, "Optimal control and pathwise nonlinear filtering of nondeg. diff.", 20th IEEE Conf. Dec. Confr., San Diego, Dec. 1981.

[7] J.S. Baras, G.L. Blankenship and W.E. Hopkins, Jr., "Existence, Uniqueness, and Asymptotic Behavior of Solutions to a Class of Zakai Equations with unbounded Coeff," to appear in IEEE Trans. Aut. Contr.

[8] W.E. Hopkins, Jr., "Nonlinear Filtering of Multivariable Diffusions with Strongly Unbounded Coefficients", to appear in Proc. of 1982 Conference on Information Sciences and Systems, Princeton Univ.

[9] W.E. Hopkins, Jr., Ph.D. Thesis, Electrical Engineering Dept., University of Maryland, in preparation.

[10] P. Besala, "Fundamental Solution and Cauchy Problem for a Parabolic System with Unbounded Coefficients" J. Diff. Eqs., 33, 1979, pp. 26-38.

[11] D.W. Stroock and S.R.S. Varadhan, Multidimensional Diffusion Processes, Springer-Verlag, New York, 1979.

[12] H.J. Sussmann, "Rigorous Results on the Cubic Sensor Problem", in [2], above.

OPTIMAL STOPPING UNDER PARTIAL OBSERVATIONS

Vaclav E. Benes

Bell Laboratories
Murray Hill, New Jersey 07974

1. Introduction

The area of stochastic control, though long ago supplied by Richard Bellman with the powerful methodology of dynamic programming, is still strewn with unsolved problems. Yes, *problems*, because although the *theory* of stochastic control is well and even elegantly formulated, the gritty task of finding or describing optimal control laws languishes, and *problems* continue to stay generically unsolved, except on a set of empty interior by (shudder!) numerical methods. In the difficult and currently active field of control with incomplete noisy information, solutions are even more sparse; for good Lie algebraic reasons they are limited with a few exceptions to linear dynamics and Gauss-Markov processes.

It is therefore important to note that a special but significant subclass of control problems with incomplete noisy information turns out to be more readily approachable than one might expect: these are the problems of optimally stopping a process (say a diffusion x_t,) while observing only a function $h(x_t)$ muddied up by white noise, so as to maximize $Ek(x_\tau)$ for some return function k.

Now stopping a process to maximize an expected return depending on the final value at the stopping time is a problem that has attracted much theoretical interest, and found applications in economics, finance, and inventory theory, *inter alia*. Here the admissible stopping times are those of the process itself, because one has perfect information. Thus in the problem's usual formulation it is assumed that the process to be stopped is Markov and that it is exactly observed; the task then becomes one of finding a *continuation region* in which the optimal return is *harmonic* (in the generalized sense appropriate to the process) while everywhere dominating the return function k, and fitting it smoothly at the (unknown) boundary. These optimality conditions can also be reformulated more analytically as a quasi-variational inequality, convenient for proofs but obscuring the "free boundary" problem.

We pose a version of the stopping problem in which only noisy incomplete information about the past of the process x_t to be stopped is available: the decision to stop must be based on a given initial distribution for x_0 and on a noisy observation $dy_t = h(x_t)dt + db_t$ consisting of a function h of x_t corrupted by white Gaussian noise. This kind of problem was posed first by Grigelionis,[1] and studied extensively by Mazziotto and Szpirglas.[2–5]

It turns out, not surprisingly, that our form of the problem can be formulated and attacked by means of nonlinear filtering, and that the resulting structure is a kind of infinite-dimensional version of the original problem: the conditional distribution of the state is used as a new state, and constitutes a measure-valued diffusion; again one looks for a function (defined now on the positive cone of a Banach space of measures) which is "harmonic" in a certain continuation region, and dominates a given function. The meaning of 'harmonic' is clarified by the interpretation in terms of filtering: we "Markovize" the problem by taking the conditional distribution of x_t as an infinite-dimensional state variable, with dynamics given by Zakai's equation. It is natural, if not almost obvious, that this conditional measure is a *sufficient* statistic in the sense that by using only it one can do as well as by using all the available data; so we can expect that a particular optimal stopping time can be defined as the hitting time of a suitable boundary by the conditional distribution. The "harmonic" functions in this setting are those which are annihilated by a natural infinite-dimensional "Laplacian" operator obtained from the Zakai equation. In this way we inject, into the problem of stopping under partial observations, a more analytical method than had been present in the previous work[3–4] of Szpirglas and Mazziotto.

We end this introduction with a summary of the rest of the work. Sections 2 and 3 describe a formulation and interpretation of the problem whose neatness owes much to Fleming and Pardoux[7] and to Davis and Kohlmann.[8–9] In Section 4 we advance some heuristics on the appropriate meaning of 'harmonic' to be used in an analytical approach, leading, in Section 5, to an Ito differential rule. In Section 6 we describe a quasi-variational inequality proposed as equivalent to the stopping problem; the inequality is stated in terms of an infinite-dimensional "Laplacian" advanced in Section 4 as expressing the appropriate meaning of 'harmonic.' There follow a verification lemma in Section 7, a discussion of the regularity of the optimal return in Section 8, and a reformulation of the whole problem as a fixed point problem in Section 9, and a characterization of the optimal return as the least superharmonic majorizer of the linear functional (k,μ). Section 10 studies the optimal return as a fixed point in a complete lattice, while Section 11 suggests and calculates some examples. The final Section 13 gives the proof of the differential formula.

2. Formulations

It is convenient to give our problem a series of formulations of increasing tractability. We start with a diffusion process observed through noise, described by stochastic DEs

$$dx_t = bdt + adw_t \qquad \text{state}$$

$$dy_t = h(x)dt + db_t \quad \text{observation}$$

and the problem is to maximize the expected return $Ek(x_t)$ over stopping times τ of y, for a given initial measure μ for x_0. The first reformulation consists in eliminating all reference to x_t. Let $\mathbf{Y}_t = \sigma\{y_s, s \leq t\}$, and for a stopping time τ of y, let $\mathbf{Y}_\tau = \sigma$-algebra of events prior to τ. Since

$$E\{k(x_\tau)|\mathbf{Y}_\tau\} = E\{k(x_t)|\mathbf{Y}_t\}_{t=\tau} \quad \text{a.s.}$$

we can write the criterion as

$$EE\{k(x_t)|\mathbf{Y}_t\}_{t=\tau} = E\hat{k}_\tau$$

where $\hat{k}_t$ is the conditional expectation: $\hat{k}_t = E\{k(x_t)|\mathbf{Y}_t\}$. The process $\hat{k}_t$ is a functional of the past of y, and our problem can be rephrased as the optimal stopping of $\hat{k}_t$ on the basis of $\mathbf{Y}_t$ itself. It is now natural to invoke nonlinear filtering to describe the local character or dynamics of $\hat{k}_t$.

For a starting measure μ, let $\sigma_T\mu$ be the unnormalized conditional measure of x_t given $\mathbf{Y}_t$ defined by the Zakai dynamics

$$\sigma_0\mu = \mu$$

$$d(f,\sigma_t\mu) = (\mathbf{A}f,\sigma_t\mu)dt + (fh,\sigma_t\mu)dy_t$$

for bounded continuous f in the domain of an elliptic generator

$$\mathbf{A} = \frac{1}{2} tr\ a\nabla^2 a' + b'\nabla$$

The notation (,) refers to the pairing between the set C_b of bounded continuous functions and the set **M** of finite regular Borel measures, with the weak topology for **M**. $\sigma_t\mu$ is a measure-valued diffusion driven by the observation process. In terms of it we can write

$$\hat{k}_t = \frac{(k,\sigma_t\mu)}{(1,\sigma_t\mu)}$$

and state a first reformulation of the original problem as the search for

$$\sup_\tau E\,\frac{(k,\sigma_\tau\mu)}{(1,\sigma_\tau\mu)} \quad , \quad \tau \text{ a s.t. of } y_t$$

subject to the Zakai dynamics for $\sigma_t\mu$, the expectation E now being purely over the y_t-process.

Finally, on the suggestion of M. H. A. Davis[6,8] we can use an idea due to W. H. Fleming and E. Pardoux[7] to eliminate the troublesome-looking normalizing denominator $(1,\sigma_\tau\mu)$ if we replace the measure induced by y_t by Wiener measure. The final formulation is then this: Find, for w_t Brownian a stopping time τ^* of w_t that achieves

$$S(\mu) = \sup_\tau E(k,\sigma_\tau\mu)$$

subject to $\sigma_0\mu = \mu$ and

$$d(f,\sigma_t\mu) = (\mathbf{A}f,\sigma_t\mu)dt + (fh,\sigma_t\mu)dw_t\ .$$

3. Interpretation via convex analysis

The following explanation of $S(\mu)$ is adapted from Davis and Kohlmann.[8] The functions of the form

$$S(\mu) = \sup_\tau E(k,\sigma_\tau\mu)$$

are convex, positive homogeneous, and bounded in the sense ($\|\cdot\|$ = variation)

$$\sup_{\|\mu\|=1} |S(\mu)| < \infty \qquad \text{for bounded } k$$

Viewing $E(k,\sigma_\tau\mu)$ dually as (k^τ,μ) for suitable k^τ, and letting

$$K = \{k^\tau\colon \tau \text{ s.t.}\}$$

be the "reachable" set, we see that $S(\mu)$ is the *support function* of K, in the language of convex analysis. $S(\mu)$ is not changed if the bounded but possibly nonconvex set K is replaced by its convex hull coK, and in terms of $S(\cdot)$, coK is given by

$$coK = \{f \in C_b\colon (f,\mu) \le S(\mu) \text{ for all } \mu\}$$

This has a nice control theoretic interpretation: the set of which S is the support function consists of those f such that for any μ, given the choice of (f,μ) pesos now or of taking the maximum expected payoff from the stopping problem, you would go for the latter.

4. Some informed guesses and heuristics

We provide here an informal motivation for the mathematical ideas used in the sequel. It is tempting to conjecture that our problem of stopping with partial information, when reformulated as a complete information problem on a new state space, will have the same kind of structure as the well-known stopping problems governed by free boundary problems or quasi-variational inequalities. This structure is revealed when we view the $\mathbf{M}^+$-valued diffusion $\sigma_t \mu$ as having a drift associated with the generator $\mathbf{A}$, and a degenerate kind of diffusion proportional to h.

To fix ideas, let us use the density formulation of the Zakai equation. For $t > 0$, the unnormalized measure $\sigma_t \mu$ has a density ρ which, according to the Zakai equation, moves with "drift" $\mathbf{A}^* \rho$, and diffusion "$h\rho$" when driven by dy_t. This suggests that our problem might also be governed analytically by a second-order operator based on $\mathbf{A}$ and h.

For guidance, let us see how a smooth functional $U(t,\rho_t)$ would develop in time under the Zakai dynamics: expanding in Taylor's series up to terms of order 2, we find

$$U(t+dt,\rho_{t+dt}) = U(t,\rho_t) + U_1(t,\rho_t)dt + U_2(t,\rho_t)[\mathbf{A}^* \rho_t dt + h\rho_t dy_t]$$
$$+ \frac{1}{2} U_{22}(t,\rho_t)[d\rho_t, d\rho_t] + \cdots$$

Here U_1, U_2, U_{22} are the ordinary and Fréchet derivatives of U; U_2 is a linear functional acting on a function f to produce $U_2(t,\rho_t)[f]$, while U_{22} is a bilinear functional acting on a pair f,g to produce $U_{22}(t,\rho_t)[f,g]$. Considerations of quadratic variation indicate that U_{22} will contribute only through a term $U_{22}(t,\rho_t)[h\rho_t, h\rho_t](dy)^2$. Thus in ρ-space the analog of a backward operator on a smooth U is

$$\mathbf{L}U(\rho) = \frac{1}{2} U_{22}(\rho)[h\rho, h\rho] + U_2(\rho)[\mathbf{A}^* \rho] .$$

We can expect that under reasonable circumstances the second term can be expressed in an "adjoint" manner: $U_2(\rho)$ is represented by a function $u = u(\rho)$ in the domain of $\mathbf{A}$, such that $\mathbf{A}u$ is a bounded continuous function, and with (,) the natural pairing,

$$U_2(\rho)[\mathbf{A}^* \rho] = (\mathbf{A}u, \rho) .$$

Let now $V: \mathbf{M}^+ \to R$ be twice continuously Fréchet differentiable with respect to μ (weak topology), and such that its first derivative V_2 is (a linear functional) representable by a function $v \in \mathbf{D}(\mathbf{A})$, with $\mathbf{A}v \in C_b$. We call such functions V smooth. Then with $h\mu$ the measure defined by $(g, h\mu) = (gh, \mu)$ it is natural to associate with the process $\sigma_t \mu$ the operator

$$\mathbf{L}V = \frac{1}{2} V_{22}(\mu)[h\mu, h\mu] + (\mathbf{A}\nu, \mu) \qquad (\nu \text{ depends on } \mu)$$

as a kind of analog of the generators used in finite-dimensional diffusion. Such an interpretation of **L** is justified by its consequences: (i) **L** is associated with an Ito differential rule for $\sigma_t \mu$; (ii) $\sigma_t \mu$ solves a martingale problem posed in terms of **L**; for suitable test functions F

$$F(\sigma_t \mu) = \int_0^t \mathbf{L}F(\sigma_s \mu) ds$$

is a martingale when w_t is Brownian; finally, (iii) use of **L** leads to rather complete analogs of classical results for optimal stopping.

5. A differential rule for $\sigma_t \mu$

In order to use the method of quasi-variational inequalities, prove a "verification" lemma, and generally, do dynamic programming for partially observed diffusions, it is convenient to have an Ito differential rule for smooth functions of $\sigma_t \mu$. In the notation of Section 4, such a rule is given in

(1) Ito lemma: Let F: $R^+ \times \mathbf{M}^+ \to R$ be C^1 in t and Fréchet C^2 in μ, such that its first μ-derivative F_2 is represented by a function f_2 in the domain of **A** such that $\mathbf{A}f_2 \in C_b$. Then

$$F(t, \sigma_t \mu) = F(0, \mu) + \int_0^t (\mathbf{A}f_2(\sigma_t \mu), \sigma_t \mu) ds + \frac{1}{2} \int_0^t F_{22}(s, \sigma_s \mu)[h\sigma_s \mu, h\sigma_s \mu] ds$$

$$+ \int_0^t (hf_2(\sigma_s \mu), \sigma_s \mu) dw_s \qquad f_2 = f_2(s, \sigma_s \mu)$$

Proof: The proof is long, not quite standard, and in the Appendix, Section 13.

6. Quasi-variational inequality

We guess, in analogy with the finite-dimensional case, that an appropriate value function for our problem will be a positive homogeneous function V: $\mathbf{M}^+ \to R$ in the domain of **L**, such that

$$V(\mu) \geq (k, \mu) = \int k \, d\mu$$

$$\mathbf{L}V(\mu) \leq 0$$

$$\mathbf{L}V(\mu)[V(\mu) - (k, \mu)] = 0$$

This quasi-variational inequality (QVI) generalizes the finite-dimensional version to our context. The first condition says that the best return is at least as large as what you get by stopping at once; the other two express optimality. Hence **L** incorporates the sense of 'harmonic' that is appropriate to the problem. Continuing the analogy we can expect that

$$\tau^* = \inf t: V(\sigma_t \mu) = (k, \sigma_t \mu)$$

is an optimal stopping time. The continuation region R is such that $V(\mu) > (k,\mu)$ and $\mathbf{L}V = 0$, and it is bounded by a surface $V(\mu) = (k,\mu)$. Starting inside R we should calculate $\sigma_t\mu$ and stop when $V(\sigma_t\mu) = (k,\sigma_t\mu)$; when starting outside R, we should stop at once.

7. Verification lemma

With all these preliminaries behind us, our first result is that a smooth solution of the QVI must be the value function of the problem.

(2) Theorem: Let V be a Fréchet-C^2 function from $\mathbf{M}^+$ to R such that

(i) V satisfies the QVI

(ii) the linear functional V_2 is representable by a function $v = v(\mu,\cdot)$ in the domain of $\mathbf{A}$, with $\mathbf{A}v \in C_b$ (V is smooth, we say.)

Then

$$V(\mu) = \sup_{\tau} E(k,\sigma_\tau\mu) \triangleq S(\mu)$$

and an optimal stopping time τ^* is given by

$$\tau^* = \inf t\colon V(\sigma_t\mu) = (k,\sigma_t\mu)\ .$$

Proof: Let τ be any stopping time of the Brownian process w_t that drives $\sigma_t\mu$. By the Ito lemma and the QVI,

$$(k,\sigma_\tau\mu) \le V(\sigma_\tau\mu)$$

$$\le V(\mu) + \int_0^\tau (\mathbf{A}v,\sigma_s\mu)ds + \frac{1}{2}\int_0^\tau V_{22}(\sigma_s\mu)[h\,\sigma_s\mu,h\,\sigma_s\mu]ds + \int_0^\tau (v,h\,\sigma_s\mu)dw_s$$

$$\le V(\mu) + \text{supermartingale at } \tau$$

Taking expectations, we find that the return $E(k,\sigma_\tau\mu)$ obtained by stopping at τ is not greater than $V(\mu)$, so $V \ge S$. For the reverse inequality, consider the special stopping time $\tau^* = \inf t\colon V(\sigma_t\mu) = (k,\sigma_t\mu)$. Writing

$$V(\sigma_{\tau^*}\mu) = V(\mu) + \int_0^{\tau^*} \mathbf{L}V(\sigma_s\mu)ds + \text{martingale at } \tau^*$$

we note that $\mathbf{L}V(\sigma_s\mu) = 0$ prior to τ^*, and that by definition of τ^* $V(\sigma_{\tau^*}\mu) = (k,\sigma_{\tau^*}\mu)$. Hence taking expectations we see that

$$\begin{aligned} E \text{ return from } \tau^* &= E(k,\sigma_{\tau^*}\mu) \\ &= EV(\sigma_{\tau^*}\mu) \\ &= V(\mu) \end{aligned}$$

so τ^* *achieves* the upper bound $V(\mu)$, and hence is optimal. Only one smooth V can satisfy the conditions of Theorem (2); if there were two, each would equal S, and so the other.

8. Regularity of S

(3) Remarks: It is of course of prime importance here, as in any other problem governed by a QVI or a Bellman equation, to determine when there exist solutions having the regularity requisite for applicability of these analytical methods. This is already a difficult task in the well-known cases, and it is correspondingly harder in the present problem, if only because of the singularity of **L**. However, some results on regularity can be obtained directly from the definition of the optimal return $S(\mu) = \sup_\tau E(k,\sigma_\tau\mu)$.

(4) Lemma: S is lower semi-continuous in the weak topology.

Proof: Let $\mu_n \to \mu$. For each τ the map $\nu - E(k,\sigma_\tau\nu)$ is a bounded continuous on **M**, so $E(k,\sigma_\tau\mu_n) \to E(k,\sigma_\tau\mu)$, and hence for any $\epsilon > 0$

$$S(\mu_n) = \sup_\tau E(k,\sigma_\tau\mu_n) > S(\mu) - \epsilon \quad \text{eventually} .$$

This result is essentially the observation that the upper envelope of a family of linear functions is l.s.c. "Stronger" continuity can be obtained by strengthening the topology to that of the variational norm $\|\cdot\|$.

(5) Lemma: $E\|\sigma_\tau\mu\| = \|\mu\|$ for all $\mu \in \mathbf{M}$

Proof: Let $\mu^+ - \mu^-$ be a Hahn decomposition of μ, so that $\|\mu\| = \|\mu^+\| + \|\mu^-\|$. By linearity and positivity of σ_t

$$\sigma_t\mu = \sigma_t\mu^+ - \sigma_t\mu^-$$

and since $\|\mu^\pm\| = (1,\mu^\pm)$ and $\mathbf{A}1 = 0$, we find

$$\|\sigma_t\mu^\pm\| = (1,\sigma_t\mu^\pm) = (1,\mu^\pm) + \int_0^t (\mathbf{A}1,\sigma_s\mu^\pm)ds + \int_0^t (h,\sigma_s\mu^\pm)dy_s$$

$$\|\sigma_\tau\mu\| = \|\mu\| + \text{ martingale at } \tau$$

$$E\|\sigma_\tau\mu\| = \|\mu\| .$$

(6) Lemma: S is Lip_κ in the $\|\cdot\|$ topology, with $\kappa = \sup|k|$.

Proof: Using Lemma ( ), we obtain

$$S(\mu_1) - S(\mu_2) \le \sup_\tau E(k,\sigma_\tau\mu_1 - \sigma_\tau\mu_2)$$

$$\le \kappa E\|\sigma_\tau(\mu_1-\mu_2)\| = \kappa\|\mu_1-\mu_2\|$$

Similarly, $-\kappa\|\mu_1-\mu_2\|$ is a lower bound. This argument is from Davis and Kohlmann.[9]

9. A fixed point formulation

Consider a positive homogeneous function $V: \mathbf{M}^+ \rightarrow R$ satisfying the two properties $V(\mu) \geq (k,\mu)$ and

(7) $$V(\mu) = \sup_\tau EV(\sigma_\tau \mu) \triangleq \Gamma V(\mu)$$

If we can interpret $V(\mu)$ as an achievable average return starting from "knowledge" μ, then the first property says that the policy leading to V is at least as good as stopping at once, while the second says that if you wait for any stopping time τ (while the initial μ moves stochastically under its driving Wiener process to $\sigma_\tau \mu$) and *then* from $\sigma_\tau \mu$ follow the policy yielding V then the average return starting from $\sigma_\tau \mu$ is no larger: you can still do as well as if you had followed the V-policy in the first place. Thus the second property expresses a kind of optimality, since the null stopping time $\tau = 0$ is admissible.

Intuitively, then, we expect (7) to act as a kind of Bellman equation for the noisy stopping problem, with the value function satisfying the two conditions $V = \Gamma V$ and $V(\mu) \geq (k,\mu)$. This guess will be borne out in part by theorems to follow. However, it turns out that these two conditions are only necessary and seemingly not sufficient: such a function V might not be achievable by a stopping policy. Still, we shall show that the smallest such function is the right one, and so characterize the value function (in a manner similar to using the Snell envelope) as the least fixed point of Γ that majorizes (k,μ).

(8) Remark: An equation (and cognate results) can be formulated for the ordinary stopping problems in R^n: $v(x) \geq k(x)$ and

$$v(x) = \sup_\tau Ev(x_\tau) \qquad \tau = \text{s.t. of } x_t$$

(the superharmonic property.)

(9) Remark: From an analytical viewpoint, introduction of Γ begs all the questions, because so much is hidden in the definition of Γ itself, and you cannot easily calculate with it, or iterate it. This is one reason for trying to use the *prima facie* more explicit operator $\mathbf{L}$. However, it will appear that the inequalities $V \geq \Gamma V$ and $\mathbf{L}V \leq 0$ are both analogous to the *superharmonic (excessive)* property. We have

(10) Theorem: If V is a Fréchet-C^2 function with V' represented by $v \in \mathbf{D}(\mathbf{A})$ with $\mathbf{A}v \in C_b$, then

$$\mathbf{L}V \leq 0 \quad iff \quad V = \Gamma V$$

Proof: For any stopping time τ

$$V(\sigma_\tau \mu) = V(\mu) + \int_0^\tau \mathbf{L}V(\sigma_s \mu) ds + \text{martingale at } \tau$$

If $\mathbf{L}V \leq 0$, taking E and sup on τ gives $V \geq \Gamma V$; the reverse inequality $\Gamma V \geq V$ is trivially true, so $V = \Gamma V$. Conversely $V \geq \Gamma V$ implies $E \int_0^\tau \mathbf{L}V(\sigma_s \mu)ds \leq 0$ for all stopping times τ. So for $\tau(\omega) = h > 0$ we find

$$\mathbf{L}V(\mu) = \lim_{h \downarrow 0} h^{-1} E \int_0^h \mathbf{L}V(\sigma_s \mu)ds \leq 0 .$$

Remark: If $\mathbf{L}V \leq 0$, then $V(\sigma_t \mu)$ is a supermartingale.

We next show directly that the optimal return $S(\mu) = \sup_\tau E(k, \sigma_\tau \mu)$ is itself the least fixed point of Γ majorizing (k, μ)

(11) Theorem: $S \geq \Gamma S$, and

$$S(\mu) = \inf V(\mu)\colon\ V(\mu) \geq (k,\mu) \ \text{ and } \ V = \Gamma V$$

Proof: $V = \Gamma V$ and $V(\cdot) \geq (k,\cdot)$ imply

$$V(\mu) \geq \sup_\tau EV(\sigma_\tau \mu) \geq \sup_\tau E(k, \sigma_\tau \mu) = S(\mu)$$

To show $S \geq \Gamma S$, it is enough to prove that for any s.t. $\bar{\tau}$

$$\sup_{\tau \geq \bar{\tau}} E(k, \sigma_\tau \mu) \geq E(S(\sigma_{\bar{\tau}} \mu)) .$$

Let $\bar{\tau}$ be a fixed s.t. The function S is the upper envelope of a family of linear functions: $S(\mu) = \sup_\tau E(k, \sigma_\tau \mu)$. So S is lower semi-continuous in the weak topology of $\mathbf{M}$, and hence measurable. Let $\epsilon > 0$ be fixed. By Lusin's theorem, there is a compact subset K of $\mathbf{M}$ such that S is continuous on K and $Pr\{\sigma_{\bar{\tau}} \mu \epsilon K\} \geq 1 - \epsilon$. For each $\mu \epsilon K$ there is a neighborhood N_μ such that

$$\nu \epsilon K \cap N_\mu \quad \text{implies} \quad S(\nu) \leq S(\mu) + \epsilon$$

This is upper semi-continuity on K. By definition of S there is a stopping time $\tau(\mu)$ such that

$$E(k, \sigma_{\tau(\mu)} \mu) > S(\mu) - \epsilon$$

On $N_\mu \cap K$ then

$$E(k, \sigma_{\tau(\mu)} \mu) > S(\nu) - 2\epsilon$$

From the relative cover $\{N_\mu \cap K, \mu \epsilon K\}$ one can pick a finite subcover $N_{\mu_i} = N_i$ with corresponding s.t.s. $\tau(\mu_i) = \tau_i$. Define a partition $\{\mathbf{A}_i\}$ of K by $A_1 = N_1 \cap K$

$$A_{n+1} = N_{n+1} \cap K - \bigcup_{i=1}^{n} A_i$$

Now we can mimic each τ_i starting at $\bar{\tau}$; that is, there is a stopping time $\bar{\tau}_i$ of $w_{\bar{\tau}+\cdot} - w_{\bar{\tau}}$ with the same law as τ_i, and of course independent of $\mathbf{F}_\tau =$ events prior to τ, such that $\bar{\tau} + \bar{\tau}_i$ is a s.t. Now set $\tau' = 0 = \tau_i$ on $\{\sigma_{\bar{\tau}} \mu \epsilon K'\}$ and

$$\tau' = \bar{\tau}_i \quad \text{on} \quad \{\sigma_{\bar{\tau}} \mu \epsilon A_i\}$$

Then $\bar{\tau}+\tau'$ is a s.t., and

$$E(k,\sigma_{\bar{\tau}+\tau'}\mu) = E(k,\sigma_{\tau'}\sigma_{\bar{\tau}}\mu)$$

$$\geq E1_{\sigma_{\bar{\tau}}\mu\epsilon K}\,(k,\sigma_{\tau'}\sigma_{\bar{\tau}}\mu)$$

$$\geq E\sum_{i=1}^{\infty} 1_{\sigma_{\bar{\tau}}\mu\epsilon A_i} E(k,\sigma_{\tau(\mu_i)}\sigma_{\bar{\tau}}\mu) - \epsilon\ sup\ |k|$$

$$\geq E1_{\sigma_{\bar{\tau}}\epsilon K}\,S(\sigma_{\bar{\tau}}\mu) - 2\epsilon P\{K\} - \epsilon\ sup\ |k|$$

$$\geq ES(\sigma_{\bar{\tau}}\mu) - 2\epsilon P\{K\} - 2\epsilon\ sup\ |k|$$

But ϵ was arbitrary. To verify that τ' is a stopping time we appeal to

(12) Lemma: Let τ be a stopping time, $\mathbf{F}_\tau$ the σ-algebra of events prior to τ, and τ_i stopping times of $w_{\tau+\cdot} - w_\tau$. Let B_i be a countable system of disjoint sets from $\mathbf{F}_\tau$, of total probability 1. Then the r.v. to equal to $\tau+\tau_i$ on B_i is a stopping time.

Proof: Similar to Meyer's T58, p. 74, Ref. 10.

(13) Theorem: If V is smooth, superharmonic ($V \geq \Gamma V$), and such that for $\tau = inf\ t$: $V(\sigma_t\mu) - (k,\sigma_t\mu)$

$$V(\sigma_{t\wedge\tau}\mu) \quad \text{is a martingale}$$

then $S \geq V$.

Proof: From

$$(k,\sigma_\tau\mu) = V(\sigma_\tau\mu) = V(\mu) + \int_0^\tau \mathrm{L}V(\sigma_s\mu)ds + \text{martingale at } \tau$$

it follows that

$$S(\mu) \geq E(k,\sigma_\tau\mu) = V(\mu) + E\int_0^\tau \mathrm{L}V(\sigma_s\mu)ds$$

Similarly

$$V(\sigma_{t\wedge\tau}\mu) = V(\mu) + \int_0^{t\wedge\tau} \mathrm{L}V(\sigma_s\mu)ds + \text{martingale at } \tau$$

so if the right hand side is a martingale then

$$\int_0^{t\wedge\tau} \mathrm{L}V(\sigma_s\mu)ds = 0 \quad a.s.$$

By monotone convergence and $\mathbf{L}V \leq 0$ we find $E \int_0^\tau \mathbf{L}V(\sigma_s \mu)ds = 0$ and so $S \geq V$. Notice that the domination condition $V(\mu) \geq (k,\mu)$ is not relevant here.

(14) Theorem: For smooth $V \ni \Gamma V = V$ and $\tau = \inf t$: $V(\sigma_t \mu) = (k,\sigma_t \mu)$, $V(\sigma_{t \wedge \tau} \mu)$ is a martingale iff $\mathbf{L}V = 0$ on $V(\mu) > (k,\mu)$

Proof:

$$V(\sigma_{t\wedge\tau}\mu) = V(\mu) + \int_0^{t\wedge\tau} \mathbf{L}V(\sigma_s \mu) + \text{martingale at } t$$

If this is a martingale we find $\int_0^{t\wedge\tau} \mathbf{L}V(\sigma_s \mu)ds \equiv 0$, and for $V(\mu) > (k,\mu)$

$$\mathbf{L}V(\mu) = \lim_{t \downarrow 0} E^{-1} t\wedge\tau \, E \int_0^t \mathbf{L}V(\sigma_s \mu)ds = 0$$

The converse is obvious. Note that the domination condition $V(\mu) \geq (k,\mu)$, required by the optimal return S, plays no role here.

We can now put together the following multiple characterization of the optimal return S:

(15) Proposition: For a smooth fixed point V of Γ, satisfying $V(\mu) \geq (k,\mu)$, the following conditions are equivalent:

(i) For $\tau^* = \inf t$: $V(\sigma_t \mu) = (k,\sigma_t \mu)$, $V(\sigma_{t\wedge\tau^*} \mu)$ is a martingale

(ii) $\mathbf{L}V(V-(k,\mu)) \equiv 0$

(iii) $V(\mu) = \inf_{U \in \mathbf{E}} U(\mu)$, $\mathbf{E}$ = excessive functions, i.e. $U \geq \Gamma U$.

(iv) $V = S = \sup_\tau E(k,\sigma_\tau \mu)$

Proof: Apply (2), (11), (13), and (14).

10. S as a fixed point in a complete lattice

The equation $V = \Gamma V$ can be studied directly in an abstract context by several methods of functional analysis. Needless to say, the degrees of difficulty of such approaches are directly related to the degree of regularity sought for the solution. We have seen that S is l.s.c. in the weak topology, and Lip_κ in the variational norm, of $\mathbf{M}$; differentiability of S remains an open question, with a negative answer in general, we suspect.

It is convenient to replace the weak topology of $\mathbf{M}^+$ by the strong topology induced by the variational norm $\|\cdot\|$, in which it is harder for functions to be continuous. We consider the Banach lattice of bounded uniformly continuous functions ϕ: $M^+ \to R$ with uniform norm, and pointwise order. The subclass Lip_a, $a > 0$, is defined by the condition

$$|\phi(\mu_1)-\phi(\mu_2)| \leq a\|\mu_1-\mu_2\| .$$

Lemma (6) established that S belongs in Lip_κ, $\kappa = \sup |k|$.

(16) Lemma: Γ carries Lip_a into itself

Proof: If $\phi \in Lip_a$

$$(\Gamma\phi)(\mu_1) - (\Gamma\phi)\mu_2 \leq \sup_\tau E\{\phi(\sigma_\tau\mu_1) - \phi(\sigma_\tau\mu_2)\}$$

$$\leq \sup_\tau E|\phi(\sigma_\tau\mu_1) - \phi(\sigma_\tau\mu_2)|$$

$$\leq a \sup_\tau E\|\sigma_\tau\mu_1 - \sigma_\tau\mu_2\|$$

$$\leq a\|\mu_1 - \mu_2\|$$

by Lemma (6). Similarly, $-a\|\mu_1 - \mu_2\|$ is a lower bound.

Remark: Since we know that S is the smallest superharmonic majorant of (k,μ), it is tempting and natural to try to construct it directly as the appropriate, i.e., smallest, fixed point of Γ in a suitable complete lattice, as follows.

We now single out the set K of functions ϕ defined for $\mu \geq 0$, with these properties

(i) ϕ is positive homogeneous and convex

(ii) $\|\mu\| \cdot \sup|k| \geq \phi(\mu) \geq (k,\mu)$

(iii) $\phi \in Lip_\kappa$, $\kappa = \sup|k|$.

It can be verified that K is a complete lattice, and that Γ preserves (i)-(iii). Then Γ is an isotone function from K to itself: $\phi \leq \psi$ implies $\Gamma\phi \leq \Gamma\psi$. By the fixed point theorem for complete lattices, there is a $\phi \in K$ such that $\phi = \Gamma\phi$. In particular, if E is the set of functions ϕ from K that are excessive or superharmonic in the sense $\phi \geq \Gamma\phi$, then

$$V(\mu) = \inf_{\phi \in E} \phi(\mu)$$

is in K and is a fixed point of Γ. The set E is not empty, because ϕ defined by $\phi(\mu) = \|\mu\| \cdot \sup|k|$ belongs to it, and by Lemma (6), is itself the largest fixed point of Γ in K! To show $V \geq \Gamma V$, we note that

$$\Gamma V = \Gamma \inf_{\phi \in E} \phi(\mu) \leq \Gamma\psi \leq \psi \quad \text{for any } \psi \in E$$

$$(\Gamma V)(\mu) \leq \inf_{\psi \in E} \psi(\mu) = V(\mu)$$

Since the value function S belongs to Lip_κ, it follows that $V = S$.

11. Examples

It is known that the infinite-dimensional Zakai equation can have a unique solution describable in finite-dimensional terms. In such cases our problem also reduces to one of ordinary analysis. For an example with nonlinear drift we pick the system of one-dimension:

(17)
$$dx_t = \tanh x_t dt + dw_t$$

$$dy_t = x_t dt + db_t$$

As in Ref. 10 we calculate explicitly that for Borel $A \subseteq R$

$$\sigma_t\mu(A) = \int_R \mu(dx) \text{sech}\, x\, E 1_{x+w_t \in A} \cosh(x+w_t)$$

$$\exp\left\{\int_0^t (x+w_s)dy_s - \frac{1}{2}\int_0^t (x+w_s)^2 ds - \frac{t}{2}\right\}$$

$$= \int_R \mu(dx) \int_A dz\, \text{sech}\, x \cosh z \exp\left\{-t-m'v + \frac{1}{2}v'Rv - \frac{x^2}{2} - \frac{(z-x\,\text{sech}\, t-\theta_t)^2}{2\tanh t}\right\}\left\{\frac{1+\tanh t}{2\pi\tanh t}\right\}^{\frac{1}{2}}$$

where

$$v' = (0,1,-1)$$

$$m' = \left\{xe^{-t},\ 0,\ \int_0^t e^{-s}ysds\right\}$$

$$d\theta_t = -(\tanh t)\theta_t dt + \tanh t\ dy_t \quad , \quad \theta_0 = 0$$

$$\dot{R} = Y + AR + RA' \quad , \quad R(0) = 0$$

$$\dot{m} = Am \quad , \quad m_0 = (x,0,0)$$

$$A = \begin{bmatrix} -1 & 0 & 0 \\ 0 & 0 & 0 \\ y_t & 0 & 0 \end{bmatrix} \quad , \quad Y = \begin{bmatrix} 1 & y_t & 0 \\ y_t & y_t^2 & 0 \\ 0 & 0 & 0 \end{bmatrix}$$

The "statistics" in m, R, and θ can be combined into a single vector ξ_t, and a function $\hat{k}$ defined, so that

$$\int_R \hat{k}(t,\xi_t,x)\mu(dx) = (k,\sigma_t\mu)$$

$$d\xi_t = F_t\xi_t dt + g_t(\xi_t)dy_t$$

where F_t is a matrix and g_t a vector, neither depending on y. Now introduce an operator $\mathbf{B}$ by

$$\mathbf{B} = \frac{\partial}{\partial t} + \frac{1}{2}\sum_{i,j} g_i g_j \frac{\partial^2}{\partial\xi_i\partial\xi_j} + \xi' F_t \nabla$$

and pose the QVI: to find a function $u(t,\xi,\mu)$ such that

$$u(t,\xi,\mu) \geq \int_R \hat{k}(t,\xi,x)\mu(dx) = (\hat{k},\mu)$$

$$\mathbf{B}u \leq 0$$

$$\mathbf{B}u[u-(\hat{k},\mu)] = 0$$

It can be verified that this QVI is exactly equivalent to finding the best admissible stopping time for (17). In particular, the time t functions here as another nonrandom statistic, and

$$S(\mu) = \sup_\tau E(k,\sigma_\tau\mu) = u(0,0,\mu)$$

with the sup achieved by $\inf t\colon u(t,\xi_t,\mu) = (k,\sigma_t\mu)$.

The reader is invited to construct a similar finite-dimensional reduction for the drift

$$z + e^{-z^2} / \int_{-\infty}^{z} e^{-u^2} du \, .$$

12. Acknowledgements

We note with pleasure the value of discussions with M.H.A. Davis, G. Mazziotto, and L. A. Shepp, and the stimulating influence of the preprint[9] by Davis and Kohlmann. W. H. Fleming suggested the invitation that led to presentation of this material at the IFIP Working Conference on Recent Advances in Filtering and Optimization, Cocoyoc, Mexico, February 1-6, 1982.

13. Appendix: Differential formula for $\sigma_t\mu$

The weak topology of $\mathbf{M}$ can be metrized as follows: let $\{\phi_n\}$ be a countable set of uniformly continuous functions dense in the bounded uniformly continuous functions on R^d, and write for a metric on $\mathbf{M}$

$$d(\mu,\nu) = \sum_{n=1}^{\infty} w_n \frac{|(\phi_n,\mu) - (\phi_n,\nu)|}{\|\phi_n\|} \quad , \quad \mu,\nu \in \mathbf{M}$$

with $\{w_n\}$ a sequence of weights to be chosen. There is no loss of generality in supposing that the ϕ_n are all C^2 functions of norm 1 in the domain of $\mathbf{A}$. Furthermore we can choose the weights w_n to decrease so fast that

$$\sum_{n=1}^{\infty} w_n 2^{4n} < \infty$$

When the coefficients in $\mathbf{A}$ are bounded, we can further adjust the w_n so that $\sum_{n=1}^{\infty} w_n |\mathbf{A}\phi_n(x)|$ defines a bounded function.

Put $\hat{h}_t = (h,\sigma_t\mu)/(1,\sigma_t\mu)$. If w_t were really the observed function y_t, then $\hat{h}$ would be $E\{h(x_t)|\mathbf{Y}_t\}$. We have the relation

$$\|\sigma_t\mu\| = (1,\sigma_t\mu) = \|\mu\| + \int_0^t q(\hat{h},w)_s \hat{h}_s dw_s$$

where q is the Girsanov functional

$$q(f,g)_t = \exp\left\{\int_0^t f_s dg_s - \frac{1}{2}\int_0^t |f_s|^2 ds\right\}$$

By the Kallianpur-Striebel formula, $\sigma_t\mu$ can be written in the form

$$\sigma_t\mu(A) = \int \mu(dx) E 1_{x+W_t \in A}\, q(a^{-1}b,W)_t q(h,w)_t$$

where E represents integration over an independent Brownian motion W, and $a^{-1}b = (a^{-1}b)(x+W)$, $h = h(x+W)$. Further, assuming $a^{-1}b$ is of at most linear growth, $q(^{-1}b,W)_t$ is in L_p for some $p > 1$, and $Eq(a^{-1},b,W)_t = 1$. Then by Hölder

$$\sigma_t\mu(A) \le \int \mu(dx) E^{(p-1)/p} 1_{x+W_t \in A} q(a^{-1}b,W)_t E^{1/p} q^p(h,w)_t q(a^{-1}b,W)_t$$

$$\le \int \mu(dx) E^{(p-1)/p^2} 1_{x+W_t \in A} E^{(p-1)/p^2} q^p(a^{-1}b,W)_t E^{1/p} q^p(h,w)_t q(a^{-1}b,W)_t$$

The first E does not involve w, the second is bounded on t-intervals, and the third has mean

$$E^w E^W q(a^{-1}b,W)_t q^p(h,w)_t = Eq(a^{-1}b,W)_t q(ph,w) e^{\frac{p^2-1}{2}\int_0^t |h|^2 ds}$$

also bounded on intervals. Hence there is an increasing sequence of w-sets A_N, of probabilities increasing to 1 as $N \to \infty$, and for each N a class of compact sets $\{K_\epsilon^N \subseteq R^d, \epsilon > 0\}$, such that on A_N

(i) $\sup_{0 \le s \le t} \|\sigma_s\mu\| \le N$

(ii) $\sup_{0 \le s \le t} Eq(a^{-1}b,W)_s q^p(h,w)_s \le N$

(iii) $\sup_{0 \le s \le t} \sigma_s\mu(R^d - K_\epsilon^N) < \epsilon$

Thus on A_N the orbit $\{\sigma_s\mu,\ 0 \le s \le t\}$ remains in a weakly compact set K^{Nt} of positive measures.

As usual, we consider a partition $\pi = \{0 = t_0 < t_1 < \ldots < t_n = t\}$ of the interval $[0,t]$; setting $mesh(\pi) = \max_i |t_{i+1} - t_i|$, we start with the identity

$$F(t,\sigma_t\mu) - F(0,\mu) = \sum_{i=0}^{n-1} [F(t_{i+1},\mu_{i+1}) - F(t_i,\mu_i)] = \sum_{i=0}^{n-1} \Delta F^i$$

where $\mu_i = \sigma_{t_i}\mu$. Next, by the ordinary mean value theorem, for some $\theta_i \in [0,1]$

$$\Delta F^i = F(t_{i+1},\mu_{i+1}) - F(t_i,\mu_{i+1}) + F(t_i,\mu_{i+1}) - F(t_i,\mu_i)$$

$$= F_1(t_i,\mu_{i+1})\Delta t_i + [F_1(t_i+\theta_i\Delta t_i,\mu_{i+1}) - F_1(t_i+\mu_{i+1})] + [F(t_i,\mu_{i+1}) - F(t_i,\mu_i)]$$

Except on sets of arbitrarily low probability the orbit s, $\sigma_s\mu$ stays in a compact set, so a standard uniform continuity argument shows that the sum over $t_i \in \pi$ of the second term on the right is $o(1)$ in pr. We turn then to the third term.

By the calculus theorem for Fréchet differentials

$$F(t_i,\mu_{+1}) - F(t_i,\mu_i) = \int_0^1 F_2(t_i,\mu_i+s\,\Delta\mu_i)[\Delta\mu_i]ds$$

$$= \int_0^1 \left\{F_2(t_i,\mu_i)[\Delta\mu_i] + \int_0^s F_{22}(t_i,\mu_i+u\,\Delta\mu_i)[\Delta\mu_i,\Delta\mu_i]du\right\}ds$$

$$= F_2(t_i,\mu_i)[\Delta\mu_i] + \frac{1}{2}F_{22}(t_i,\mu_i)[\Delta\mu_i,\Delta\mu_i]$$

$$+ \int_0^1\int_0^s \left\{F_{22}(t_i,\mu_i+u\,\Delta\mu_i) - F_{22}(t_i,\mu_i)\right\}[\Delta\mu_i,\Delta\mu_i]duds$$

For some $s_i \in [0,1]$ the last term is

$$\int_0^{s_i} \left\{F_{22}(t_i,\mu_i+u\,\Delta\mu_i) - F_{22}(t_i,\mu_i)\right\}[\Delta\mu_i,\Delta\mu_i]du$$

which in turn is equal, for some $u_i \in [0,s_i]$, to

$$s_i\left\{F_{22}(t_i,\mu_i+u_i\Delta\mu_i) - F_{22}(t_i,\mu_i)\right\}[\Delta\mu_i,\Delta\mu_i]\,.$$

Thus we have

$$F(t_i,\mu_{i+1}) - F(t_i,\mu_i) = F_2(t_i,\mu_i)[\Delta\mu_i] + \frac{1}{2}F_{22}(t_i,\mu_i)[\Delta\mu_i,\Delta\mu_i]$$

$$+ s_i\{F_{22}(t_i,\mu_i+u_i\Delta\mu_i) - F_{22}(t_i,\mu_i)\}[\Delta\mu_i,\Delta\mu_i] \tag{18}$$

We claim that the sum over $t_i \in \pi$ of the third term on the right is $o(1)$ in pr. Thinking of F_{22} as a map of $[0,1]\times M^+$ into the linear vector space of bilinear forms B on $M\times M$, such that

$$|B[\mu,\nu]| \leq \|B\| d(\mu,0)d(\nu,0)$$

$$\|B\| = \sup_{\mu,\nu} \frac{|B[\mu,\nu]|}{d(\mu,0)d(\nu,0)}$$

we find that with $\Delta F_{22}^i = F_{22}(t_i,\mu_i+u_i\Delta_i) - F_{22}(t_i,\mu_i)$

$$\sum_{t_i\in\pi} s_i F_{22}^i[\Delta\mu_i,\Delta\mu_i] \leq \max_{t_i\in\pi} \|\Delta F_{22}^i\| \sum_{t_i\in\pi} d^2(0,\Delta\mu_i)$$

For $a \in [0,1]$ the metric d has the homogeneity property

(19) $$d(\mu, a\mu + (1-a)\nu) = (1-a)d(\mu,\nu)$$

To proceed further we need a series of technical results:

(20) Lemma: $E\left(\int_s^{s+\epsilon} (\phi_n h, \sigma_u \mu) dw_u\right)^4 \leq 0(\epsilon^2)$ uniformly in n and $0 \leq s \leq t$, in an interval.

Proof: First estimate $\|\sigma_s \mu\|^4$ by noting that

$$\|\sigma_s \mu\| = \|\mu\| + \int_0^s q(\hat{h}, w)\hat{h} dw$$

$$\|\sigma_s \mu\|^4 \leq 8\|\mu\|^4 + 8\left(\int_0^s q(\hat{h}, w)\hat{h} dw\right)^4$$

$$q^4(\hat{h}, w)_s = q(4\hat{h}, w)_s \exp\left\{\frac{15}{2}\int_0^s \hat{h}^2 du\right\}$$

$$Eq^4(\hat{h}, w)_s \leq e^{Ks} \quad , \quad K = \frac{15}{2} \sup |h|^2$$

We can now use an inequality of Skorokhod[11] to conclude that

$$E\left(\int_0^s q(\hat{h}, w)\hat{h} dw\right)^4 \leq 36s \int_0^s Eq^4(\hat{h}, w)_u \hat{h}_u^4 du$$

$$\leq const.\ s(e^{Ks} - 1)$$

Hence

$$E\|\sigma_s \mu\|^4 \leq 8\|\mu\|^4 + const.\ s(e^{Ks} - 1)$$

Then using Skorokhod's result again, we see that

$$E\left(\int_s^{s+\epsilon} (\phi_n h, \sigma_s \mu) dw_s\right)^4 \leq const.\ \epsilon \int_s^{s+\epsilon} E\|\sigma_s \mu\|^4 ds$$

$$\leq const.\ \epsilon \int_s^{s+\epsilon} [4\|\mu\|^4 + const.\ u(e^{Ku} - 1)] du$$

$$\leq 0(\epsilon^2) \text{ uniformly in } n \text{ and } s \in [0,t].$$

In the next two lemmas we use the abbreviations $\Delta\mu_i = \sigma_{t_{i+1}}\mu - \sigma_{t_i}\mu$, and

$$d_i = d(0, \Delta\mu_i) = \sum_{n=1}^{\infty} w_n \left| \int_{t_i}^{t_{i+1}} (A\phi_n, \sigma_s \mu) ds + \int_{t_i}^{t_{i+1}} (\phi_n h, \sigma_s \mu) dw_s \right|$$

Thus d_i is the distance, in our special metric for the weak topology of measures on R^d, from $\sigma_{t_{i+1}}\mu$ to $\sigma_{t_i}\mu$, $t_{i+1} > t_i$.

(21) Lemma: $E \sum_{t_i \in p} d_i^2 \leq o(1) + 0(t)$ as $mesh(\pi) \to 0$

Proof: We have

$$d_i = \sum_{n=1}^{\infty} w_n |(\phi_n, \Delta\mu_i)|$$

$$\leq \int_{t_i}^{t_{i+1}} \sum_{n=1}^{\infty} w_n |(A\phi_n, \sigma_s\mu)| ds + \sum_{n=1}^{\infty} w_n \left| \int_{t_i}^{t_{i+1}} (\phi_n h, \sigma_s\mu) dw_s \right|$$

Use $(a+b)^2 \leq 2(a^2+b^2)$ and Schwarz's inequality to get

$$\sum_{t_i \in \pi} \leq 2 \left(sup \sum_{n=1}^{\infty} w_n \; |A\phi_n| \right)^2 \sum_{t_i \in \pi} \left(\int_{t_i}^{t_{i+1}} \| \sigma_s \mu \| ds \right)^2 + 2 \left(\sum_{n=1}^{\infty} w_n \right) \sum_{m=1}^{\infty} w_m \sum_{t \in \pi} \left| \int_{t_i}^{t_{i+1}} (\phi_m h, \sigma_s \mu) dws \right|^2$$

The Lebesgue integral has quadratic variation zero, and the last sum has mean at most $t \; sup \, |h|^2$.

(22) Lemma: $\max_{t_i \in \pi} d_i \to 0$ in probability, as $mesh(\pi) \to 0$

Proof: Write $d_i = \lambda_i + M_i$ where λ_i is the Lebesgue integral part, and M_i the sum of stochastic integrals. Evidently, under our choice of ϕ_n, $E\lambda_i^2 \leq const. \; (t_{i+1}-t_i)^2$, and

$$Pr\{|\lambda_i| > \varrho\} \leq \varrho^{-2} \, const. \; (t_{i+1}-t_i)^2$$

$$Pr\{\max_i |\lambda_i| > \varrho\} \leq \varrho^{-2} \, const. \sum_i (t_{i+1}-t_i)^2$$

$$\leq \varrho^{-2} \, const. \; t \; mesh(\pi)$$

Hence $\max_i |\lambda_i|$ goes to zero in pr, as π refines. Similarly, using

$$|M_i| \leq \sum_{n=1}^{\infty} w_n \left| \int_{t_i}^{t_{i+1}} (\phi_n h, \sigma_s \mu) dw_s \right| \triangleq \sum_{n=1}^{\infty} w_n x_{in}$$

we find that by Lemma (20)

$$Pr\{x_{in} > \varrho\} \leq \varrho^{-4} E |x_{in}|^4$$

$$Pr\{w_n x_n > 2^{-n}\varrho\} \leq \varrho^{-4} 2^{4n} w_n^4 E |x_{in}|^4$$

$$Pr\left\{ \sum_{n=1}^{\infty} w_n x_n > \varrho \right\} \leq \varrho^{-4} \sum_{n=1}^{\infty} E |x_{in}|^4 2^{4n} w_n^4$$

$$Pr\{\max_i |M_i| > \varrho\} \leq \varrho^{-4} \sum \sum_{n=1}^{\infty} E |x_{in}|^4 2^{4n} w_n^4$$

$$\leq \varrho^{-4} \, const. \sum_i (t_{i+1}-t_i)^2 \sum_{n=1}^{\infty} w_n^4 2^{4n}$$

$$\leq \varrho^{-4} \, const. \; t \; mesh(\pi)$$

so that $\max_i |M_i|$ goes to zero in pr.

On $[0,t] \times K^{N_t}$, F_{22} is uniformly continuous, and there is a modulus ω_N, a function such that using (19)

$$\|F_{22}(t_i,\mu_i+u_i\Delta\mu_i) - F(t_i,\mu_i)\| \leq \omega_N(u_i d_i)$$

whence on A_N

$$|\Delta F^i_{22}[\Delta\mu_i,\Delta\mu_i]| \leq \omega_N(d_i)d_i^2$$

and

$$\sum_{t_i \in \pi} |\Delta F^i_{22}[\Delta\mu_i,\Delta\mu_i]| \leq \omega_N\left(\max_{t_i\in\pi} d_i\right) \sum_{t_i\in\pi} d_i^2$$

By (22) and (21) resp., the maximum converges to zero in probability, and the sum $\Sigma\, d_i^2$ is bounded in the mean *uniformly* in π. Since $\omega_N(u)$ decreases to zero with u it follows that

$$1_{A_N} \sum_{t_i\in\pi} |\Delta F^i_{22}[\Delta\mu_i,\Delta\mu_i]|$$

goes to zero in pr.; but $P\{A_N\} \to 1$, so we can drop the 1_{A_N}. This completes proof that the third term on the right contributes $o(1)$ in pr. as $mesh(\pi) \to 0$.

It remains to deal with the first two terms on the right of (18). If F_2 is represented by the function f_2 in the domain of A with $Af_2 \in C_b$ as assumed then with $f_2^s = f_2(s,\sigma_s\mu)$, $f_2^i = f_2^{t_i}$, the first of these terms can be written as

$$F_2(t_i,\mu_i)[\Delta\mu_i] = \int_{t_i}^{t_{i+1}} (\mathrm{A}f_2^s,\sigma_s\mu)ds + \int_{t_i}^{t_{i+1}} (hf_2^s,\sigma_s\mu)dw_s$$

$$- \int_{t_i}^{t_{i+1}} (\mathrm{A}f_2^s - Af_2^i,\sigma_s\mu)ds - \int_{t_i}^{t_{i+1}} (hf_2^s - hf_2^i,\sigma_s\mu)dw_s$$

Methods exactly like those already used show that the sums on $t_i \in \pi$ of the absolute values of the third and fourth terms on the right go to zero in pr. as $mesh(\pi) \to 0$. The second term can be written out in the form:

$$
\begin{aligned}
F_{22}(t_i\mu_i)[\Delta\mu_i,\Delta\mu_i] &= F_{22}(t_i,\mu_i)[h\,\sigma_{t_i}\mu, h\,\sigma_{t_i}\mu]\Delta t_i \\
&+ F_{22}(t_i,\mu_i)[h\,\sigma_{t_i}\mu, h\,\sigma_{t_i}\mu]((\Delta w_{t_i})^2 - \Delta t_i) \\
&+ 2F_{22}(t_i,\mu_i)\left[\int_{t_i}^{t_{i+1}} h(\sigma_s\mu - \sigma_{t_i}\mu)dw_s, h\,\sigma_{t_i}\mu\right]\Delta w_{t_i} \\
&+ F_{22}(t_i,\mu_i)\left[\int_{t_i}^{t_{i+1}} h(\sigma_s\mu - \sigma_{t_i}\mu)dw_s\, ,\, \int_{t_i}^{t_{i+1}} h(\sigma_s\mu - \sigma_{t_i}\mu)dws\right] \\
&+ 2F_{22}(t_i,\mu_i)\left[\int_{t_i}^{t_{i+1}} h\,\sigma_s\mu dw, \alpha_i\right] + F_{22}(t,\mu_i)[\alpha_i,\alpha_i]
\end{aligned}
\tag{23}
$$

where α_i is the signed measure defined by

(24) $$(\alpha_i,\phi_n) = \int_{t_i}^{t_{i+1}} (A\phi_n,\sigma_s\mu)ds$$

and $\int_{t_i}^{t_{i+1}} h\,\sigma_s\mu dw_s$ is that whose value at a set A is

$$\int_{t_i}^{t_{i+1}} \int_{x\in A} \sum_{j=1}^{n} h_j(x)(\sigma_s\mu)(dx)dw_j(s) \qquad n = \dim h$$

That (24) defines α_i as a signed measure rather than just a linear functional on a dense subspace can be seen from

$$\Delta\mu_i = \alpha + \int_{t_i}^{t_{i+1}} h\,\sigma_s\mu dw_s\ .$$

Again, methods like those used before show that the sums on $t_i \in \pi$ of the absolute values of all the terms of $F_{22}(t_i,\mu_i)[\Delta\mu_i,\Delta\mu_i]$ in (23) except the first go to zero in pr. as $mesh(\pi) \to 0$.

References

1. B. Grigelionis, On sufficient statistics for optimal stopping problems of stochastic processes, Lietuvos Math. Rink., Vol. XI, No. 3, (1971), pp. 529-533.
2. J. Szpirglas and G. Mazziotto, Modèle général de filtrage non linéaire et équations differentielles stochastiques associées, Ann. Inst. H. Poincarée, Vol. XV, No. 2, (1979), pp. 147-173.
3. _____, Théorème de séparation pour l'arrêt optimal, Sém. Prob. XIII, Lecture Notes in Maths. No. 721. Springer Verlag 1979.
4. _____, Théoreme de séparation pour le contrôle impulsionnel, Note technique NT/PAA/ATR/MTI/246, CNET, September 1980.
5. _____, Separation theorems for optimal stopping and impulse control, ibid.
6. M. H. A. Davis, private communication.
7. W. H. Fleming and E. Pardoux, Optimal control for partially observed diffusions, SIAM J. on Control and Optimization, Vol. 20, No. 2, March 1982, pp. 261-285.
8. M. H. A. Davis, Some current issues in stochastic control theory, preprint of paper for *Stochastics,* Special Issue on Stochastic Optimization (ed., M. H. A. Dempster), based on a lecture at the Taskforce Workshop on Stochastic Optimization, IIASA, December, 1980.
9. M. H. A. Davis, and M. Kohlmann, On the nonlinear semigroup of stochastic control under partial observations, unfinished manuscript, 1981.
10. P. A. Meyer, Probability and potentials, Blaisdell, Waltham, 1966.
11. V. E. Beneš, Exact finite dimensional filters for certain diffusions with nonlinear drift, Stochastics, Vol. 5, (1981), pp. 65-92.
12. A. V. Skorokhod, Studies in the theory of random processes, Addision-Wesley, Reading, Mass., 1965, p. 23, Theorem 4.

OPTIMAL CONTROL OF PARTIALLY OBSERVED DIFFUSIONS

A. Bensoussan
University Paris-Dauphine and INRIA

INTRODUCTION.

We discuss in this paper the stochastic maximum principle and the dynamic programming approaches of the problem of stochastic control under partial observation. We formulate the problem as a problem of stochastic control for a stochastic partial differential equation, with full observation. The stochastic PDE is the Zakaï equation. A formal approach of the stochastic maximum principle has been already given by H. Kwakernaak [7]. Our approach is rigorous. We also present an analytic problem which is a surrogate of Dynamic Programming. The solution of the analytic problem coïncides with the value function of the problem. We present here the results and some ideas of the proofs. A detailed version of the paper will appear elsewhere.

1. STUDY OF A STOCHASTIC PARTIAL DIFFERENTIAL EQUATION.

1.1. Notation.

Let :

(1.1) $\quad g(x,v) : R^n \times R^k \to R^n$, bounded continuous

$\mathcal{U}_{ad}$ will be a compact non empty subset of R^k, (here after the set of control values)

(1.2) $\quad \sigma(x) : R^n \to \mathcal{L}(R^n;R^n)$ bounded uniformly continuous ;

Setting $a_{ij} = \frac{1}{2}(\sigma\sigma^*)_{ij}$ we assume that :

$$\sum_{i,j} a_{ij}\,\xi_i\xi_j \geq \alpha|\xi|^2 \ , \quad \forall\xi \in R^n, \quad \alpha > 0$$

We will assume the additional regularity :

(1.3) $$\frac{\partial g_i}{\partial x_i} \in L^\infty \ , \quad \frac{\partial a_{ij}}{\partial x_j} \in L^\infty \ , \quad \frac{\partial^2 a_{ij}}{\partial x_i \partial x_j} \in L^\infty$$

We define the operator :

$$(1.4) \qquad A = - \sum_{i,j} a_{ij} \frac{\partial^2}{\partial x_i \partial x_j} - \sum_i g_i \frac{\partial}{\partial x_i}$$

which we may written under the divergence form

$$A = - \sum_{i,j} \frac{\partial}{\partial x_i} a_{ij} \frac{\partial}{\partial x_j} + \sum_i a_i \frac{\partial}{\partial x_i}$$

where

$$(1.5) \qquad a_i = -g_i + \sum_j \frac{\partial}{\partial x_j} a_{ij}$$

We also consider the adjoint :

$$(1.6) \qquad A^* = - \sum_{i,j} \frac{\partial^2}{\partial x_i \partial x_j}(a_{ij}\cdot) + \sum_i \frac{\partial}{\partial x_i} (g_i\cdot)$$

$$= - \sum_{i,j} \frac{\partial}{\partial x_i} a_{ij} \frac{\partial}{\partial x_j} - \sum_i a_i \frac{\partial}{\partial x_i} - \sum_i \frac{\partial a_i}{\partial x_i}$$

In fact, since g_i, hence a_i depends on a parameter w, we will index the operators A and A^* by v, and write

$$A^v \qquad A^{*v} .$$

Let h(x) such that :

$$(1.7) \qquad h : R^n \to R^d \text{ , bounded, } h \in W^{2,\infty}(R^n)$$

Let (Ω,M,P) be a probability space, on which we can construct a standardised d dimensional Wiener process, denoted by y(t). We will define

$$Y^t = (y(s), s \leq t)$$

We write :

$$(1.8) \qquad H = L^2(R^n), \quad V = H^1(R^n) , \quad V' = H^{-1}(R^n)$$

(1.9) $L^2_Y(o,T;V)$ = subspace of

$L^2((o,T) \times \Omega ; dt \times dP;V)$ of processes z(t) such that a.e.t,
$z(t) \in L^2(\Omega,Y^t;P;V)$.

In (1.9) we can take $T = \infty$, and replace V by any Hilbert space (in particular we will use it with V replaced by R^k)

1.2. Admissible controls. State equation.

The set of admissible controls is defined as follows

$$(1.10) \qquad v(.) \in L^2_Y(o,T;R^k) \forall T \text{ finite},$$

$$\text{a.e} \qquad v(t) \in \mathcal{U}_{ad}, \text{ a.s.}$$

Let

$$(1.11) \qquad \pi \in L^2(R^n), \pi \geq 0$$

For a given control we want to solve the stochastic PDE

$$(1.12) \qquad \begin{vmatrix} dp + A^{*v(.)} p \, dt = p \, h \, . \, dy \\ p(o) = \pi \quad , \end{vmatrix}$$

which we will call the Zakaï equation, following the common practice. It is convenient to define :

$$(1.13) \qquad A_o = - \sum_{i,j} \frac{\partial}{\partial x_i} a_{ij} \frac{\partial}{\partial x_j}$$

which does not depend on the control, and

$$(1.14) \qquad B^v = + \sum_i a_i(x,v) \frac{\partial}{\partial x_i} + \sum_i \frac{\partial}{\partial x_i} a_i(x,v)$$

Therefore (1.12) can be written as follows :

$$(1.15) \qquad \begin{vmatrix} dp + A_o p \, dt = B^{v(.)} p \, dt + p \, h \, . \, dy \\ p(o) = \pi \end{vmatrix}$$

We state the following result, which is a variant of the general results of E. Pardoux [9] (see also A. Bensoussan [2]).

Theorem 1.1. : We assume (1.1), (1.2), (1.3), (1.11), then for any admissible control v(.) (cf. (1.10)), there exists one and only one solution of (1.15) in the following functional space :

$$(1.16) \qquad p \in L^2_Y(o,T;V) \cap L^2(\Omega,M,P;C(o,T;H)), \qquad \forall T \text{ finite.}$$

Moreover one has :

$$(1.17) \qquad p \geq 0$$

□

1.3. Additional properties.

Let us notice that for $\phi \in H^1(R^n)$, we have :

$$(B^v\phi,\phi) = \frac{1}{2}\int_{R^n} \phi^2 \sum_i \frac{\partial a_i}{\partial x_i}$$

Setting :

(1.18) $$\bar{a}(x,v) = \sum_i \frac{\partial a_i}{\partial x_i}(x,v) + |h(x)|^2$$

we can write the following energy equality :

(1.19) $$E\,|p(T)|^2 + 2E\int_0^T \langle A_0 p,p\rangle dt = E\int_0^T\int_{R^n} \bar{a}(x,v(t))p^2(x,t)dxdt + |\pi|^2$$

also

(1.20) $$E\,e^{-2\gamma T}\,|p(T)|^2 + 2E\int_0^T e^{-2\gamma t}\,\langle A_0 p,p\rangle\,dt$$
$$+ E\int_0^T\int_{R^n} (2\gamma - \bar{a}(x,v(t)))p^2\,e^{-2\gamma t}dx\,dt = |\pi|^2$$

For a convenient choice of γ, we deduce :

(1.21) $$E\int_0^\infty e^{-2\gamma t}\,\|p(t)\|^2\,dt < \infty\ .$$

Consider next the case of a constant control $v(.) \equiv v$. Let us write $p^v(t)$ to be the solution of (1.15) at time t, emphazing the dependence with respect to v and to the initial condition. The map :

(1.22) $$\pi \to p_\pi^v(t) \in \mathcal{L}(H\ ;\ L^2(\Omega,M,P;H))$$

and from (1.20) we have :

(1.23) $$E\,|p_\pi^v(t)|^2 \le e^{2\gamma t}\,|\pi|^2$$

Let us consider the semi-group $T^v(t)$ from H into itself, which is defined by solving the Cauchy problem

(1.24) $$\left|\begin{array}{l} \frac{dz}{dt} + (A_0 - B^v)z = 0 \\ z(o) = \pi \\ z(t) = T^v(t)\,\pi \end{array}\right.$$

We have the following :

Lemma 1.1. : The following representation formula holds :

(1.25) $$p_\pi^v(t) = T^v(t)\pi + \int_0^t T^v(t - s)(p(s)h) \, . \, dy(s)$$

□

We can consider (1.25) as a linear integral equation, whose solution is searched in $C(o,T;L^2(\Omega,M,P;H))$, for T fixed. It has a unique solution, since taking $\pi = 0$, we deduce the estimate :

$$E\,|p(t)|^2 \le C_T\, E \int_0^t |p(s)|^2 ds$$

and by Gronwall's inequality, it follows p=0.

Let us set :

$$y_\theta(s) = y(s + \theta) - y(\theta)$$

which is a standardised Wiener process with respect to $Y^{s+\theta} = \tilde{Y}^s$. In addition, the process $y_\theta(s)$ is independent of Y^θ.

Consider now the integral equation analogue to (1.25) (1)

(1.26) $$q_\nu(t) = T(t)\nu + \int_0^t T(t-s)(q(s)h).dy_\theta(s)$$

which has a unique solution in say $C(o,T;L^2(\Omega,M,P;H))$. We can claim that the random variable $q_\nu(t\,;\omega)$ with values in H is Y^θ independent. Moreover we have the following

Proposition 1.1. : We have the property :

(1.27) $$E\,[F(p_\pi^v(t + \theta)) \mid Y^\theta] = E\,[F(p_\nu^v(t))]_{\nu=p_\pi(\theta)}$$

and the process $p_\pi^v(t)$ is a Markov process with values in H.

□

1.4. Linear semi group.

Since $p_\pi^v(t)$ is a Markov process in H, we can define a linear semi-group operating on the Banach spaces :

B = space of Borel bounded functionals on H

C = space of uniformly continuous bounded functionals on H.

Setting :

(1.28) $$\Phi^v(t)(F)(\pi) = E\,[F(p_\pi^v(t))]$$

(1) We omit to write the index v.

for $F \in B$, then we have from Proposition 1.1, the semi group property :

(1.29) $\quad \Phi^v(t + s) = \Phi^v(t)\, \Phi^v(s)$

In addition we can assert that :

(1.30) $\quad \Phi^v(t) : C \to C.$

This follows easily from (1.22), (1.23). Now, since $p_\pi^v(t)$ depends linearly on π, it it useful to consider functionals F, which are not bounded, and rather have a linear growth.

To that extent, we introduce :

B_1 = space of Borel functionals on H, which have linear growth.

We put on B_1 the norm :

$$||F|| = \sup \frac{|F(\pi)|}{1 + |\pi|}$$

and B_1 is a Banach space.

Similarly, we define :

C_1 = subspace of B_1 of functionals F such that $\frac{F(\pi)}{1 + |\pi|} \in C$.

For $F \in B_1$ or C_1, we have :

$$|F(p_\pi^v(t))| \le ||F||\,(1 + |p_\pi^v(t)|)$$

$$|E\, F(p_\pi^v(t))| \le ||F||\,(1 + |\pi|\, e^{\gamma t})$$

therefore $\Phi^v(t) \in \mathscr{L}(B_1, B_1)$ or $\mathscr{L}(C_1, C_1)$ with norm

(1.31) $\quad ||\Phi^v(t)||_{(B_1;B_1)} \le e^{\gamma t}$

Hence $\Phi^v(t)$ is not a contraction on B_1 (we recall that γ does not depend on v). The semi group $\Phi^v(t)$ has also the following regularity property :

<u>Proposition 1.2.</u> : <u>If</u> F <u>satisfies</u>

(1.32) $\quad |F(\pi_1) - F(\pi_2)| \le ||F||_\delta\, |\pi_1 - \pi_2|^\delta \qquad 0 \le \delta \le 1$

<u>then one has</u> :

(1.33) $\quad |\Phi^v(t)(F)(\pi_1) - \Phi^v(t)(F)(\pi_2)| \le ||F||_\delta\, e^{\gamma\delta t}\, |\pi_1 - \pi_2|^\delta$

□

2. A STOCHASTIC CONTROL PROBLEM.

2.1. *Setting of the problem.*

Let for any $v \in \mathcal{U}_{ad}$

(2.1) $$f(x,v) \equiv f_v(x) \in H$$

$$|f_v|_H \leq C \text{ , independent of } v \text{ in } \mathcal{U}_{ad}$$

(2.2) $$\ell \in H$$

Consider an admissible control v(.) (cf. (1.10)) and let $p_\pi^{v(.)}(t)$ be the corresponding solution of (1.15). We define :

(2.3) $$J_\pi(v(.)) = E\,[\int_0^T (f_v(t), p_\pi^{v(.)}(t))dt + (\ell, p_\pi^{v(.)}(T))]$$

The integral to the right hand side of (2.3) is well defined.

2.2. *Preliminaries.*

We want to minimize the functional (2.3). The existence of an optimal control for (2.3) seems to be a very difficult problem. Introducing a concept of relaxed control, W. Fleming and E. Pardoux [6] were able to obtain an existence theorem. Our objective here is to derive a Maximum Principle. We follow an idea of H. Kwakernaak [7] who derived a Maximum Principle, working with the Kushner - Stratonovitch equation of the conditional probability. The treatment of Kwakernaak is formal, whereas our derivation is rigorous.

We will assume that :

(2.4) g(x,v) is C^2 in v and

$$\frac{\partial g_i}{\partial v_j}, \frac{\partial^2 g_i}{\partial v_j \partial v_\ell} \text{ are bounded}$$

$$i=1,\dots,n \qquad j,\ell=1-k$$

(2.5) f(x,v) is C^2 in v and

$$\int_{R^n} \left(\frac{\partial^2 f}{\partial v_i \partial v_j}(x,v)\right)^2 dx \leq \text{Constant independent of } v$$

$$\int_{R^n} \left(\frac{\partial f}{\partial v_i}(x,v)\right)^2 dx \leq \text{Constant independent of } v$$

We will set

$$f'_v \equiv \frac{\partial f}{\partial v}(x,v)$$

and f'_v can be considered as an element of H^k. We start with some preliminaries. Let u(.) be an optimal control. We denote by p(.) the corresponding state, which is the solution of

$$(2.6)\quad \left|\begin{array}{l} dp + A_o p\, dt = B^{u(.)} p\, dt + p\, h\, .\, dy \\ p(o) = \pi \\ p \in L^2_{Y,\gamma}(o,\infty;V) \cap L^2(\Omega,M,P;C(o,T;H)) \qquad \forall T \end{array}\right.$$

where we have denoted

$$L^2_{Y,\gamma}(o,\infty;V) = \{z \mid E \int_0^\infty e^{-2\gamma t} \|z(t)\|^2 dt < \infty \quad ,$$

$$z(t) \in L^2(\Omega, Y^t, P; V), \text{ a.e.} t\}$$

The fact that $p \in L^2_{Y,\gamma}(o,\infty;V)$ follows from the energy inequality (derived from (1.20))

$$2\, E \int_0^\infty e^{-2\gamma t} \langle A_o p, p\rangle\, dt + E \int_0^\infty \int_{R^n} (2\gamma - \bar{a}(x,u(t)))p^2 dx\, dt \leq |\pi|^2$$

Since π will not vary in this context, we write J(v(.)) for the functional (2.3). We consider it as a functional on the Hilbert space $L^2_Y(o,T;R^k)$, and we want to show that it is Gateaux-differentiable. Let $v(.) \in L^2_Y(o,T;R^k)$, we consider the equation :

$$(2.7)\quad dz + A_o z\, dt = B^{u(.)} z\, dt + \sum_i \frac{\partial}{\partial x_i} (a_{i,v}(x,u(t))v(t)p)dt + z\, h\, .\, dy$$

$$z(o) = 0$$

$$z \in L^2_Y(o,T;V) \cap L^2(\Omega,M,P;C(o,T;H)).$$

Provided that v(t) is bounded by a deterministic constant, then (2.7) has one and only one solution. In fact, we have the energy equality :

$$(2.8) \qquad E\, e^{-2\gamma T}\, |z(T)|^2 + 2E \int_0^T e^{-2\gamma t} \langle A_0 z, z\rangle \, dt +$$

$$+ E \int_0^T \int_{R^n} (2\gamma - \bar{a}(x,u(t)))z^2\, e^{-2\gamma t}\, dx\, dt =$$

$$= -2E \int_0^T \int_{R^n} e^{-2\gamma t} \sum_i a_{i,v}(x,u(t))v(t)p \frac{\partial z}{\partial x_i}\, dx\, dt$$

We then state the following :

Lemma 2.1 : We have the formula

$$(2.9) \qquad \frac{d}{d\theta} J(u(.) + \theta v(.))|_{\theta=o} = E \{ \int_0^T [(f'_{u(t)}\, v(t), p(t)) +$$

$$+ (f_{u(t)}, z(t))]\, dt + (\ell, z(T))\}$$

□

2.3. Maximum Principle.

Let $\phi \in L^2_\gamma(o,T;V')$. Consider ρ to be the solution of :

$$(2.10) \qquad d\rho + A_0\rho\, dt = B^{u(.)}\rho\, dt + \phi(t)dt + \rho h \,.\, dy$$

$$\rho(o) = 0$$

$$\rho \in L^2_\gamma(o,T;V) \cap C(o,T;L^2(\Omega,M;\mathring{P}))$$

From the energy equality :

$$(2.11) \qquad E\, e^{-2\gamma T}\, |\rho(T)|^2 + 2E \int_0^T e^{-2\gamma t} \langle A_0\rho, \rho\rangle\, dt +$$

$$+ E\int_0^T \int_{R^n} (2\gamma - \bar{a}(x,u(t)))\rho^2\, e^{-2\gamma t}\, dx\, dt =$$

$$= 2E \int_0^T \langle \rho(t), \phi(t)\rangle\, e^{-2\gamma t} dt$$

It follows that the map : $\phi \to \rho$

(2.12) is linear and continuous from

$L^2_\gamma(o,T;V')$ in $L^2_\gamma(o,T;V) \cap C(o,T;L^2(\Omega,M,\mathring{P}))$

Consider then the functional

$$\phi \to E\,\{\int_0^T (f_{u(t)},\rho(t))dt + (\ell,\rho(T))\}$$

it is linear and continuous on $L^2_Y(o,T;V')$.

Therefore, there exists one and only one element $\lambda \in L^2_Y(o,T;V)$ such that :

$$E\{\int_0^T (f_{u(t)},\rho(t))dt + (\ell,\rho(T))\} = E\int_0^T <\lambda(t),\phi(t)>dt \tag{2.13}$$

Applying this result to (2.7), we can write (2.9) as follows :

$$\frac{d}{d\theta} J(u(.) + \theta v(.))|_{\theta=0} = E\int_0^T [(f'_{u(t)}v(t),p(t)) + \tag{2.14}$$

$$+ (\lambda(t), \sum_i \frac{\partial}{\partial x_i} (a_{i,v}(x,u(t))v(t)p))]dt =$$

$$= E\int_0^T \sum_j v_j(t)\,[\int_{R^n} (\frac{\partial f}{\partial v_j}(x,u(t)) +$$

$$+ \sum_i \frac{\partial\lambda(x;t)}{\partial x_i}\frac{\partial g_i}{\partial v_j}(x,u(t)))p(t,x)dx]\,dt$$

where we have used the fact that :

$$\frac{\partial a_i}{\partial v_j} = -\frac{\partial g_i}{\partial v_j}$$

We can then state the following

Theorem 2.1 : We make the assumptions of Theorem 1.1 as well as (2.1), (2.2), (2.4), (2.5). We also assume that

$$\mathscr{U}_{ad} \text{ convex.} \tag{2.15}$$

Let u(.) be an optimal control for problem (1.15), (2.3) and p the corresponding trajectory, solution of (2.6). Then, there exists λ in $L^2_Y(o,T;V)$ such that the following condition holds :

$$\sum_{j=1}^{k}(v_j - u_j(t))\,[\int_{R^n} (\frac{\partial f}{\partial v_j}(x,u(t)) + \sum_{i=1}^{n}\frac{\partial\lambda(x;t)}{\partial x_i}\frac{\partial g_i}{\partial v_j}(x,u(t)))p(t;x)dx] \geq 0 \quad \forall v \in \mathscr{U}_{ad}, \quad \text{a.e.t, a.s.} \tag{2.16}$$

2.4. Equation for the adjoint process.

We derive here an equation for λ, the adjoint process. Let us introduce μ to be the solution of

$$(2.17)\qquad -\frac{\partial\mu}{\partial t} + A_o\mu + \sum_i \frac{\partial\mu}{\partial x_i}[a_i(x,u(t)) + 2\sum_j a_{ij}\frac{\partial}{\partial x_j}(y(t)\,.\,h)] +$$

$$+ \mu\,[\tfrac{1}{2}|h|^2 - \sum_{i,j} a_{ij}\frac{\partial}{\partial x_j}(y(t)\,.\,h)\frac{\partial}{\partial x_i}(y(t)\,.\,h) +$$

$$+ \sum_{i,j}\frac{\partial}{\partial x_i}(a_{ij}\frac{\partial}{\partial x_j}(y(t)\,.\,h)) - \sum_i a_i(x,u(t))\frac{\partial}{\partial x_i}(y(t)\,.\,h)] =$$

$$= f(x,u(t))\exp y(t)\,.\,h(x)$$

$$\mu(x,T) = \ell(x)\exp + y(T)\,.\,h(x)$$

For a.s. ω , (2.17) has one and only one solution in the functional space :

$$(2.18)\qquad \mu \in L^2(o,T;V)\ ,\ \frac{\partial\mu}{\partial t} \in L^2(o,T;V').$$

We will also set :

$$(2.19)\qquad \nu(x,t) = \mu(x,t)\exp - y(t)\,.\,h(x).$$

Of course, we can assert that

$$(2.20)\qquad \nu \in L^2(o,T;V) \cap C(o,T;H)\ \text{a.s.}$$

One has to be careful in taking the mathematical expectation. However we have the following result :

Lemma 2.2. : $\forall\phi \in H$ (deterministic), $\forall s \in [o,T]$ $(\phi,\nu(s)) \in L^2(\Omega,M,P)$ and there exists $\hat\nu(s) \in L^2(\Omega,Y^s,P;H)$ such that

$$(2.21)\qquad (\phi,\hat\nu(s)) = E\,[(\phi,\nu(s))\ |\ Y^s]\quad .$$

Moreover the adjoint process λ satisfies

$$(2.22)\qquad \lambda(s) = \hat\nu(s)\qquad \text{a.e. } s.\ \text{a.s.}$$

□

To proceed we will need an additional regularity property of $\hat\nu(s)$, namely we have :

Lemma 2.3. : The process $\hat\nu(s)$ satisfies

$$(2.23)\qquad E\,|\hat\nu(s)|_H^4 \le C\quad ,\ \forall s \in [o,T].$$

□

The previous result can be strenghtened as follows :

Lemma 2.4. : Assume that

(2.24) $$\frac{\partial a_{ij}}{\partial x_k} \in L^\infty$$

and

(2.25) $\ell \in V$.

Then the process $\hat{\nu}(s)$ satisfies

(2.26) $$E \|\hat{\nu}(s)\|_V^4 \leq C$$

□

We then deduce that :

Lemma 2.5.: Set

$$\hat{u}(x,t) = \hat{\nu}(x,t) \exp y(t) . h(x)$$

then one has

(2.27) $$E \|\hat{\mu}(t)\|_V^2 \leq C$$

(2.28) $$E^{Y^t} \frac{\partial \mu}{\partial t} \in L^\infty(o,T;L^2(\Omega,T;L^2(\Omega,M,P;V'))).$$

Lemma 2.6. : There exists $r \in L^2_Y(o,T;V'^d)$ such that

(2.29) $$m(t) = \hat{\mu}(t) - E\mu(o) - \int_0^t E^{Y^s} \frac{\partial \mu}{\partial s} ds = \int_0^t r(s)dy$$

We can now state the following □

Theorem 2.2. : We make the assumptions of Theorem 2.1, (2.24), (2.25). Then the adjoint process λ defined in Theorem 2.1 satisfies :

(2.30) $$\lambda \in L^2_Y(o,T;V) \cap L^\infty(o,T;L^4(\Omega,M,P;V))$$

$$\lambda(t) \exp y(t) . h \in L^2(\Omega,M,P;C(o,T;V')) -$$

$$- d\lambda + (A_0\lambda + \sum_i a_i (x,u(t)) \frac{\partial \lambda}{\partial x_i} +$$

$$+ \lambda|h|^2)dt = \lambda h . dy - \exp - y(t) . h\, r(t) . dy +$$

$$+ (f(x,u(t)) + \exp - y(t) . h\, r(t) . h)dt$$

$$\lambda(x,T) = \ell(x)$$

Moreover there exists one and only one pair (λ,r) with $r \in L_Y^2(o,T;V')$ such that (2.30) holds.

□

3. SEMI-GROUP ENVELOPE AND APPLICATIONS TO STOCHASTIC CONTROL WITH PARTIAL INFORMATION.

3.1. *Setting of the problem.*

We go back to the notation of section 1 and consider the family of semi-groups $\Phi^V(t)(F)$ on B_1 and C_1 defined in (1.28). Let f_v be as in (2.1). We identify it with the element of C_1,

$$f_v(\pi) = (f_v,\pi) \qquad \forall \pi \in H. \tag{3.1}$$

We take

$$\beta > \gamma \tag{3.2}$$

where γ has been chosen in §1.3.

We consider the following problem, called the problem of semi-group envelope. This problem considered in A. Bensoussan - M. Robin [3], is closely connected to the approach of M. Nisio [8], who introduced a non linear semi-group associated with stochastic control. As we shall see, the framework fits perfectly with the control problem for the Zakaï equation considered in section 2 (although, we consider here an infinite horizon version of the problem). For different semi-group approaches we refer to W. Fleming [5] and M.H.A. Davis, M. Kohlmann [4].

We introduce the set of functions $S(\pi)$,

$$\begin{aligned} & S \in C_1 \\ & S \leq \int_0^t e^{-\beta s}\,\Phi^V(s)f_v\,ds + e^{-\beta t}\,\Phi^V(t)S \qquad \forall t \geq 0 \end{aligned} \tag{3.3}$$

Our objective is to study the structure of the set (3.3).

3.2. *Preliminaries.*

We give here some useful additional properties of the semi-group $\Phi^V(t)(F)$.

Lemma 3.1. : We have the property

$$t \to \Phi^V(t)(F)(\pi) \quad C\,[o,\infty)\ , \qquad \forall \pi \in H, \quad \forall F \in C_1. \tag{3.4}$$

□

Lemma 3.1 and (3.2) imply in particular that the integral

(3.5) $$\int_0^\infty e^{-\beta t}\, \Phi^v(t) f_v \, dt \in C_1$$

Let h be a parameter which will tend to 0. We will define the family of operators

(3.6) $$T_h(F) = \underset{v\in\mathcal{U}_{ad}}{\text{Min}} \left[\int_0^h e^{-\beta t}\Phi^v(t) f_v \, dt + e^{-\beta h}\, \Phi^v(h)(F)\right] , \qquad F \in C_1.$$

We assume

(3.7) $v \to f_v$ is continuous from $\mathcal{U}_{ad}$ into H, and bounded.

Lemma 3.2. : The operator T_h maps C_1 into itself.

□

3.3. Approximation.

We solve the following equation :

(3.8) $$S_h = T_h(S_h) \quad , \qquad S_h \in C_1.$$

Lemma 3.3. : Equation (3.8) has one and only one solution.

□

Lemma 3.4. : The solution S_h is uniformly Lipschitz and

(3.9) $$|S_h(\pi) - S_h(\pi')| \leq \frac{C}{\beta - \gamma}|\pi - \pi'|$$

where C is the bound on $|f_v|_H$, $\forall v$.

□

We can then state the following :

Theorem 3.1. : We assume (1.1), (1.2), (1.3), (1.7), (3.2), (3.7). Then the set (3.3) is not empty and has a maximum element, which is moreover uniformly Lipschitz.

□

3.4. Interpretation of the maximum element.

Let us consider the functional (2.3) over an infinite horizon, with discount β, namely :

(3.10) $$J_\pi(v(.)) = E \int_0^\infty e^{-\beta t}(f_{v(t)}, p_\pi^{v(.)}(t))dt$$

where v(.) is an admissible control and $p_\pi^{v(.)}(t)$ is the solution of (1.15).

We denote by W the class of step processes adapted to Y^t, with values in $\mathcal{U}_{ad}$. More precisely if $v(.) \in W$, there exists a sequence :

$$\tau_o = 0 < \tau_1 < \ldots < \tau_n < \ldots$$

which is deterministic, increasing and convergent to $+\infty$, such that :

$$v(t;\omega) = v_n(\omega) \qquad , \qquad t \in [\tau_n, \tau_{n+1}).$$

and v_n is $Y^{\,n}$ measurable.

We have the following :

Lemma 3.5. : Let $\phi \in B$ and $v(.) \in W$. Then for $\tau_n \leq t < \tau_{n+1}$, one has

$$\text{(3.11)} \qquad E\,[\phi(p_\pi^{v(.)}(t)) \mid Y^{\tau_n}] = \Phi^{v_n}(t - \tau_n)\, \phi(p_\pi^{v(.)}(\tau_n)).$$

□

Define for $v(.) \in W$

$$\text{(3.12)} \qquad p_n = p_n^{v(.)} = p_\pi^{v(.)}(\tau_n)$$

$$\text{(3.13)} \qquad f_{h,v} = \int_0^h e^{-\beta t}\, \Phi^v(t) f_v \, dt$$

From Lemma 3.5 which can be extended to $\phi \in C_1$ we deduce

$$E \int_{\tau_n}^{\tau_{n+1}} e^{-\beta(t-\tau_n)} (f_{v_n}, p_\pi^{v(.)}(t))dt = E \int_0^{\tau_{n+1}-\tau_n} e^{-\beta t}\, \Phi^{v_n}(t) f_{v_n}(p_n)dt$$

$$= E(f_{\tau_{n+1}-\tau_n, v_n}, p_n)$$

hence finally the formula :

$$\text{(3.14)} \qquad J_\pi(v(.)) = E \sum_{n=0}^{\infty} e^{-\alpha\tau_n} (f_{\tau_{n+1}-\tau_n, v_n}, p_n) .$$

Let us consider the subset of W

$$W_h = \{v(.) \in W \mid \tau_n = nh\}$$

then we have the following :

Theorem 3.2. : The assumptions are those of Theorem 3.1. We then have :

(3.15) $$S_h(\pi) = \operatorname*{Inf}_{v(.)\in W_h} J_\pi(v(.)).$$

□

We finally obtain the interpretation of the maximum element still denoted by S, of the set (3.3).

We have :

Theorem 3.3. : The assumptions are those of Theorem 3.1 and (2.15). Then we have :

(3.16) $$S(\pi) = \operatorname*{Inf}_{v(.)} J_\pi(v(.)).$$

□

REFERENCES.

[1] A. Bensoussan, Filtrage Optimal des Systèmes Linéaires, Dunod, Paris, 1971.

[2] A. Bensoussan, Control of Stochastic Partial Differential Equations, in Distributed Parameter Systems, edited by W.H. Ray and D.G. Lainotis, Marcel Dekker, N.Y., 1978.

[3] A. Bensoussan ; M. Robin, On the Convergence of the Discrete Time Dynamic Programming Equation for General Semi groups, SIAM Control, to be published.

[4] M.H.A. Davis; M. Kohlmann, On the Non linear Semi-group of Stochastic Control under Partial Observations, submitted to SIAM Control.

[5] W.H. Fleming, Non linear Semi group for Controlled Partially Observed Diffusions, preprint.

[6] W.H. Fleming ; E. Pardoux, Optimal Control for Partially Observed Diffusions, submitted to SIAM Control.

[7] H. Kwakernaak, A Minimum Principle for Stochastic Control Problems with Output Feedback, Systems and Control Letters, Vol. 1, n°1 (July 1981).

[8] M. Nisio, On a Non linear Semi-group Attached to Stochastic Optimal Control, RIMS, Kyoto University, 13 (1976), 513-537.

[9] E. Pardoux, Stochastic Partial Differential Equations and Filtering of Diffusion Processes, Stochastics, 1979, Vol. 3, pp. 127-167.

ACCURATE EVALUATION OF CONDITIONAL DENSITIES IN NONLINEAR FILTERING

W.E. Hopkins, Jr., G.L. Blankenship and J.S. Baras
Electrical Engineering Department
University of Maryland
College Park, Maryland 20742

ABSTRACT

Using some approximation formulas for stochastic Wiener function space integrals, it is possible to approximate the conditional densities which arise in the nonlinear filtering of diffusion processes to within $O(n^{-2})$, with $n \geq 1$ arbitrary, by n-fold ordinary integrals. The latter have the simple form of a "rectangular rule", but their accuracy is an order of magnitude better. The n-fold integral can be further decomposed into a recursion involving n one dimensional integrals. The sequence is recursive in the increments of the observation process in the filtering problem. It is not, however, recursive in time. The one dimensional integrals are naturally treated by an m-step Gaussian quadrature which has an error proportional to $nm!/[2^m(2m)!]$. (The proportionality constant can be estimated and optimized.) The computation of these individual integrals can be reduced further by exploiting certain inherent symmetries of the problem, and by doing some preliminary, "off-line" computing. The end result is a highly accurate, computationally efficient numerical algorithm for evaluating conditional densities for a substantial class of nonlinear filtering problems. By accepting slight reductions in accuracy, one can obtain an algorithm (apparently) fast enough, when efficiently coded, for "on-line," recursive filtering in real time.

1. THE PROBLEM

Consider the stochastic dynamical system

$$dx(t) = f(x(t))dt + g(x(t))dv(t) \tag{1.1a}$$

$$dy(t) = h(x(t))dt + dw(t) \tag{1.1b}$$

$$x(0) = x_0 \sim p_0(x), \quad 0 \le t \le T$$

written in the Itô calculus. Here f,g,h are smooth functions mapping R into R; v(t), w(t) are independent, R-valued standard Wiener processes mutually independent of the initial condition x_0 which has a smooth, bounded density $p_0(x)$. We call x the signal and y the observation process. The filtering problem is to determine an estimate of x(t) given observations y(s), $0 \le s \le t$, or, more precisely, the σ-algebra Y_t generated by $\{y(s), 0 \le s \le t\}$. For the estimate one usually takes the conditional mean $\hat{x}(t) = E[x(t)|Y_t]$ since this produces the minimum mean square error.

To compute $\hat{x}(t)$, it suffices to know $p(t,x|Y_t)$, the conditional density of x(t) given Y_t, if it exists. This satisfies a complex, nonlinear stochastic partial differential equation which is difficult to analyze or to treat numerically [1] [2]. Alternately, one can write

$$p(t,x|Y_t) = u(t,x)/[\int_{-\infty}^{\infty} u(t,z)dz] \tag{1.2}$$

where the unnormalized conditional density u satisfies

$$du(t,x) = [\tfrac{1}{2}\partial_{xx}(g^2u) - \partial_x(fu) - \tfrac{1}{2}h^2 u]dt + h(x)u\,\eth y(t) \tag{1.3}$$

$$u(0,x) = p_0(x), \quad 0 \le t \le T$$

(written in the Stratonovich calculus*) a linear stochastic PDE discovered by M. Zakai [3] (and independently by R. Mortensen and T. Duncan - see the remarks in [3]). The solution to this equation may be written in terms of a function space integral. Specifically,

$$u(t,x) = E_x\{\exp[\int_0^t h(x(s))dy(s) - \tfrac{1}{2}\int_0^t h^2(x(s))ds]p_0(x(t))\} \tag{1.4}$$

where $E_x\{\cdot\}$ is expectation over the paths of (1.1) starting at x(0)=x. That is, (1.4) is formally the solution of (1.3) as a short calculation using the Stratonovich calculus shows. (Existence and uniqueness of solutions to (1.3) are discussed in [4], among other papers.) Our objective here is the evaluation of the function space integral (1.4) by quadrature type approximations.

---

* We use đu, đy, etc. to signify the Stratonovich calculus.

The first step of the approximation is carried out by regarding (a transformed version of) (1.4) as a "stochastic Wiener integral" and applying the formulas in [5]. This leads to an approximation of the function space integral by an n-fold ordinary integral with an error $0(n^{-2})$. The second step of the approximation is reduction of the n-fold integral to a recursive sequence of one dimensional integrals which may be evaluated by Gaussian quadrature.

To describe the formulas in [5], we will use the following setup: Let $C_0([0,t])$ be the space of R-valued, continuous functions x(s) on [0,t] with x(0)=0 and let W be Wiener measure on C_0. If $F:C_0 \to R$ is a smooth functional, the Wiener integral

$$I = \int_{C_0} F(x)dW(x) \tag{1.5}$$

is defined as the sequential limit [6]

$$I = \lim_{\substack{\max|t_j - t_{j-1}| \to 0 \\ 1 \le j \le n \\ n \to \infty}} \int_R \overset{n}{\cdots} \int_R da_1 \ldots da_n F(z_{sx}) \cdot \prod_{j=1}^{n} \frac{\exp[-(a_j - a_{j-1})^2/2(t_j - t_{j-1})]}{[2\pi(t_j - t_{j-1})]^{n-2}} \tag{1.6}$$

where $0<t_1<t_2<\ldots<t_n = t$ and z_{sx} is a polynomial function on [0,t] passing through x at s=0 and a_j at t_j, j=1,2,...,n.

Except for a few simple cases, it is impossible to evaluate Wiener integrals explicitly. Approximation formulas suitable for numerical computations are, therefore, of considerable interest in applications. In [7] A.J Chorin presented some formulas of this type. His results were based on the use of parabolas to interpolate the Wiener paths and on expansions of the nonlinear functional F in a Taylor series with the quadrature formula adjusted to optimize the approximation of the first two terms. Chorin's formulas were of the form

$$\int_C F(x)dW(x) = \pi^{-n/2} \int_{R^n} F_n(u_1,\ldots,u_n) e^{-u_1^2 - \ldots - u_n^2} du_1 \ldots du_n + 0(n^{-2}) \tag{1.7}$$

where F_n had the simple form of a rectangle rule in specific cases.

The function space integral in (1.4) involves a random (Ito) process $\{y(s), 0 \le s \le t\}$, and the formulas of Chorin do not apply to it. In [5] these formulas were adapted and extended to cover this case. The result we need is the following:

Lemma. *Let* $\{w(s),\ 0\le s\le t\}$ *be an* R-*valued standard Wiener process on* $(\Omega,\mathcal{F},P)$ *and let* $\{f(s),\ g(s),\ 0\le s\le t\}$ *be* R-*valued random processes non-anticipative with respect to* w *which have continuous paths almost surely and second moments uniformly bounded in* $s\varepsilon[0,t]$. *Let*

$$\begin{aligned} dy(s) &= f(s)ds + g(s)dw(s) \\ y(0) &= 0,\quad 0\le s\le t. \end{aligned} \tag{1.8}$$

Suppose $V:R\to R$ *has derivatives up to order 4 with*

$$\int_t^{t+1/n}|V''''(z(s))|^2 ds = O(n^{-1}) \tag{1.9}$$

for all $t\varepsilon[0,T]$ *and any continuous* $z:[0,T]\to R$. *Then for any* $t\varepsilon[0,T]$

$$\begin{aligned} I &= \int_C \exp\left[\int_0^t V(x(s))dy(s)\right]dW(x) \\ &= (2\pi)^{-n/2}\int_{R^n}\left\{\exp\left[\sum_{i=1}^n V\left(x_{i-1}+\frac{u_n t}{\sqrt{2n}}\right)\Delta y_{i-1}\right]\right\} \\ &\quad\cdot \exp\left[-\frac{1}{2}(u_1^2+\ldots+u_n^2)\right]du_1\ldots du_n + e_n \end{aligned} \tag{1.10}$$

where

$$\begin{aligned} x_i &= t(u_1+\ldots+u_i)/n,\quad i=1,2,\ldots,n \\ t_i &= it/n,\quad \Delta y_{i-1} = y(t_i)-y(t_{i-1}) \end{aligned} \tag{1.11}$$

The approximation error is

$$(Ee_n^2)^{1/2} = O(n^{-2}) \tag{1.12}$$

where E *is expectation with respect to the distribution of* $\{y(s),0\le s\}$.

Remarks 1. The "ordinary" integral which appears in (1.4) admits a similar approximation with $\Delta t = n^{-1}$ replacing Δy_{i-1} in (1.10). The error, which is deterministic, is also $O(n^{-2})$.

2. The remarkable feature of formula (1.10), as noted in [7], is that it is no more complicated in structure nor does it require more computing effort than the standard "rectangle rule" [8] which has accuracy $O(n^{-1})$.

3. The evaluation of the n-fold integral in (1.10) may be reduced to a sequence of one dimensional integrals which are recursive in the increments Δy_{i-1}. This has some important implications in the filtering problem.

4. The simple form of (1.10), the error estimate, and the recursive evaluation depend on the fact that the underlying measure W is Wiener measure. Since the process x(t) in (1.1) is not a Wiener

process, it is necessary to make a change of coordinates or a change of measure (i.e., a Girsanov transformation) in (1.1) to take advantage of this structure.

2. COORDINATE TRANSFORMATIONS

If the coefficients in (1.1a) are sufficiently smooth, we can change coordinates (pathwise) so that the resulting diffusion is a Wiener process. When the coefficients are not smooth or when $x(t)$ is a multivariable process, this procedure may not work, and a change of measure, i.e., a Girsanov transformation, may be required to implement our computational algorithms.

Suppose

(A1) $g(x) \geq g_0 > 0$ for some g_0 and all $x \in R$

$$\int_0^\infty dx/g(x) = \int_{-\infty}^0 dx/g(x) = +\infty$$

(A2) $p_0(x) = \exp[\mu_0(x)]$

(A3) $f \in C^1(R)$, $g \in C^2(R)$

Consider the change of coordinates in (1.1)

$$z(t) = \phi[x(t)] = \int_0^{x(t)} dx/g(x) \tag{2.1}$$

Using Itô's formula

$$\begin{aligned} dz(t) &= (f/g - \tfrac{1}{2}g')(\phi^{-1}[z(t)])dt + dv(t) \\ dy(t) &= h(\phi^{-1}[z(t)])dt + dw(t) \end{aligned} \tag{2.2}$$

The associated Zakai equation is

$$\begin{aligned} \bar{d}u(t,z) = \{&\tfrac{1}{2}\partial_{zz}u + [\tfrac{1}{2}g' - f/g](\phi^{-1}(z))\partial_z u \\ &+[g(\tfrac{1}{2}g'' - (f/g)') - \tfrac{1}{2}h^2](\phi^{-1}(z))u\}dt \\ &+ h(\phi^{-1}(z))u\bar{d}y(t) \end{aligned} \tag{2.3}$$

(written in the Stratonovich calculus). We can eliminate the first order term in (2.3) by using the exponential transformation

$$\begin{aligned} v(t,z) &= u(t,z)\exp[-\psi(z)] \\ \psi(z) &= \int_0^z (f/g - \tfrac{1}{2}g')(\phi^{-1}(x))dx \end{aligned} \tag{2.4}$$

The equation for $v(t,z)$ is

$$\bar{d}v(t,z) = [\tfrac{1}{2}\partial_{zz}v(t,z) - V(z)v(t,z)]dt + H(z)v\bar{d}y(t) \tag{2.5}$$

where

$$V(z) = \tfrac{1}{2}[h^2+(f/g-\tfrac{1}{2}g')^2+g(f/g-\tfrac{1}{2}g')'](\phi^{-1}(z))$$

(2.6)

$$H(z) = h(\phi^{-1}(z))$$

Since the Laplacean in (2.5) is "isolated," the fundamental solution of (2.5) involves Wiener measure, and we can apply the formulas of [5] to evaluate it. This is done in the next section.

Instead of changing coordinates in (1.1) we could have changed the coordinates in the Zakai equation (1.3) and then used an exponential transformation to eliminate the drift term. This would also lead to an equation like (2.5), but involving different functions $\psi(z)$, $V(z)$. While this procedure is perfectly valid in the context of numerical studies, the resulting equation cannot, in general, be associated with a nonlinear filtering problem like (1.1), and we shall not pursue it further.

In situations where f and g are not smooth, i.e., (A3) does not hold, then we must consider some more general type of weak transformation to accomplish the reduction of the Zakai equation. Also, when x(t) is an R^d-valued process with $d\geq 2$, it will be necessary to have the integrand in (2.4) be a gradient. That is, with g(x) dxd and non-singular $g^{-1}(x)f(x) - \frac{1}{2}g'(x)$ will have to be a gradient for the exponential transformation (2.4) to have a meaning.

3. APPROXIMATE EVALUATION OF THE CONDITIONAL DENSITY

Using the Feynman-Kac formula or the Kallianpur-Striebel formula as it is called in filtering theory, we can write the solution to (2.5) as

$$v(t,z) = E_z\{\exp[\int_0^t H(z(s))dy(s) - \frac{1}{2}\int_0^t H^2(z(s))ds] \qquad (3.1)$$
$$\exp[-\int_0^t V(z(s))ds+\mu_0(z(t))]\}$$

where E_z is expectation over Brownian paths starting at $z(0) = z$. Applying the formulas of Chorin and the Lemma, we can write

$$v(t,z) = I_n(t,z) + O(n^{-2}) \qquad (3.2)$$

where

$$I_n(t,z) = (2\pi)^{-n/2}\int_{R^n} K_n(t,z,\underline{r},\underline{y})d\underline{r} \qquad (3.3a)$$

$$\underline{y} = (y_0, y_1, \ldots, y_{n-1}),\quad y_i = y(t_{i+1}) - y(t_i)$$

$$\underline{r} = (r_0, r_1, \ldots, r_{n-1}),\quad t_i = it/n,\quad i = 0,1,\ldots,n-1 \tag{3.3b}$$

$$\begin{aligned} K_n(t,z,\underline{r},\underline{y}) = \exp\{\sum_{i=0}^{n-1} & [H(z+\langle\alpha_i^n(t),\underline{r}\rangle)y_i \\ & - \tfrac{1}{2}H^2(z+\langle\alpha_i^n(t),\underline{r}\rangle)t/n \\ & - V(z+\langle\alpha_i^n(t),\underline{r}\rangle)t/n] \\ & +(\mu_0-\psi)(\phi^{-1}(z+\langle\alpha_{n-1}^n(t),\underline{r}\rangle)) \\ & - \tfrac{1}{2}\langle\underline{r},\underline{r}\rangle\} \end{aligned} \tag{3.3c}$$

where for $i = 0,1,\ldots,n-1$

$$\alpha_i^n(t) = (t/\sqrt{2n},\ t/\sqrt{n},\ldots,t/\sqrt{n},\ 0,\ldots,0)^T \varepsilon R^n \tag{3.4}$$

0th entry $\quad (i-1)^{st}$ entry $\quad (n-1)^{st}$ entry

and $\langle\alpha,\underline{r}\rangle$ is the Euclidean inner product. The error in (3.2) is interpreted as in the Lemma (even though it has a deterministic component which is $O(n^{-2})$). Efficient evaluation of the n-fold integral (3.3a) is our main objective here.

3.1 A Recursive Evaluation of $I_n(t,z)$

Let

$$w_i = z + \langle\alpha_i^n(t),\ \underline{r}\rangle,\quad i = 0,1,\ldots,n-1 \tag{3.5}$$

$$\underline{w} = [w_0,\ w_1,\ldots,w_{n-1}]^T$$

and note $dw_0 = (t/\sqrt{2n})dr_0$, $dw_i = (t/\sqrt{n})dr_i$, $i=1,\ldots,n-1$. So

$$I_n(t,z) = (\tfrac{n}{2\pi})^{n/2}\ \frac{\sqrt{2}}{t^n}\int_{R^n} \hat{K}_n(t,z,\underline{w},\underline{y})d\underline{w} \tag{3.6}$$

where

$$\begin{aligned} \hat{K}_n(t,z,\underline{w},\underline{y}) = \exp\{\sum_{i=0}^{n-1} & [H(w_i)y_i - \tfrac{1}{2}H^2(w_i)t/n \\ & - V(w_i)t/n] + (\mu_0-\psi)(\phi^{-1}(w_{n-1})) \\ & - \tfrac{1}{2}\ \tfrac{n}{t^2}[2(w_0-z)^2 + (w_1-w_0)^2 + \ldots + (w_{n-1}-w_{n-2})^2]\} \end{aligned} \tag{3.7}$$

If we now define

$$g(t,w,y) = H(w)y - \tfrac{1}{2}H^2(w)t/n - V(w)t/n \tag{3.8}$$

and the sequence of integrals

$$I_1(t,z,w_1) = c_n(t)\int_{-\infty}^{\infty}\exp\{-\tfrac{1}{2}\tfrac{n}{t^2}[2(w_0-z)^2+(w_1-w_0)^2]+g(t,w_0,y_0)\}dw_0 \qquad (3.9)_1$$

$$\vdots$$

$$I_k(t,z,w_k) = c_n(t)\int_{-\infty}^{\infty}\exp\{-\tfrac{1}{2}\tfrac{n}{t^2}(w_k-w_{k-1})^2+g(t,w_{k-1},y_{k-1})\} \qquad (3.9)_k$$

$$\cdot\, I_{k-1}(t,z,w_{k-1})dw_{k-1},\ k = 2,3,\ldots,n-1,$$

then

$$I_n(t,z) = c_n(t)\int_{-\infty}^{\infty}\exp\ (\mu_0-\psi)(\phi^{-1}(w_{n-1})) + g(t,w_{n-1},y_{n-1})\}$$

$$\cdot I_{n-1}(t,z,w_{n-1})dw_{n-1} \qquad (3.9)_n$$

where $c_n(t) = (2^{1/2n}\sqrt{n})/(t\sqrt{2\pi})$.

<u>Remarks</u> 1. The integrals $I_1, I_2,\ldots,I_{n-1}$ are independent of the initial data; and so, $(3.9)_n$ has the form of a Green's representation

$$I_n(t,z) = \int_{-\infty}^{\infty}q_0(w)G(t,z;0,w)dw \qquad (3.10)$$

That is, $c_n(t)\exp[g(t,w_{n-1},y_{n-1})]I_{n-1}(t,z,w_{n-1})$ approximates the Green's function of the partial differential equation (2.5) for the (transformed) unnormalized conditional density.

2. The system (3.9) is <u>recursive</u> in the observations $y_0,y_1,\ldots,y_{n-1}$; that is, I_k depends on $y_0,y_1,\ldots,y_{k-1}$, and it is computed from I_{k-1} and y_{k-1}. Unfortunately, it is not recursive in n or in t. This last means that (3.9) does not constitute a recursive filter in the usual sense of the term.

At this point there is considerable flexibility in designing a numerical implementation of (3.9). In discussing possible designs we will show that the n-step recursive evaluation (3.9) has a significant computational advantage over (most) direct evaluations of the n-fold integral in (3.6).

3.2 Implementation

Suppose we require the evaluation of $v(t,z) = I_n(t,z) + O(n^{-2})$ at points $z = a_1,a_2,\ldots,a_m$ for some $m\geq 1$, and $t=t_1, t_2,\ldots,t_N = T$. The t_j are prespecified observation times. Let $n_1,n_2,\ldots,n_N$ be integers with $n_j\geq 1$, $j = 1,2,\ldots,N$. Consider the discretization of the time interval shown below

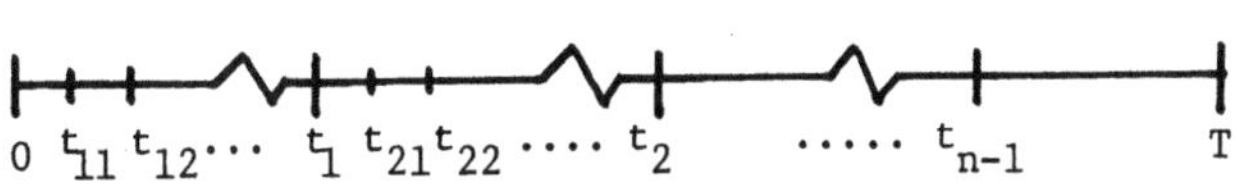

Here $t_{11} = t_1/n_1$, $t_{12} = 2t_1/n_1,\ldots,t_{21} = (t_2-t_1)/n_2$, $t_{22}=2(t_2-t_1)/n_2,\ldots$. If we only want $v(t,a_i)$ at $t_1,t_2,\ldots,t_N$, then the mesh points t_{ij} can be chosen arbitrarily; that is, the integers $n_1,n_2,\ldots,n_N$ can be

different. The recursion (3.9) is then used to evaluate $I_n(t,z)$ at $t=t_1,t_2,\ldots,t_N$. If $v(t,z)$ is required at the intermediate times t_{ij}, then it is advantageous to take $n_1,n_2=\ldots=n_N=n$, since this permits the use of parallel processors (n of them) to compute the intermediate values.

The operational count for this procedure is as follows: Each integral $I_k(t,z,w)$ is evaluated for each t at $(z,w) = (a_i,a_j), i,j=1,2,\ldots,m$, a total of m^2 evaluations of I_k. For this purpose we can use <u>any</u> quadrature rule for one-dimensional integrals of the form $I(z,w) = \int_{-\infty}^{\infty} e^{-\phi(v)} F(z,w,v)dv$ with ϕ a given weight function; that is,

$$I(z,w) = \sum_{i=1}^{m} F(z,w,a_i)A_i \tag{3.11}$$

Here A_i are weights associated with the rule, and the a_i are chosen in accordance with the prescription of the quadrature rule. By using the same rule for all the I_k no interpolation is needed to evaluate I_{k+1} in terms of I_k. At the final stage we require only m values of $v(t_1,z)$. Thus, on the initial interval $(0,t_1)$ we make $(n-1)m^2 + m$ evaluations of the integral (3.11) and $(n-1)m^3 + m^2$ function evaluations and the same number of multiplications and additions in (3.11) to compute (3.9). For example, if $m = 20$, $n = 20$, we must compute about 7500 integrals, implying about 150,000 additions and multiplications. While this is not insubstantial, it compares very favorably with product formulas for (3.6) that do not use the special structure in (3.9). By introducing a particular quadrature representation, we can obtain a more precise count of the elementary operations required in the evaluation.

4. FURTHER COMPUTATIONAL ASPECTS OF THE RECURSIVE FORMULAS

By modifying the notation we can streamline the representation of the recursion (3.9) and, in this way, make the underlying structure of the computation more evident. Referring to (3.9) a natural choice for the weight functions ϕ in (3.11) are the diagonal terms of the quadratic form in w_i in (3.7). Unfortunately, it will be seen in section 5 that the error associated with the resulting Hermite-Gauss quadrature rule is an increasing function of bounds on derivatives of the integrand in (3.11); these may not be finite because of the remaining cross terms in the quadratic form. This suggests selecting as weight functions a portion of each diagonal element, and choosing the proportionality constants to preserve the negative definiteness of the quadratic form in (3.7) while simultaneously optimizing the error bounds for the quadrature rule. To this end let $\alpha_i \varepsilon (0,1), 0 \le i \le n-1$, and define

$$J^1 = (t/\sqrt{n})\mathrm{diag}[\sqrt{2/3},\ 1,\ldots,1,\sqrt{2}]$$
$$J^2 = \mathrm{diag}[(1-\alpha_i)^{-\frac{1}{2}}] \qquad (4.1)$$

(diag = diagonal matrix), and

$$\underline{w} = J\underline{\xi},\quad J=J^1J^2 \triangleq \mathrm{diag}[J_i] \qquad (4.2)$$

Then

$$I_n(t,z) = c_n(t,z)\int_{R^n} [\prod_{i=0}^{n-1} e^{-\xi_i^2}/\sqrt{\pi}]\tilde{K}_n(t,z,\underline{\xi},\underline{y})d \qquad (4.3)$$

where

$$c_n(t,z) = 2\sqrt{2/3}\ \prod_{i=0}^{n-1} (1-\alpha_i)^{\frac{1}{2}}\ e^{-nz^2/t^2}/2^{n/2}$$

$$\tilde{K}(t,z,\underline{\xi},\underline{y}) = \exp\{[\mu_0-\psi)(\phi^{-1}(w_{n-1})) + \sum_{i=0}^{n-1} g(t,w_i,y_i)]_{w=J\underline{\xi}}$$
$$- \sum_{i=0}^{n-1} a_{ii}\xi_i^2 - 2\sum_{i=0}^{n-2} a_{i,i+1}\xi_i\xi_{i+1} - b_0\xi_0\}$$

$$b_0 = -\frac{2\sqrt{2/3}\sqrt{n}}{t\sqrt{1-\alpha_0}}\,z \qquad (4.4)$$

$$a_{ii} = \alpha_i/(1-\alpha_i)$$

$$a_{i,i+1} = \begin{cases} -(1/\sqrt{6})/\sqrt{(1-\alpha_0)(1-\alpha_1)} & ,\ i = 0 \\ -(1/2)/\sqrt{(1-\alpha_i)(1-\alpha_{i+1})} & ,\ 1\leq i\leq n-3 \\ -(1/\sqrt{2})/\sqrt{(1-\alpha_{n-2})(1-\alpha_{n-1})} & ,\ i=n-2 \end{cases}$$

The transformation J^1 isolates a factor $\exp-\Sigma(1-\alpha_i)\xi_i^2$, which is then normalized by J^2. To express (4.3) recursively, define

$$r_i^1(\xi_i,t) = \exp[H(J_i\xi_i)]$$

$$r_i^2(\xi_i,t) = \exp[-\tfrac{1}{2}H^2(J_i\xi_i)t/n - V(J_i\xi_i)t/n]$$

$$q_0(\xi_0,z,t) = \exp[(2\sqrt{2n/3}/t\sqrt{1-\alpha_0})z\xi_0] \qquad (4.5)$$

$$s_k(\xi_k,\xi_{k+1}) = \exp[a_{kk+1}\xi_k\xi_{k+1} - a_{kk}\xi_k^2],\ 0\leq k\leq n-2$$

$$s_{n-1}(\xi_{n-1}) = \exp[(\mu_0-\psi)(\phi^{-1}(J_{n-1}\xi_{n-1}))]$$

Here $w_i = J_i\xi_i$ and the dependence on α_i has been suppressed. Then (4.3) becomes

$$I_n(t,z) = c_n(t,z)\int [\prod_{i=0}^{n-1} e^{-\xi_i^2}/\sqrt{\pi}]\ [\prod_{i=0}^{n-1} (r_i^1(\xi_i,t))^{y_i}]$$
$$\cdot[\prod_{i=0}^{n-1} r_i^2(\xi_1,t)s_i(\xi_i,\xi_{i+1})]q(\xi_0,z,t)d\underline{\xi} \qquad (4.6)$$

To express this recursively in terms of one dimensional integrals, let $\rho(\xi) = e^{-\xi^2}/\sqrt{\pi}$, and consider the sequence

$$\begin{aligned}
I_1(\xi_1,z) &= \tilde{c}_n(t,z)\int_{-\infty}^{\infty}\rho(\xi_0)(r_0^1(\xi_0,t))^{y_0} r_0^2(\xi_0,t) \\
&\quad \cdot s_0(\xi_0,\xi_1)q(\xi_0,z,t)d\xi_0 \\
&\vdots \\
I_{k+1}(\xi_{k+1},z) &= \tilde{c}_n(t,z)\int_{-\infty}^{\infty}\rho(\xi_k)(r_k^1(\xi_k,t))^{y_k} r_k^2(\xi_k,t) \\
&\vdots \\
&\quad \cdot s_k(\xi_k,\xi_{k+1})I_k(\xi_k,z)d\xi_k, \quad k=1,2,\ldots,n-2 \\
I_n(t,z) &= \tilde{c}_n(t,z)\int_{-\infty}^{\infty}\rho(\xi_{n-1})(r_{n-1}^1(\xi_{n-1},t))^{y_{n-1}} r_{n-1}^2(\xi_{n-1},t) \\
&\quad \cdot s_{n-1}(\xi_{n-1}\xi_n)\, I_{n-1}(\xi_{n-1},z)d\xi_{n-1}
\end{aligned} \tag{4.7}$$

where

$$\tilde{c}_n(t,z) = [c_n(t,z)]^{1/n} \tag{4.8}$$

Note that the functions r_k^1, r_k^2, s_k do not depend on k, for $k=1,\ldots,n-2$.

For fixed α_i, we can integrate $I = \int_{-\infty}^{\infty}\rho(\xi)f(\xi)d\xi$ with an M-point Gaussian formula (see, e.g., [9]):

$$\begin{aligned}
I &= S + e_M \\
S &= \sum_{i=1}^{M} A_i f(a_i)
\end{aligned} \tag{4.9}$$

where $A_i>0$ are weights and a_i are the (real) zeros of the Mth Hermite polynominal

$$H_M(x) = 2^M x^M + \ldots \tag{4.10}$$

and are symmetric about $x = 0$. Using this, the kth term in (4.7), $k = 1,2,\ldots,n-2$, becomes

$$I_{k+1}(a_j,a_i) = \tilde{c}_n(t,a_i)[S_k(a_j,a_i) + e_k] \tag{4.11}$$

where

$$\begin{aligned}
S_k(a_j,a_i) &= \sum_{\ell=1}^{M} A_\ell[r^1(a_\ell,t)]^{y_k} r^2(a_\ell,t) \\
&\quad \cdot s(a_\ell,a_j)I_k(a_\ell,a_i) \\
&\approx \sum_{\ell=1}^{M} A_\ell[r^1(a_\ell,t)]^{y_k} r^2(a_\ell,t) \\
&\quad \cdot s(a_\ell,a_j)\tilde{c}_n(t,a_i)S_{k-1}(a_\ell,a_j)
\end{aligned} \tag{4.12}$$

(Note: we have dropped the k-dependence on r^1, r^2, s.)

The error in the second expression should be clear from (4.11). Now let $\underline{S}_k$ be the MxM matrix with elements $S_k(a_j,a_i), i,j=1,2,\dots,M$.

$$\underline{S}_k = \{S_k(a_j,a_i)\}_{i,j=1}^{M} \tag{4.13}$$

and let

$$R(t,i,j,\ell) = r^2(a_\ell,t)s(a_\ell,a_j)\tilde{c}_n(t,a_i) \tag{4.14}$$

Then

$$S_k(a_j,a_i) = \sum_{\ell=1}^{M} A_\ell[r^1(a_\ell,t)]^{y_k}R(t,i,j,\ell) \cdot S_{k-1}(a_\ell,a_i) \tag{4.15}$$

The terms A_ℓ, $r^1(a_\ell,t)$, $R(t,i,j,\ell)$ can be precomputed "off-line". In this case, calculation of the matrix $\underline{S}_k$ consists of

(i) raising the elements of the vector $r^1(a_\ell,t)$ $\ell = 1,2,\dots,M$ to the y_k power;

(ii) performing M^2 vector products of $\underline{x}_{ij} = \{A_\ell[r^1(a_\ell,t)]^{y_k} R(t,i,j,\ell),\ \ell = 1,\dots,M\}$ and $\underline{v}_{k-1}^{i} = \{S_{k-1}(a_\ell,a_i),\ \ell = 1,\dots,M\}$ for $i,j = 1,2,\dots,M$.

To compute $I_{k+1}(a_j,a_i)$, it is necessary to multiply $S_k(a_j,a_i)$ by $\tilde{c}_n(t,a_i)$,

Thus, a total of $2M^3+M^2+M$ elementary operations is required to compute the M^2 entries $I_{k+1}(a_\ell,a_i)$, $i,j=1,\dots,M$. This is for $k=1,2,\dots,n-2$, a total of $(n-2)(2M^3+M^2+M)$ operations. The operation count for $I_1(a_\ell,a_i)$ is the same, adding $(2M^3+M^2+M)$ operations. Only $2M^2 + 2M$ operations are required to compute the M-vector

$I_n(t,a_\ell)$, $\ell=1,\dots,M$. Thus, the total number of elementary operations N_{oper} required to compute the approximation $I_n(t,z)$ of $v(t,z)$ at the points $z = a_\ell$, $\ell=1,\dots,M$ is

$$N_{oper} = 2(n-1)M^3 + (n+1)(M^2+M) \tag{4.16}$$

For $n=20$, $M=20$ we have $N_{oper} = 312820$.

Note that it is possible to do the multiplications of $\underline{x}_{ij}$ and $\underline{v}_{k-1}^{i}$ and that of $S_k(a_j,a_i)$ and $\tilde{c}_n(t,a_i)$ using "parallel processing" which leads to considerable time savings.

Finally, referring back to (2.4) the evaluation of the unnormalized density $u(t,a_i)$ from $v(t,a_i)$, $i = 1,2,\dots,M$, requires an additional M function evaluations and M multiplications.

5. SOME COMMENTS ON AN ERROR ANALYSIS

Our final formula (4.11) for the approximation of v(t,z) depends on the weight parameters α_i and involves two basic errors: (i) the error $O(n^{-2})$ in the approximation (1.10) of the function space integral by an n-fold "ordinary" integral; and (ii) the error e_M in the M-point Gaussian quadrature formula (4.9). (The error in the substitution (4.12) is clearly a linear multiple of e_M.) For obvious reasons we would like to have sharp estimates of these quantities, the last of which should be minimized with respect to $\underline{\alpha}$.

The analysis in [5] which led to the estimate $O(n^{-2})$ in (1.10) is so complex that a precise evaluation of the order expression appears to be out of the question. In specific cases one can probably emulate the analysis of Cameron [8], sec. 6, which shows that in some cases the $O(n^{-2})$ estimate is actually very conservative. (E.g., "Simpson's rule," included in our formulas, evaluates the Wiener integral of the functional $F[x] = [\int_0^1 x^2(s)dx]^2$ to within an error of $1/8\pi^4 n^{-3}$.)

We can say somewhat more about the error $e_M = e_M(t,z,\underline{\alpha})$. Natural bounds are of the form

$$|e_M(t,z,\underline{\alpha})| \leq C(n,M)N(t,z,\underline{y},n,\underline{\alpha}) \tag{5.1}$$

where C is the error coefficient and N is some bound on a derivative of the integrand in (4.6). There are several possible procedures, see, e.g., [10][11]. We shall follow the approach used by Lether in [11]. Since the recursive evaluation (4.7) does not reduce or increase the error relative to direct evaluation of the n-fold integral (4.3), we may as well apply Lether's procedure to the latter.

Because the weight function in (4.3) is normalized, and the error coefficient for each coordinate is (independent of the coordinate)

$$\varepsilon(M) = M!/2^M(2M)! \tag{5.2}$$

the error for fixed $(t,z,\underline{\alpha})$ is bounded by (from [11], equ.(7))

$$E_M(t,z,\underline{\alpha}) = \varepsilon(M) \sum_{i=0}^{n-1} \frac{\partial^{2M}}{\partial \tilde{\xi}_i^{2M}} \tilde{K}_n(\tilde{\xi}_0,\ldots,\tilde{\xi}_{n-1}) \tag{5.3}$$

for some point $(\tilde{\xi}_0,\ldots,\tilde{\xi}_{n-1})$ in R^n. Here $\tilde{K}_n$ is the integrand in (4.3). (We have supressed the $t,z,\underline{y},\underline{\alpha}$ variables in the argument of $\tilde{K}$.) Of course, we assume that $\tilde{K}$ and, therefore ρ_0, f,g,h, are sufficiently smooth for (5.3) to make sense.

The dependence of E_M on α may be found by computing upper bounds for the derivatives of $\tilde{K}_n$. Writing

$$\tilde{K}_n(\underline{\xi},\underline{\alpha}) = F(\underline{J\xi})G(\underline{\xi},\underline{\alpha}), F(\underline{w}) = \exp[(\mu_0-\psi)(\phi^{-1}(w_{n-1})) + \sum_{i=0}^{n-1} g(w_i)] \tag{5.4}$$

$$G(\underline{\xi},\underline{\alpha})=\exp[-(\sum_{i=0}^{n-1} a_{ii}(\alpha_i)\xi_i^2 + \sum_{i=0}^{n-2} a_{ii+1}(\alpha_i,\alpha_{i+1})\xi_i\xi_{i+1} + b_0(\alpha_0)\xi_0)]$$

assume F has continuous derivatives up to order 2M+1, bounded by a constant F_0 possibly depending on the observation path $y(t)$. By completing the square with respect to ξ_i in $G(\underline{\xi},\underline{\alpha})$ and employing a bound for Hermite polynomials [12],

$$H_j(x)e^{-x^2} \le C_0 2^{j/2}\sqrt{j!}\,e^{-x^2/2} \quad , \quad C_0 \cong 1.086$$

and $d^i e^{-x^2}/dx^i = e^{-x^2} H_i(x)\cdot(-1)^i$, it may be shown that

$$E_M(t,z,\underline{\alpha}) \le C_1 \varepsilon(M) \sum_{i=0}^{n-1} \underline{P}_i(\underline{\alpha}) \sup_{\xi} e^{-Q_i(\underline{\xi},\underline{\alpha})} \tag{5.6}$$

$$C_1(n,t,z) = C_0 F_0^n \sqrt{2}\ (\tfrac{n}{2})^{-n/2}\ t^{-n} \exp(-nz^2/t^2)$$

$$P_i(\underline{\alpha}) = (\prod_{j=0}^{n-1}(1-\alpha_j)^{-1/2})(J_i^1)^{-m}(1-\alpha_i)^{-m} \sum_{j=0}^{2m} \binom{2m}{j}\sqrt{j!}(2J_i^1\ \alpha_i)^{j/2}$$

$$\cdot\ \exp[\tfrac{1}{8}\ b_i^2/a_{ii} + \tfrac{1}{4}\ B_i^T A_i^{-1} B_i](\underline{\alpha})$$

$$Q_i(\underline{\xi},\underline{\alpha}) = [(\xi + \tfrac{1}{2}\ A_i^{-1}B_i)^T A_i(\xi + \tfrac{1}{2}\ A_i^{-1}B_i)](\underline{\alpha})$$

where

$$A_i(\underline{\alpha}) = M_i(\alpha)A(\alpha),$$

$$A = (a^{ij}),\ M_i = I_{n\times n} - \frac{1}{2}\begin{bmatrix} & 0 & \\ & \vdots & \\ & 0 & \\ & a_{i-1,i}/a_{ii} & \\ 0 & 1 & 0 \\ & a_{ii+1}/a_{ii} & \\ & 0 & \\ & \vdots & \\ & 0 & \end{bmatrix}_{n\times n}$$

$$B_0 = \begin{bmatrix} \frac{1}{2}b_0 \\ 0 \\ \\ 0 \end{bmatrix} \quad , \quad B_i = \begin{bmatrix} b_0 \\ 0 \\ \\ 0 \end{bmatrix}$$

Here a_{ij}, b_i are defined in (4.4) and A_i is quint-diagonal. By continuity arguments it may be shown that the matrices A_i are all positive definite in some nonempty subset S of the cube $\mathcal{D} = \prod_{i=0}^{n-1}(0,1)$. In this region the factors $\exp{-Q_i}$ in (5.6) may be bounded by unity. Also, the function $\Sigma\,\underline{P}_i(\underline{\alpha})$ is continuous in $\mathcal{D}$ and diverges to $+\infty$ as $\underline{\alpha} \to \partial\mathcal{D}$ Therefore, $\Sigma\,\underline{P}_i(\underline{\alpha})$ attains a minimum in S at some point $\underline{\alpha}^*(m)$.

This value α^* may be computed numerically and used to define the approximations in section 4, giving rise to the error bound

$$|e_M(t,z,\alpha^*)| \le E_M(t,z,\alpha^*) \le C_1 \varepsilon(M) \sum_{i=0}^{n-1} \underline{P}_i(\alpha^*) \tag{5.7}$$

It is robust in the sense that α^* does not depend on the data f,g,h, or the observation process y(t). Of course, the dependence of (5.7) on (m,n) is crucial. At present, the asymptotic behavior of α^* as $m \to \infty$ has not been determined, but analysis of similar integrals with A_i diagonal suggests that the performance should be good.

Remark : The assumption that $F(\underline{w})$ have bounded derivatives is not very restrictive. From (2.6) (3.8), (5.4) it is clear that we include the cases

(i) f,g,h and their derivatives bounded

(ii) g constant , f,h together with their derivatives of "polynomial" growth, and $\lim_{|x| \to \infty} \operatorname{sgn}(x) f(x) < \infty$.

More general classes of f,g,h may be identified from (2.6),(3.8),(5.4).

REFERENCES

1. H. Kushner, "Dynamical equations for optimal nonlinear filtering", J. Differential Equations, 2 (1967), pp. 179-190.
2. F. Levieux, "Conception d'algorithms parallelisables et convergents de filtrage récursif non-linéare", Appl. Math. Opt., 4 (1977), pp. 61-95.
3. M. Zakai, "On the optimal filtering of diffusion processes", Z. Wahrsch. Verw. Geb., 11 (1969), pp. 230-243.
4. J. S. Baras, G. L. Blankenship, W. E. Hopkins, Jr., "Existence, uniqueness, and asymptotic behavior of solutions to a class of Zakai equations with unbounded coefficients", to appear.
5. G. L. Blankenship, J. S. Baras, "Accurate evaluation of stochastic Wiener integrals with applications to scattering in random media and to nonlinear filtering", SIAM J. Appl. Math., 41 (1981), pp. 518-552.
6. R. H. Cameron, "A family of integrals serving to connect the Wiener and Feynman integrals", J. Math. Physics, 39 (1960), pp. 126-140.
7. A. J. Chorin, "Accurate evaluation of Wiener integrals", Math. Comp., 27 (1973), pp. 1-15.
8. R. H. Cameron, "A Simpson's rule for the numerical evaluation of Wiener's integrals in function space", Duke Math. J., 18 (1951) pp. 111-130.
9. A. H. Stroud, D. Secrest, Gaussian Quadrature Formulas, Prentice-Hall, Englewood Cliffs, N.J., 1966.
10. F. G. Lether, "Cross-product cubature error bounds", Math. Comp., 24 (1970), pp. 583-592.
11. F. G. Lether, "An error representation for product cubature rules" SIAM J. Numer. Anal., 7 (1970), pp. 363-365.
12. A. Erdelyi, ed., Bateman Manuscript Project, Higher Transcendental Functions, vol. II, Calif. Inst. Tech., McGraw-Hill, New York, 1955.

AN EFFICIENT APPROXIMATION SCHEME FOR A CLASS OF STOCHASTIC DIFFERENTIAL EQUATIONS

J.M.C. Clark

Department of Electrical Engineering
Imperial College
London SW 7 2BT U.K.

INTRODUCTION

A feature that is characteristic of many of the multidimensional stochastic differential equations that arise in nonlinear filtering theory is that they are forced by a single, "drifting", Brownian motion ; that is, when written in Stratonovich form, they look like

$$dX_t(x,w) = f(X_t(x,w)dt + g(X_t(x,w))_{\circ}dw(t), \quad X_0(x,w) = x \tag{1}$$

where (X_t) is a d-dimensional continuous process on $[0,T]$ and $(w(t))$ is a scalar continuous process with a distribution P that is absolutely continuous with respect to Wiener measure P_0. In what follows $(w(t))$ is taken to be the coordinate process of the canonical space $C[0,T]$, topologised with the uniform norm, and P and P_0 are distributions on its Borel field B_T. $X_T(x,w)$ is the natural solution of (1) that is known to be continuous in t, x and w for sufficiently smooth f and g ([4], p. 215, [2], [10]) .

This paper is concerned with the asymptotic behaviour of a scheme for approximating the solutions of such equations in which each approximant depends on the values of the forcing process only at the points of a regular partition; that is, the scheme produces for each n a P_n^0-measurable approximation, where P_n^0 is the partition σ-field $\sigma\{w(iT/n) : i = 0,1,\dots,n\} \subset B_T$.

Suppose the accuracy of approximation is measured in terms of the L_2-norm $(E[X_T-\hat{X}_T]^2)^{\frac{1}{2}}$ of the error of approximation of X_T; then the best rate of convergence among all (P_n^0)-adapted schemes is provided by the sequence of conditional means $E(X_T|P_n^0)$. For scalar equations of the type (1) with a single forcing Brownian motion it is known [1] that for such a sequence the convergence is of order n^{-1} and that a number of difference schemes such as those of McShane ([4], p. 205) and Mil'stein [5], and the second order Runge-Kutta schemes of Rumelin [8], have the same order of convergence. If the norm $\sup_{0\le t\le T} (E[X_t-\hat{X}_t]^2)^{\frac{1}{2}}$ were to be used instead the best order would drop to $n^{-\frac{1}{2}}$; so the choice of norm is crucial here. If there are two or more forcing Brownian motions driving non-commutative vector fields then the best order is again $n^{-\frac{1}{2}}$ [1].

Newton [6], in an analysis of approximations for bilinear equations with Brownian motions driving commuting vector fields, has shown that a truncated Taylor-series scheme is efficient, in the sense of possessing the best rate of convergence, only if most terms of third order are retained. This suggests that the schemes just mentioned, which are basically second-order methods, would generally be inefficient when applied to equations of the type (1).

This paper considers the efficiency of a "prototypical" approximation scheme for (1). It is shown that the solution of a sequence of relatively simple differential equations driven by piecewise linear approximations of the process w form a scheme that is efficient for all sufficiently regular distributions of w. This is done by means of a central limit theorem for the conditional distributions of normalized error. The technique of representing the the solution of a differential equation as the composition of the solutions of two simpler equations will be used repeatedly in the proofs; this technique is at the heart of the work of Doss [2] and Sussmann [10] and has also been used by Kunita [3] in a more general context.

AN ODE APPROXIMATION SCHEME

Let $\bar{w}_n$ denote the piecewise linear approximation of $w \in C[0,T]$:

$$\bar{w}_n(t) = \frac{1}{h}((i+1)h-t)w(ih) + \frac{1}{h}(t-ih)w((i+1)h)$$

$$\text{for } ih \leq t \leq (i+1)h, \qquad i = 0,\dots,n-1$$

where $h = \frac{T}{n}$. Let ${}^n\hat{X}_t(x,w)$ be the solution of the ordinary differential equation

$$\dot{\hat{X}}_t = f(\hat{X}_t) + \frac{T}{12n}[g,[g,f]](\hat{X}_t) + g(\hat{X}_t)\dot{\bar{w}}_n(t), \hat{X}_0(x,w) = x \tag{2}$$

for $0 \leq t \leq T$. Here the vector fields f and g on $\mathbb{R}^d$ are of class C_b^2 (continuous bounded derivatives up to second order) and $[g,f]$ is the Lie bracket vector field $\left\{\sum_j (g^j \frac{\partial f^i}{\partial x^j} - f^j \frac{\partial g^i}{\partial x^j}\right\}$. Then $({}^n\hat{X}_T)$ is our scheme of approximations for X_T. The presence of the double Lie bracket in (2) distinguishes it from the familiar Wong - Zakai piecewise linear scheme [11],[4]. It also has an incidental geometric virtue not possessed by the simpler difference schemes : suppose f and g are C_b^∞ and N is a maximal integral manifold generated by f and g with a dimension d_1 possibly less than d. It follows from the special structure of (1) ([2], [10], also Lemmas 1 and 2) that if $x \in N$ then $X_T(x,\cdot)$ stays in N for all t. However the vector field generating ${}^n\hat{X}_t$ depends only on f and g and their brackets and so ${}^n\hat{X}_t$ also stays in N.

CONDITIONAL CENTRAL LIMIT THEOREMS

It turns out that the most natural way of studying the asymptotic behaviour of ${}^n\hat{X}_T$ $(= {}^n\hat{X}_T(\cdot,w)$ is by the action of its inverse map; roughly speaking if x lies in a maximal integral manifold N of f and g, then in the limit $n({}^n\hat{X}_T^{-1} \circ X_T(x) - x)$ behaves like a normal random vector in the tangent space of N at x. Let $\mathcal{P}_n$ be the completion of $\mathcal{P}_n^0$ by the P-null sets in $\mathcal{B}_T$.

Theorem 1: Suppose f and g are of class C_b^3, and $[g,f]$ and $[g,[g,f]]$ are of class C_b^2.

i) If $P = P_0$, Wiener measure, then for almost all w, any version of the P_n-conditional distribution of $n(\hat{X}_T^{-1} \circ X_T(x)-x)$ converges weakly to a normal distribution in $\mathbb{R}^d$ with mean at the origin and covariance matrix

$$V_0(x,w) := \frac{T^2}{12} \int_0^T \{X_{t*}^{-1}[g,f](X_t(x,w))\}\{X_{t*}^{-1}[g,f](X_t(x,w)\}'dt. \tag{3}$$

Furthermore the P_n-conditional covariance matrix converges a.s. to V_0.

ii) If $P \ll P_0$ and $E[\frac{dP}{dP_0}] := \int \frac{dP}{dP_0} dP < \infty$,

then for almost all w (P), the conditional distribution and conditional covariance matrix converge to the same limits along any infinite subsequence I for which $(P_n : n \in I)$ is increasing.

Here $X_{t*}^{-1}[g,f]$ denotes the vector field obtained by "pulling back" $[g,f]$ through the diffeomorphism $X_t(\cdot,w)$; that is,

$$X_{t*}^{-1}[g,f](x) := DX_t^{-1}(X_t(x))\cdot[g,f](X_t(x)) \tag{4}$$

where DF denotes the Jacobian $\frac{\partial F^i}{\partial x^j}$ of a diffeomorphism F. The proof, which is rather long, is given in the final section.

Rootzén [7] has a comparable "Renyi-stable" functional limit theorem for approximations of Stratonovich integrals. In the present context a functional limit theorem is not appropriate, for the following reason. In the proof of Theorem 1, a "ripple free" approximation ${}^n\bar{X}_t$ of X_t is introduced that equals X_t at the points of a partition and that is differentiable between these points. It is to this process that ${}^n\hat{X}_t$ converges uniformly with rate $1/n$; it converges uniformly to X_t only with rate $1/n^{\frac{1}{2}}$.

In the theorem corresponding to Theorem 1 for the usual piecewise linear approximations ((2) without the bracket term) the limit distribution would have a non-zero mean.

The corresponding theorem for $n({}^n\hat{X}_T - X_T)$ takes a more complicated form. Intuitively, for fixed x, the pair $\{\hat{X}_T(x,\cdot),\ n({}^n\hat{X}_T(x,\cdot)-X_T(x,\cdot))\}$ behaves in the limit like a random element of the tangent bundle of the maximal integral manifold containing x.

Theorem 2: Suppose f and g are as in Theorem 1. Under either of the conditions (i), (ii) on P, the P_n-conditional distribution of $n({}^n\hat{X}_T - X_T)$ and its covariance matrix converge in the same way as those of $n({}^n\hat{X}_T^{-1} \circ X_T(x)-x)$, except that the covariance matrix limit now has the form

$$V_T(x,w) := E[U_T U_T' | w] \tag{5}$$

where (U_t), $0 \le t \le T$, is a continuous process in $\mathbb{R}^d$ that solves the Stratonovich equation

$$dU_t = Df(X_t)U_t dt + Dg(X_t)U_t \circ dw(t) + \frac{T}{(12)^{\frac{1}{2}}}[g,f](X_t)\circ dv(t),\ U_0 = 0 \tag{6}$$

on a probability space augmented to carry a Brownian motion $(v(t))$ that is independent of $(w(t))$.

Again, the proof is given in the final section.

EFFICIENCY

An immediate consequence of Theorem 2 is the following (Var[•] denotes a variance) :

<u>Corollary</u>: Suppose $E\left(\frac{dP}{dP_0}\right) < \infty$ and $x, c \in \mathbb{R}^d$. Then $(c'{}^n\hat{X}_T)$ is an efficient set of approximations of $c'X_T$ in the sense that for all infinite subsequences of integers I for which $(P_n : n \in I)$ is increasing,

$$\lim_{n\in I} \frac{E[(c'{}^n\hat{X}_T(x) - c'X_T(x))^2 | P_n]}{Var[c'X_T | P_n]} = 1.$$

for almost all w in the set $\{w : c'V_T(x,w)c > 0\}$

It is of interest to determine conditions under which the covariance matrix $V_T(x,w)$ is a.s. positive definite. It is instructive to look at the elementary Gaussian case where

$$dX_t = AX_t dt + bdw(t), \quad X_0(x,w) = x.$$

Then (6) reduces to

$$dU_t = AU_t dt + \frac{T}{\sqrt{12}} Abdv(t) \qquad U_0 = 0.$$

and the $V_T(x,w)$ becomes the "controllability" Grammian

$$V_T = \int_0^T (e^{A(T-t)}Ab)(e^{A(T-t)}Ab)' dt \cdot \frac{T^2}{12}$$

which is positive definite if and only if the matrix

$$[Ab, A^2b, \ldots A^d b] = A[b, Ab, \ldots, A^{d-1}b] =: AR$$

is of full rank. If there exists a vector c that is orthogonal to the range of A or, more generally, to the image under A of the "controllable" subspace of $\dot{x} = Ax + bu$, then $c'Vc = 0$ and the corollary is uninformative about the efficiency of $(c'{}^n\hat{X}_T)$. It so happens that in this case $c'{}^n\hat{X}_T = c'X_T$, and the scheme is still "efficient" in a trivial sense, but what is of more significance is that the positive definiteness of V_T is not guaranteed by the full rank of R, which is equivalent to the condition that the transition densities of (X_t) are non-degenerate in $\mathbb{R}^d$.

For the general case a promising line of argument(suggested to me by D. W. Stroock) would seem to be to make use of the exact-time reachability theorems of Sussmann and Jurdjevic [9] and others to establish the positive definitness of V_T, but I have not yet looked at this in detail.

PROOFS

We begin with some lemmas. In what follows, rather than saying that a sequence of continuous functions (q_n) converges uniformly on bounded sets to a continuous limit g_∞, we shall say that $(x,n) \to q_n(x)$ is "continuous on $\mathbb{R}^d \times \bar{\mathbb{N}}$", where

$\bar{\mathbb{N}}$ is the usual one-point compactification $\mathbb{N} \cup \{\infty\}$ of the integers. $|\cdot|$ will denote Euclidean norms in various finite dimensional spaces, and also the uniform norm in $C[0,T]$.

Suppose $g:\mathbb{R}^d \to \mathbb{R}^d$ is a vector field of class C_b^3 and $q:\mathbb{R}^d \times \bar{\mathbb{N}} \to \mathbb{R}^d$ an indexed vector field with bounded first and second derivatives continuous in (x,n). Let $\gamma_t(x)$ solve

$$\dot{\gamma}_t = g(\gamma_t), \quad \gamma_0(x) = x \in \mathbb{R}^d, \quad t \in \mathbb{R} \tag{7}$$

and $Z_t(x,w,n)$ solve, for $w \in C[0,T]$

$$\overset{\circ}{Z}_t = D\gamma_{-w(t)} \circ \gamma_{w(t)}(Z_t) \cdot q_n \circ \gamma_{w(t)}(Z_t), \; Z_0(x,w,n) = x \tag{8}$$
$$=: \gamma^{-1}_{w(t)*} q_n(Z_t)$$

Lemma 1: i) (Doss [2], Sussman [10]) If the distribution P of w is absolutely continuous with respect to Wiener measure P_0, the composite function:

$$Y_t(x,w,n) := \gamma_{w(t)} \circ Z_t(x,w,n) \tag{9}$$

is continuous in t,x,w and n on $\mathbb{D} := [0,T] \times \mathbb{R}^d \times C[0,T] \times \bar{\mathbb{N}}$, solves the Stratonovich equation

$$dY_t = q_n(Y_t)dt + g(Y_t) \circ dw(t) \qquad Y_0 = x$$

and solves the corresponding ordinary differential equation for all w that are piecewise C^1.

ii) The x-derivatives of $\gamma_t(x)$ up to third order are continuous in t and x and bounded by ke^{kt} for some constant k.

iii) The first and second x-derivatives of $Y_t := Y_t(\cdot,w,n)$ and those of its inverse Y_t^{-1} are all continuous on $\mathbb{D}$ and bounded uniformly in t,x, and n for bounded w.

Proof: (i) is simply an amalgam of theorems in [2] and [10]. A standard theorem of ODE's shows that γ_t and its x-derivatives exist and are continuous in t and x. Furthermore, for $d = 1$,

$$D\dot{\gamma}_t = Dg \cdot D\gamma_t, \; D\gamma_0 = 1,$$
$$D^2\dot{\gamma}_t = Dg \cdot D^2\gamma_t + D^2 g(D\gamma_t)^2, \qquad D^2\gamma_0 = 0,$$

and boundedness of the derivatives of g and Gronwall's inequality establishes (ii). The same argument applies for the third order and vector cases. For w in a bounded region the vector field in (8) has bounded x-derivatives up to second order that are continuous on $\mathbb{D}$. The same ODE theorem implies that Z_t has the same properties. Application of the chain rule of calculus to (9) shows that Y_t also has these properties. Now consider the solutions $\bar{Z}_{s,t}$ and $\bar{Y}_{s,t}$ defined by equations (8) and (9) with $w(t)$ replaced by $w(t)-w(s)$ and with an initial condition $\bar{Z}_{s,s}(x,w,n) = x$. Then it is clear that the arguments that have just been applied to Y_t also show that $\bar{Y}_{s,t}(x,w,n)$ and its derivatives are continuous in s,t,x,w and n, and are bounded for bounded w. But it is easy to verify that $\bar{Y}_{t,0} = Y_t^{-1}$ and so the proof of (iii) is complete.

To simplify the notation, from now on we shall generally suppress the arguments n and w.

Lemma 2: Suppose $\hat{X}_t$ is defined by (2) and f,g and [g,[g,f]] are of class C_b^2. The Jacobian matrices $D\hat{X}_t$, $D\hat{X}_t^{-1}$ and $D(\hat{X}_{t*}^{-1}k)$, where k is a C_b^1 vector field on $\mathbb{R}^d$, are continuous on $\mathcal{D}$, bounded for bounded w and $\mathcal{P}_n$-measurable for fixed t, x and n.

Proof: Set in Lemma 1 $q_n := f + \frac{T}{12n}[g[g,f]]$. Then $\hat{X}_T(x,w) = Y_t(x,\bar{w}_n,n)$. Since $|\bar{w}_n| \le |w|$ and $(w,n) \to \bar{w}_n$ is continuous it follows from Lemma 1 (iii) that $D\hat{X}_t$ and $D\hat{X}_t^{-1}$ have the required properties. That the Jacobian of

$$\hat{X}_{t*}^{-1} k(t) := (D\hat{X}_t^{-1} k) \circ \hat{X}_t(x)$$

also has these properties follows from the chain rule and the continuity and boundedness of the first and second derivatives of Y_t and Y_t^{-1}. $\mathcal{P}_n$-measurability is obvious.

Now let $\tilde{w}(t)$ be the process $w(t)-\bar{w}_n(t)$; then on $(C[0,T],\mathcal{B}_T,P_0)$ $\tilde{w}$ is a "Brownian bridge" process. The following lemma summarizes its properties. The proof is elementary and is omitted.

Lemma 3: i) $\tilde{w}$ is a continuous Gaussian process independent of $\bar{w}_n$ that is pinned to zero at $t = ih$, $i = 0,1,\dots,n$, where $h = T/n$.

ii) $E\tilde{w}(t) = 0$ for all $t \in [0,T]$

iii) $E[\tilde{w}(t)\tilde{w}(s)] = h^{-1}(s-ih)((i+1)h-t)$, for $ih \le s \le t \le (i+1)h$
$= 0$ otherwise

iv) $\int_0^h E\tilde{w}(t)^2 dt = \frac{h^2}{6}$

v) $\int_0^h \int_0^h E[\tilde{w}(t)\tilde{w}(s)] ds\, dt = \frac{h^3}{12}$

Proof of Theorem 1 (i): As before, we set $h = T/n$ and we generally suppress the arguments n and w. Now $P = P_0$; so w is a Brownian motion and $\tilde{w}$ a Brownian bridge process. First we introduce a piecewise smooth approximation of (X_t); let $\bar{X}_t(x,w,n)$ be the solution of the ordinary differential equation

$$\dot{\bar{X}}_t = D\gamma_{-\tilde{w}(t)} \cdot f \circ \gamma_{\tilde{w}(t)}(\bar{X}_t) + g(\bar{X}_t)\dot{\bar{w}}(t), \quad \bar{X}_0 = z . \tag{10}$$

Then it follows from the decomposition-of-solution arguments used in the proof of Lemma 1 that $X_t = \gamma_{\tilde{w}(t)} \circ \bar{X}_t$ and $\bar{X}_t = \gamma_{-\tilde{w}(t)} \circ X_t$. Notice that $\bar{X}_{ih} = X_{ih}$ for $i=0,1,\dots,n$, because $\tilde{w}(ih) = 0$. Now let $L_g f$ denote $[g,f]$ regarded as the Lie derivative of the vector field f in the direction of g. The first term of (10) can be expanded as a Lie series. So

$$\begin{aligned}\dot{\bar{X}}_t &= f(\bar{X}_t) + L_g f(\bar{X}_t)\tilde{w}(t) + \tfrac{1}{2} L_g^2 f(\bar{X}_t)\tilde{w}(t)^2 + R_1(t) + g(\bar{X}_t)\dot{\bar{w}}(t) \\ &= f(\bar{X}_t) + L_g f(\bar{X}_t)\tilde{w}(t) + \frac{T}{12n} L_g^2 f(\bar{X}_t) + g(\bar{X}_t)\dot{\bar{w}}(t) + R_1(t) + R_2(t),\end{aligned}$$

where $R_1(t)$ is the integral remainder

$$R_1(t) := \tfrac{1}{2}\tilde{w}(t)^3 \int_0^1 (1-s)^2 \gamma_{s\tilde{w}(t)*}^{-1} L_g^3 f(\bar{X}_t) ds$$

and $R_2(t) := \frac{1}{2}L_g^2 f(\bar{X}_t)w(t)^2 - \frac{T}{12n} L_g^2 f(\bar{X}_t)$.

By well-known moment inequalities for stochastic differential equations we have that for all $r \geq 1$, $\sup_t E|X_t|^r < \infty$; that is, X_t is bounded in L_r-norm uniformly in t. Now $\tilde{w}$ is Gaussian, and it follows from Lemma 3 that $\exp K|\tilde{w}(t)|$ is similarly bounded. But by Lemma 1 (ii),

$|\bar{X}_t| \leq K \exp K|\tilde{w}(t)|\cdot|X_t|$, and $\sup_t E|\bar{X}_t|^r < \infty$ also for all $r \geq 1$. Now it follows from Lemma 1 (ii) and Lipschitz nature of $L_g^3 f$ that $|R_1(t)| \leq k_1 \exp k_2|\tilde{w}(t)|\cdot|X_t|\cdot|\tilde{w}(t)|^3$. Since $|\tilde{w}(t)|$ is $O(h^{\frac{1}{2}})$ in L_r for all $r \geq 1$, uniformly in t, Hölder's inequality implies that $|R_1(t)|$ is $O(h^{3/2})$ in L_r for all r uniformly in t. Similarly by Lemma 3, $|R_2(t)|$ is uniformly $O(h)$.

It is convenient at this point to introduce a special notation for orders. We shall say a parameterized process $U(t,x,n,w)$ is "$O_c(h^\delta)$" if for all integer $r \geq 1$, $\sup_{t,x,n}(E|\frac{U}{b_n h^\delta}|^r) < \infty$ where, as before, $h = T/n$, and where $(b_n(w))$ is a positive (P_n)-adapted sequence bounded uniformly in n. It is clear from Lemma 2 that DX_t, $D\hat{X}_t^{-1}$ and $D\hat{X}_{t*}^{-1}[g,f]$ are all bounded by such a (P_n)-adapted sequence b_n and so are $O_c(1)$ in this sense. Now consider the "error" process $Z_t := \hat{X}_t^{-1}\circ\bar{X}_t$. We have $Z_0(x) = x$. Ordinary calculus yields

$$\begin{aligned}\dot{Z}_t &= \hat{X}_{t*}^{-1}[g,f](Z_t)\tilde{w}(t) + D\hat{X}_t^{-1}(\bar{X}_t)(R_1(t) + R_2(t)) \\ &=: \quad J_1(t) \qquad\qquad + J_2(t) \qquad + J_3(t)\end{aligned} \tag{11}$$

The term $J_1(t)$ has the alternative expression $(D\hat{X}_t^{-1}\cdot[g,f](\bar{X}_t)\tilde{w}(t)$.
Hence $|J_1(t)| \leq k_3 b_n|\bar{X}_t||\tilde{w}(t)|$. An appeal to the previous moment bounds then shows that $J_1(t)$ is $O_c(h^{\frac{1}{2}})$.

Similarly $J_2(t)$ is $O_c(h^{3/2})$ and $J_3(t)$ is $O_c(h)$.

Now an application of Gronwall's inequality to (11) shows that $Z_t(x)-x$ is $O_c(h^{\frac{1}{2}})$, (with b_n being replaced here by $k_4\exp(k_4 b_n T)$ for some k_4) and that for $ih \leq t \leq (i+1)h$ $Z_t - Z_{ih}$ is $O_c(h^{3/2})$. Furthermore it follows from (2) and the equations for its derivatives, and the Lévy modulus theorem, that there is a (P_n)-adapted sequence c_n, uniformly bounded in n, such that

$$|X_{t*}^{-1} L_g^i f(t) - X_{s*}^{-1} L_g^i f(t)| \leq c_n|x|h^{\frac{1}{2}-\delta} \qquad , \quad i = 1 \text{ or } 2$$

for all $|t-s| \leq h$, $0 \leq s \leq t \leq T$, and for $0 < \delta < \frac{1}{2}$. These bounds and the bounds on the derivatives of the vector fields of (11) given by Lemma 2 show that it can be transformed into the discrete form

$$\begin{aligned}Z_{ih+h} - Z_{ih} &= \hat{X}_{ih*}^{-1}[g,f](Z_{ih})\Delta_i\tilde{w} + \tfrac{1}{2}\hat{X}_{ih*}^{-1} L_g f(Z_{ih})\Delta_i u + R_{3,i} \\ &=: \quad J_3(i)\Delta_i\tilde{w} \qquad + J_4(i)\Delta_i u \qquad\qquad + R_{3,i}\end{aligned}$$

where $\Delta_i\tilde{w} = \int_{ih}^{ih+h} \tilde{w}(t)dt$ and $\Delta_i u = \int_{ih}^{ih+h}\tilde{w}(t)^2 dt - \frac{1}{6}h^2$.

The remainder $R_{3,i}$ is $0_c(h^{5/2-\delta})$. Both $J_3(i)$ and $J_4(i)$ are $P_n \vee B_{ih}$-measurable. $\Delta_i\tilde{w}$ and $\Delta_i u$ are independent of this σ-field and have zero mean. It follows from the Brownian bridge properties of $\tilde{w}$ and the usual moment inequalities for martingale transforms that for some (P_n)-adapted sequence d_n, bounded in n,

$$(E|d_n^{-1}\sum_{i=0}^{n-1} J_4(i)\Delta_i u|^2)^{\frac{1}{2}} = 0(h^{3/2})$$

and, more generally, that $\sum_i J_4(i)\Delta_i u$ is $0_c(h^{3/2})$. Clearly

$\sum_i R_{3i} = 0_c(h^{3/2-\delta})$. A further expansion of the coefficients about x gives

$$Z_T(x)-x = \sum_{i=0}^{n-1} \hat{X}_{ih*}^{-1}[g,f](x)\Delta_i\tilde{w} + (\sum_i J_5(i)\cdot(Z_{ih}-x)\cdot\Delta_i\tilde{w} + R_4 \tag{12}$$

$$=: \quad J_6 \quad + \quad R_5$$

where R_4 is $0_c(h^{3/2-\delta})$. Since $J_5(i) := D\hat{X}_{ih*}^{-1}[g,f](y)$ for some $y = x-\theta x+\theta Z_{ih}$, $0\leq\theta\leq 1$, $J_5(i)$ is $0_c(1)$; so $J_5(i)(Z_{ih}-x)$ is $0_c(h^{\frac{1}{2}})$ and is $P_n \vee B_{ih}$ measurable and independent of $\Delta_i\tilde{w}$. It follows, again by martingale arguments, that $\sum_i J_5(i)(Z_{ih}-x)$ $\Delta_i\tilde{w}$ is $0_c(h^{3/2})$.

We now consider the behaviour of the normalized error $\xi_n := n(Z_T(x)-x)$. If R_5 is the sum of the remainder terms in (12), then $nR_5 = 0_c(h^{\frac{1}{2}-\delta})$. So, for all $r > \frac{2}{1-2\delta}$, it follows that for some positive n-bounded (P_n)-adapted sequence (b_n), (remembering that $h = T/n$)

$$\sum_{n=1}^{\infty} E[b_n^{-1} E[n^r R_5^r|P_n]] = \sum_1^{\infty} E\, b_n^{-1}|nR_5|^r < \infty$$

and the moment form of the Borel-Cantelli lemma, and Jensen's inequality, then show that for all $r \geq 1$, $E[n^r R_5^r|P_n]$ converges to zero a.s. .

Now consider nJ_6. This possesses a P_n-conditional normal distribution on $\mathbb{R}^d$ which the same as that of $\sum_i \hat{X}_{ih*}^{-1}[g,f](x)\Delta_i v$, where $(v(t))$ is some Brownian motion on an augmented probability space that is independent of w, and where $\Delta_i v := v(ih+h) - v(ih)$. Given the continuity properties of $\hat{X}_{ih*}^{-1}[g,f]$, it is clear that this distribution converges weakly to the limit normal distribution of the theorem, and that its moments converge correspondingly. The a.s. convergence to zero of all the conditional moments of the remainder terms nR_5 then implies that the distribution of ξ_n behaves in the same way. This completes the proof of (i).

Now consider (ii). Let E and E_0 denote expectations for P and P_0 respectively. Let $M := \frac{dP}{dP_0}$ and assume for the moment that $M > 0$ a.s. (P_0). For $n\in I$, (P_n) is increasing and $\bigvee_n P_n = B_T$; and so $(M_n, P_n, B_T, P_0)_{n\in I}$ is a martingale with $M_n \to M$ a.s. (P_0). Since $E_0[M^2] = E[M] < \infty$, we have that the tail of the quadratic variation

$$E_0[(M-M_n)^2|P_n] = E_0[M^2|P_n] - M_n^2$$
$$\to 0 \quad \text{a.s.} \quad (P_0).$$

To prove (ii) it is sufficient to establish convergence to the corresponding limits of the moments $E[|\xi_n|^2|P_n]$ and the values of the characteristic function $E[\exp ic'\xi_n|P_n]$ for rational $c \in \mathbb{R}^d$. We consider just the first; the argument for the others is the same. Now $M > 0$ a.s. (P_0) implies that $M_n > 0$ a.s. (P_0). We have, by routine arguments,

$$E[|\xi_n|^2|P_n] = \frac{1}{M_n} E_0[M|\xi_n|^2|P_n]$$

$$= E_0[|\xi_n|^2|P_n] + \frac{1}{M_n} E_0[(M-M_n)\cdot|\xi_n|^2|P_n]$$

The second term is bounded by

$$\frac{1}{M_n} E_0[M-M_n)^2|P_n]^{\frac{1}{2}} E_0[|\xi_n|^4|P_n]^{\frac{1}{2}} ;$$

so it follows from (i) and the properties just established for M_n that this converges to zero a.s. (P_0) and that $E[|\xi_n|^2|P_n]$ converges to the correct limit. For cases where $P_0(M = 0) > 0$, we restrict our attention to those w for which $M > 0$; since $\{M > 0\}$ is P_n-measurable and $P(M > 0) = 1$, the results still hold a.s. (P). Similarly, we have that

$$P(E[\exp ic'\xi|P_n] \to \exp(-\tfrac{1}{2}c'V_0c), \text{ all rational } c \in \mathbb{R}^d) = 1$$

and, given the continuity properties of characteristic functions, this suffices to prove the a.s. weak convergence of the conditional distributions. The proof of Theorem 1 is complete.

Proof of Theorem 2: This follows from Theorem 1 and an application of the "δ-method". First notice that $X_T(x) = \hat{X}_T \circ Z_T(x)$; it follows from the smoothness properties of $\hat{X}_T$ that

$$X_T(x) = \hat{X}_T(x) + D\hat{X}_T(Z_T(x)-x) + 0_c(n^{-2}).$$

The limits of the conditional distribution and moments of $n(X_T-\hat{X}_T)$ can be established in exactly the same way as in the proof of Theorem 1 and it is easy to see that these correspond to the conditionally normal distribution of the Ito (or Stratonovich) integral

$$U_T := DX_T(x)\cdot \frac{T}{(12)^{\frac{1}{2}}} \int_0^T X_{t*}^{-1} [g,f](x)dv(t) \tag{13}$$

where v(t) is a Brownian motion independent of w(t)). But it follows from Lemma 2 that

$$d(DX_t) = Df(X_t)\cdot DX_t\, dt + Dg(X_t)\cdot DX_t \circ dw(t),$$

and an application of "Stratonovich" stochastic calculus to (13) then yields the alternative form (6).

REFERENCES

[1] J.M.C. Clark, R.J. Cameron. The maximum rate of convergence of discrete approximations for stochastic differential equations. B. Grigelionis (Ed.) Stochastic Differential Systems, Proc. IFIP-WG 7/1 Working Conference Vilnius 1978, Springer-Verlag Berlin 1980, pp. 162-171

[2] Halim Doss. Liens entre équations différentielles stochastiques et ordinaires. Ann. Inst. Henri Poincaré XIII(2) 1977, p. 99-125

[3] Hiroshi Kunita. On the decomposition of solutions of stochastic differential equations. D. Willems (Ed.) Stochastic Integrals, Proc. Durham Synposium 1980. Lect. Notes in Maths. 851. Springer-Verlag Berlin 1981

[4] E.J. McShane. Stochastic Calculus and Stochastic Models. Academic Press NY 1974.

[5] G.N. Mil'stein. Approximate integration of stochastic differential equations. Theory Prob. Appli. 19 1974, pp. 557-562

[6] N.J. Newton. PhD. Thesis, E.E. Dept. Imperial College, Univ. of London 1982

[7] Holger Rootzén. Limit distributions for the error in approximations of stochastic integrals. Ann. Prob. 1980, 8(2), pp. 241-251

[8] W. Rümelin, Numerical treatment of stochastic differential equations. To appear in SIAM J. Num. Anal. (1982)

[9] Hector J. Sussmann, V. Jurdjevic. Controllability of nonlinear systems. J. Diff. Equ. 12(1) 1972, pp. 95-116

[10] Hector J. Sussmann. On the gap between deterministic and stochastic ordinary differential equations. Ann. Prob. 1978, 6(1), pp. 19-41.

[11] E. Wong, M. Zakai. On the convergence of ordinary integrals to Stochastic integrals. Ann. Math. Statist. 36 (1965)

STOCHASTIC CONTROL WITH NOISY OBSERVATIONS

M.H.A. Davis

Department of Electrical Engineering
Imperial College, London,
ENGLAND

The last few years have seen considerable progress in nonlinear filtering theory; the proceedings [23] of the 1980 Les Arcs summer school can be consulted for an up-to-date account. It is natural to ask what the impact of these developments might be on control theory for stochastic systems with noisy observations, since, as indicated by the "separation principle", filtering plays an essential part in the optimal control of such systems. In my talk at the Cocoyoc meeting I discussed the general problem of control with incomplete observations and outlined some recent approaches based on nonlinear filtering theory. Most of this material is covered in a survey [14] written for a special issue of *Stochastics*. In this paper I aim to provide a very brief summary of recent work together with an updated list of references.

I. PROBLEM FORMULATION

Let us first consider control of a partially-observed diffusion process of the form

$$dx_t = b(x_t, u_t)dt + g(x_t)dv_t \tag{1}$$

$$dy_t = h(x_t)dt + dw_t \tag{2}$$

Here the state process (x_t) takes values in R^d and is governed by equation (1), in which (v_t) is a vector Brownian motion (BM) and (u_t) is the control process. The observation process (y_t) is supposed, for notational convenience, to be scalar, and is given by (2), (w_t) being a BM independent of (v_t). The control u_t should in some sense be a function of the observations $(y_s, 0 \leq s \leq t)$. Precisely how this dependence should be formulated is a major question in the theory. The objective will be to minimize the cost

$$J(u) = E[\Phi(x_T)] \tag{3}$$

where Φ is a non-negative real-valued function and T a fixed terminal time. Some apparently more general cost functions can be put in this form; see for example [15]. Standing assumptions will be that b,g are Lipschitz continuous and have at most linear growth in x_t (uniformly in u in the case of b.) The control process (u_t) takes values in some compact Borel space U.

If $u_t = u(t,\{y_s, 0\leq s\leq t\})$ depends sufficiently smoothly on y then (1), (2) can be considered as a stochastic differential equation in the normal Ito sense. However, one does not wish to be restricted to smooth functions since discontinuous controls of the bang-bang type arise in many situations. One way around this is to use the Girsanov measure transformation; for this it is necessary that the diffusion matrix g(x) be square and non-singular for each x, with, say, bounded inverse. Let (v_t^0, y_t) be independent BMs on some probability space (Ω,F,P) and let x_t be the solution of the Ito equation

$$dx_t = g(x_t)dv_t^0$$

Now let u_t be any process adapted to $Y_t = \sigma\{y_s, 0\leq s\leq t\}$ and define a new measure P^u by the Radon-Nikodym derivative

$$\frac{dP^u}{dP} = \exp(\int_0^T g^{-1}(x_s)b(x_s,u_s)dv_s^0 - \int_0^T |g^{-1}(x_s)b(x_s,u_s)|^2 ds$$

$$+ \int_0^T h(x_s)dy_s - \frac{1}{2}\int_0^T h^2(x_s)ds$$

Under the standing assumptions P^u is a probability measure, i.e. $P^u(\Omega) = 1$, and under P^u, (x_t,y_t) are a weak solution of (1), (2) in that the processes (v_t), (w_t) *defined by* (1), (2) are independent BMs. (u_t) is then a *feedback control* in the direct sense of being adapted to Y_t.

An alternative method of defining a solution, which does not require non-singularity of g, was introduced by Fleming and Pardoux [22]. For a given BM (v_t), equation (1) has a strong (Ito) solution for any given measurable *function* $u : [0,T] \to U$. Denote by $\bar{P}^u$ the measure induced by this solution on the canonical space $C[0,T;R^d]$ and let (x_t) be the coordinate process on $C[0,T;R^d]$. Define

$\Omega = C[0,T;R^d] \times C[0,T;R^1]$ and let (y_t) be the coordinate functions on the second factor, generating σ-fields Y_t, and define a measure $\mathring{P}^u$ on Ω by

$$\mathring{P}^u(dx,dy) = \bar{P}^{u(y)}(dx)\mu_w(dy) \tag{4}$$

where μ_w is Wiener measure. Now define P^u by taking

$$\frac{dP^u}{d\overset{\circ}{P}{}^u} = \exp\left(\int_0^T h(x_s)dy_s - \frac{1}{2}\int_0^T h^2(x_s)ds\right) =: \Lambda_T \tag{5}$$

It is not hard to show that this definition of P^u coincides with that obtained previously via the Girsanov transformation.

A somewhat wider class of controls can be introduced in this framework, following [22]. Let G be the set $L_2[0,T;U]\times C[0,T;R^1]$ with coordinate functions (u_t,y_t) and let (G_t) be the filtration

$$G_t = \sigma\{y_s, \int_0^s u(r)dr, 0\leq s\leq t\}$$

Let A be the set of probability measures on (G,G_t) such that (y_t) is a G_t-BM. This is the set of *wide-sense* controls. For any $\nu \in A$ there is a regular conditional distribution $\gamma(du;y)$ of u given y, so that

$$\nu(du,dy) = \gamma(du;y)\mu_w(dy)$$

The *strict sense* controls A_s are those for which γ is a Dirac measure: $\gamma(du;y) = \delta_{\bar{u}(y)}(du)$. Then the process $\bar{u}_t = \bar{u}(y)(t)$ is adapted to Y_t, so that strict sense controls coincide with feedback controls as previously introduced. The sample space measure corresponding to $\nu \in A$ is given not by (4) but by

$$\overset{\circ}{P}{}^\nu(dx,dy,du) = \bar{P}^u(dx)\nu(du,dy)$$

and finally the measure P^ν is given by the exponential transformation (5) as before. The cost corresponding to $\nu \in A$ is

$$J(\nu) = E^\nu[\Phi(x_T)] = \int_\Omega \Phi(x_1)dP^\nu$$

We now wish to choose $\nu \in A$ so as to minimize $J(\nu)$.

II. NONLINEAR FILTERING AND THE "SEPARATED" PROBLEM

Since the state (x_t) cannot be observed directly we reformulate the problem in terms of observed (i.e. Y_T-measurable) quantities. First, the cost can be written

$$J(u) = E^u[E^u[\Phi(x_T)|Y_T]] \tag{6}$$

It is advantageous to calculate the conditional expectation in terms of measure $\overset{\circ}{P}{}^u$ rather than P^u. The relation between the two is

$$E^u[\Phi(x_T)|Y_T] = \frac{\mathring{E}^u[\Phi(x_T)\Lambda_T|Y_T]}{\mathring{E}^u[\Lambda_T|Y_T]} =: \frac{\sigma_T(\Phi)}{\sigma_T(1)}$$

where $\mathring{E}^u$ denotes integration with respect to measure $\mathring{P}^u$ and "1" denotes the function $1(x) \equiv 1$. Since the denominator does not depend on the function Φ, σ_T can be regarded as an unnormalized conditional distribution of x_T given Y_T, i.e.

$$\sigma_T(\Phi) = \int_{R^d} \Phi(x)\sigma_T(dx)$$

Returning now to (6) we have

$$\begin{aligned} J(u) &= E^u[\sigma_T(\Phi)/\sigma_T(1)] \\ &= \mathring{E}^u[\Lambda_T\sigma_T(\Phi)/\sigma_T(1)] \\ &= \mathring{E}^u[\mathring{E}^u[\Lambda_T|Y_T]\sigma_T(\Phi)/\sigma_T(1)] \\ &= \mathring{E}^u[\sigma_T(\Phi)] \end{aligned} \tag{7}$$

A small extension of standard results in filtering theory shows that the conditional distribution σ_t satisfies the "Zakai equation"

$$d\sigma_t(f) = \sigma_t(A(u_t)f)dt + \sigma_t(hf)dy_t \tag{8}$$

where $A(u)$ is the differential generator

$$A(u)f(x) = \frac{1}{2}\sum_{i,j}(g(x)g'(x))_{ij}\frac{\partial^2 f}{\partial x_i \partial x_j} + \sum_i b_i(x,u)\frac{\partial f}{\partial x_i}(x)$$

Recall that under measure $\mathring{P}^u$, (y_t) is BM. Equations (7), (8) give the control problem in "separated" form. The new "state" is the unnormalized conditional distribution σ_t whose "dynamics" are given by the Zakai equation (8), and the cost is the linear functional at (7). The initial condition for (6) is $\sigma_0 = \pi$ where π is the given probability distribution of the initial state x_0. If this has a density $p_0(x)$, in a suitable function space then σ_t has a density $p(t,x)$ which satisfies the "forward" version of equation (6), namely

$$\begin{aligned} \frac{\partial}{\partial t} p(t,x) &= A^*(u_t)p(t,x)dt + h(x)p(t,x)dy_t \\ P(0,x) &= p_0(x) \end{aligned} \tag{9}$$

See Pardoux [23][31] for details of the Zakai equation in this form .

III CONTROL OF FINITE-STATE MARKOV PROCESSES WITH NOISY OBSERVATIONS

An exactly analogous formulation can be given for control of a Markov process on the state space $\{1, 2, \dots, N\}$ with controlled transition intensities $a_{ij}(u)$:

$$P[x_{t+\delta} = i \mid x_t = j] = a_{ij}(u)\delta + o(\delta)$$

See [16], [8] for details. The observations are as before, i.e.

$$dy_t = h(x_t) + dw_t$$

where (w_t) is a BM independent of (x_t). The unnormalized conditional distribution is given by

$$P[x_t = i \mid \mathcal{Y}_t] = q_t^i / \sum_{j=1}^{N} q_t^j$$

where $q_t' = (q_t^1, \dots, q_t^N)$ satisfies the Zakai equation

$$dq_t = A'(u_t)q_t dt + Hq_t dy_t \tag{10}$$

Here $A(u)$ is the matrix with i, j^{th} entry $a_{ij}(u)$ and H is the diagonal matrix with i, i^{th} entry $h(i)$. This problem is formulated for $\mathcal{Y}_t$-adapted or wide-sense controls in a way exactly similar to the diffusion case. By conditioning on $\mathcal{Y}_t$ the cost function (3) becomes

$$J(u) = E\langle \Phi, q_T \rangle \tag{11}$$

where $\Phi' := (\Phi(1), \dots, \Phi(N))$ and $\langle \cdot \cdot \rangle$ denotes the inner product in R^d. Thus the "separated" problem (10) - (11) concerns control of the degenerate process (q_t). This is an instructive case to study as it is a problem of partial observations where the conditional distribution is finite-dimensional. It has been studied in detail by Bismut [8] but his methods do not generalize to the diffusion case.

IV EXISTENCE OF OPTIMAL CONTROLS

This has been a long-standing open question. Even in the linear case satisfactory existence results are available only under very restrictive conditions [15][35]. For nonlinear systems Christopeit [10] and Elliott and Kohlmann [20] have given existence results for systems with rather specific information patterns, using arguments not involving nonlinear filtering. The result of [20] has also been proved by Cutland [13] by a method involving non-standard analysis.

The best recent contribution in this area is undoubtedly that of Fleming and Pardoux [22] who demonstrate the existence of an optimal control in the class $\mathcal{A}$ for the problem described in §I above. Their argument takes its simplest form when the cost is a terminal cost as in (3). The cost is re-expressed as in (7), but with σ_t now being the un-normalized conditional distribution given G_t (rather than Y_t as would be appropriate for strict-sense controls). $\sigma_T(\Phi)$ is then a G_T-measurable random variable, and $\mathring{P}^{\nu}$ restricted to G_T is just the control measure ν, so that

$$J(\nu) = \int_G \sigma_T(\Phi)d\nu$$

The class of control measures A is compact under weak convergence, so it remains to show that $J(\nu)$ is lower semi-continuous, and this is done in [22] under some additional assumptions on the system coefficients. The argument is more involved if $J(\nu)$ includes a "running cost". There is some evidence, adduced in [22], that the optimal control is not a strict sense one, so that the generalized controls cannot be dispensed with. Their status is not entirely clear :they are not randomized or relaxed controls in the conventional sense of the term (this would involve selecting a distribution over U at each time t in a Y_t-adapted way) although they can be approximated by sequences of piecewise-constant relaxed controls. These remarks are relevant to the question of conditions for optimality, discussed in §VII below.

V DYNAMIC PROGRAMMING

A formal "Bellman equation" for a problem similar to the "separated" problem (7) (8) was derived many years ago by Mortensen [28] and the corresponding equation for the Markov chain problem was given even earlier by Shiryaev [33]. Little, however, could be said about the solution of these equations. This subject has been revisited by Beneš and Karatzas [2] [5]. The "Mortensen equation" is to be solved for the function V(t,p) representing the minimal cost starting at time t with the state x_t having density p. The unnormalized density then evolves according to the Zakai equation (9) with p(t,x)=p. In [2] the densities live in an L_1 space and the Mortensen equation involves first and second Fréchet derivatives in the p variable. A change-of variables formula is proved for L_1-valued processes and this is used to prove a "verification theorem", to the effect that any solution of the Mortensen equation gives a lower bound to the achievable cost. The class of controls that arises naturally from the dynamic programming formalism is that of functions $u(t,p_t)$, i.e. feedback of the current "state", the state being in this case the unnormalized conditional density. One is now faced with the problem of solving the

the state equation (9) with $u_t = u(t,p_t)$. The existence of strong solutions is shown in [5] for sufficiently smooth dependence of $u(t,p)$ on p, but no general conditions have so far been stated, to my knowledge, under which the control constructed in the usual way from the solution of the (Bellman-)Mortensen equation would have these smoothness properties. In view of the remarks in the previous section it seems unlikely that such conditions can be given in any generality since this would imply the existence of optimal strict-sense controls.

VI. NONLINEAR SEMIGROUPS

Let $W(t,p,\Phi)$ be the value function for the control with initial density p, *time-to-go* t and terminal cost function Φ. Thus $W(t,p,\Phi) = V(T-t,p)$ in the notation of the previous section. A natural way of formulating Bellman's "principle of optimality" is to say that W has the semigroup property

$$W(t+s,p,\Phi) = W(t,p,W(s,\cdot,\Phi)) \tag{12}$$

If we define $S_t\Phi(p) = W(t,p,\Phi)$ then S_t is a family of (non-linear) operators acting on real-valued functions and (12) becomes

$$S_{t+s}\Phi(p) = S_t(S_s\Phi)(p)$$

If W is obtained by solving the Bellman-Mortensen equation the semi-group property is automatic since the equation is of "evolution" type, evolving forwards from initial data. Since, however, little informat-ion is available about the Bellman equation, there is some interest in examining the semigroup S_t directly. This idea was initiated by Nisio[29], [30] for a variety of control problems with complete obser-vations. Her method was to construct S_t as the lower envelope of a sequence $S_t^{(n)}$ of operators corresponding to piecewise-constant con-trols. Bensoussan and Lions [7] took a different approach, *defining* $W(t,p,\Phi)$ as the infimal cost and then establishing directly its semi-group and other properties.

For the Markov chain problem with noisy observations the Zakai equation (10) for the un-normalized conditional density is finite-dimensional, and hence the "separated" problem falls within the scope of Nisio's original theory; the details were worked out in Davis [16]. The diffusion case has been studied by Fleming [21] following the Bensoussan/Lions "direct" approach. In his formulation the Zakai equation (8) is taken to define (for fixed control value u) a process σ_t taking values in the set M of positive measures on R^d with the weak* topology. The semigroup S_t is then constructed, acting on functions $\Phi : M \to R$ satisfying growth properties of the form $\Phi(\mu) \leq (1+\|\mu\|)^k$ for some k.

Davis and Kohlmann [19] noted - but have so far failed to exploit successfully - some very strong convexity properties in this problem. Details will be found in [14]. The Zakai equation (8) is a bilinear equation and hence linear in the initial condition $\sigma_0 = \mu$. The cost $\mathring{E}^u[\sigma_T(\Phi)]$ is a linear functional of σ_T. Using these facts it is easy to show that the infimal cost over *constant controls* regarded as a function $S(\mu)$ of the initial measure is a *support function*, i.e. it satisfies conditions of

- positive homogeneity : $S(\lambda\mu) = \lambda S(\mu)$, $\lambda \in R^+$, $\mu \in M$
- concavity : $S(\mu_1+\mu_2) \geq S(\mu_1) + S(\mu_2)$

It is argued in [19] that support functions form the natural class of functions on which the semigroup S_t should act. However, certain topological questions arise: if one takes the space M with any weak topology then (in contrast to the situation in finite-dimensional spaces) concave functions are not automatically continuous, and the Nisio construction depends on some continuity properties. There seem to be two ways around this obstacle. N. El Karoui [25] has outlined a semigroup construction which does not depend on continuity properties. It would, however, be disappointing if continuity had to be thrown away altogether since one major purpose of constructing S_t is to establish smoothness properties of the value function. An alternative approach is to take the Zakai equation in its density function form (9) and prove that the solution evolves in some barrelled space (e.g. Hilbert space). Then the required continuity is automatic [32]. See Bensoussan [6].

VII. NECESSARY CONDITIONS FOR OPTIMALITY

Consider first the separated Markov chain problem (10), (11). One notes that the observation process (y_t) appears in the rôle of "noise" in (10) so that this problem looks like a "stochastic open-loop" problem if Y_t-adapted controls are being considered. Thus, the stochastic maximum principles given by Bismut [9] and Kushner [26] are applicable. Bismut's result states that if u^o is optimal and (q_t) is the corresponding solution of (10) then

$$\langle A(u_t^o)p_t, q_t\rangle = \max_{a\in U} \langle A(a)p_t, q_t\rangle \qquad (13)$$

where λ_t is the adjoint process which is characterized in the following way : there is a unique pair of processes (λ_t, η_t) adapted to Y_t such that $\lambda_T = -\Phi$ and

$$d\lambda_t = A'(u_t^o)\lambda_t dt + H\eta_t dt + \eta_t dy_t \tag{14}$$

For the diffusion problem of §II it is appropriate to consider the dynamics in density function form (9). Then reasoning in a purely formal manner one obtains the same maximum principle (13)(14), A now being the differential generator and H the operator of multiplication by h(x). This idea will be found in Kwakernaak [27] and a rigorous version will be found in Bensoussan [6]. However, such results require caution : as mentioned in §III above it seems highly likely that in many cases no strict sense optimal control exists, in which case necessary conditions are, of course, vacuous. Necessary conditions should be given in a framework for which good existence results are available. In the present context, this means the wide-sense controls of Fleming-Pardoux [22]. But as mentioned above, it is not currently understood in what way these controls relate to "local" choice of control value and this is of course just what one needs to know in order to formulate local conditions for optimality.

VIII. PREDICTED MISS PROBLEMS

In 1977 V.E. Beneš started a minor industry with the publication of his paper [1] on bang-bang stochastic control. The problem is to steer a linear-system to a hyperplane in fixed time using bounded controls, and the solution is to use bang-bang controls of the form $u_t = -\text{sgn}(s'(t)x_t)$ where x_t is the state and s(t) a deterministic vector-valued function. The problem is interesting because it has an explicit solution but this solution cannot be obtained by standard application of dynamic programming since the control function is discontinuous. The result was re-proved and extended by several authors : Ikeda and Watanabe [24], Davis and Clark [18], Shreve [34], Christopeit and Helmes [11]. Recently Beneš and Karatzas [3] have shown that the optimal control using partial observations is just $u_t = -\text{sgn}(s'(t)\hat{x}_t)$ where $\hat{x}_t = E[x_t|Y_t]$. This is one of the very few partially observable problems which can be solved explicitly, and, interestingly, although it is a nonlinear problem it exhibits the "certainty-equivalence" principle - the optimal control is obtained by estimating the state and then using this estimate as though it were the true state. Christopeit and Helmes [11] have further information on this problem.

IX. CONCLUDING REMARKS

As will be apparent from the above brief survey, much interesting work has been done on the partially-observable stochastic control problem, but some aspects of the theory are still in a rudimentary state. The nonlinear semigroup side is perhaps the best-developed, with satisfactory treatments existing in [21] and [6]. Further substantial progress with the Bellman-Mortensen equation seems to me very hard, while progress on a necessary conditions must await satisfactory resolution of some questions in existence theory which remains - shall we say - a challenging area. Certainly study of particular problems admitting explicit solutions should continue.

X. REFERENCES

ZW = Zeitschrift für Wahrscheinlichkeitstheorie und verwandte Gebiete
SICON = SIAM Journal on Control (and Optimization)

[1] V.E. Beneš, Full "bang" to reduce predicted miss is optimal, SICON 15(1976)62-83

[2] V.E. Beneš & I. Karatzas, On the relation of Zakai's and Mortensen's equations, submitted to SICON

[3] V.E. Beneš and I. Karatzas, Examples of optimal control for partially-observable systems: comparison, classical and martingale methods, Stochastics 5(1980)43-64

[4] V.E. Beneš and I. Karatzas, Estimation and control for linear partially-observable systems with non-gaussian initial distribution, 10th Conference on Stochastic Processes and their Application, Montreal, Canada, 1981

[5] V.E. Beneš and I. Karatzas, Filtering of diffusions controlled through their conditional measures, 20th IEEE Conference on Dicision and Control, San Diego, Cal. 1981

[6] A. Bensoussan, this volume

[7] A. Bensoussan and J.L. Lions, Applications des inequations variationelles en contrôle stochastique, Dunod, Paris, 1978.

[8] J.M. Bismut, Un problème de contrôle stochastique avec observation partielle, ZW 49(1979)63-95

[9] J.M. Bismut, An introductory approach to duality in optimal stochastic control, SIAM Review 20(1978)62-78

[10] N. Christopeit, Existence of optimal stochastic control under partial observation, ZW 51(1980)201-213

[11] N. Christopeit and K. Helmes, Optimal control for a class of partially-observable systems, Stochastics, to appear

[12] N. Christopeit and K. Helmes, On Beneš' bang-bang control problem, preprint, University of Bonn, 1981

[13] N.J. Cutland, Optimal controls for partially observed stochastic systems : an infinitesimal approach, submitted to Stochastics.

[14] M.H.A. Davis, Some current issues in stochastic control theory, Stochastics, to appear

[15] M.H.A. Davis, Martingale methods in stochastic controls, in Stochastic Control and Stochastic Differential Systems, Lecture Notes in Control and Information Sciences 16, Springer-Verlag, Berlin

[16] M.H.A. Davis, Nonlinear semigroup in the control of partially-observable stochastic systems, in Measure Theory and Applications to Stochastic Analysis, Lecture Notes in Mathematics 695, Springer-Verlag, Berlin 1978

[17] M.H.A. Davis, The separation principle in stochastic control via Girsanov solutions, SICON 14(1976)176-188

[18] M.H.A. Davis and J.M.C. Clark, On "predicted-miss" stochastic control problems, Stochastics 2(1979)197-210

[19] M.H.A. Davis and M. Kohlmann, On the nonlinear semigroup of stochastic control under partial observations, preprint, Imperial College, London, 1981

[20] R.J. Elliott and M. Kohlmann, On the existence of optimal partially observed controls, preprint, Hull University, 1981

[21] W.H. Fleming, Nonlinear semigroup for controlled partially-observed diffusions, SICON , 20(1982)286-301

[22] W.H. Fleming and E. Pardoux, Existence of optimal controls for partially-observed diffusions, SICON , 20(1982)261-285

[23] M. Hazewinkel and J.C. Willems (eds), Stochastic Systems; the mathematics of filtering and identification and applications, D. Reidel, Dordrecht, 1981

[24] N. Ikeda and S. Watanabe, A comparison theorem for solutions of stochastic differential equations and its applications, Osaka J. Math. 14(1977)619-633

[25] N. El Karoui, personal communication

[26] H.J. Kushner, Necessary conditions for continuous time stochastic optimization problems, SICON 10(1972)550-565

[27] K. Kwakernaak, A minimum principle for stochastic control problems with output feedback, Systems & Control Letters 1(1981) 74-77

[28] R.E. Mortensen, Stochastic optimal control with noisy observations, International J. Control 4(1966)455-464

[29] M. Nisio, *Stochastic Control Theory*, Indian Statistical Institute Lecture Notes No.9, MacMillan, Delhi, 1981

[30] M. Nisio, On stochastic optimal controls and envelope of Markovian semigroups, in *Stochastic Differential Equations*, ed. K. Ito, Wiley, New York, 1977.

[31] E. Pardoux, Equations du filtrage non linéaire, de la prediction et du lissage, Stochastics 6(1982)193-232

[32] R.T. Rockafellar, Level sets and continuity of conjugate convex functions, Trans. Am. Math. Soc. 123(1966)46-63

[33] A.N. Shiryaev, Some new results in the theory of controlled stochastic processes[in Russian] Trans. 4th Prague Conference on Information Theory, Statistical Decision Functions and Random Processes, Czech Academy of Sciences, Praha, 1967

[34] S.E. Shreve, Reflected Brownian motion in the "bang-bang" control of Brownian drift, SICON 19(1981)469-478

[35] W.M. Wonham, On the separation theorem of stochastic control, SICON 6(1968)312-326

APPLICATIONS OF DUALITY TO MEASURE-VALUED DIFFUSION PROCESSES

Donald A. Dawson,
Department of Mathematics and Statistics,
Carleton University,
Ottawa, Canada K1S 5B6.

Thomas G. Kurtz,
Department of Mathematics,
University of Wisconsin-Madison,
Madison WI 53706, U.S.A.

1. INTRODUCTION TO MEASURE-VALUED DIFFUSION PROCESSES

Examples of measure-valued diffusion processes arise in nonlinear filtering theory as solutions of stochastic partial differential equations which describe the evolution of the normalized and unnormalized conditional distributions. In other applications such as population genetics examples of measure-valued diffusions have arisen which cannot be obtained as strong solutions of stochastic partial differential equations. In these examples the probability law of the measure-valued process is characterized as the unique solution of a martingale problem. In this approach to measure-valued diffusion processes a key role is played by the notion of a dual process which was originally introduced in the study of infinite particle systems. The purpose of this paper is to describe the martingale problem formulation of measure-valued diffusion processes with special emphasis on the role of duality and to present a number of examples.

2. MARTINGALE PROBLEM FORMULATION OF MEASURE-VALUED DIFFUSION PROCESSES

Let S denote a complete separable metric space and $M(S)$ denote the space of bounded Radon measures on S. With an appropriate choice of metric $M(S)$ is also a complete separable metric space whose topology is equivalent to that of weak convergence of measures. Let E denote a closed subset of $M(S)$, $\mathcal{B}(E)$ the σ-algebra of Borel subsets of E and $L_\infty(E)$ the space of bounded measurable functions on E. Let $\Omega_C^E := C([0,\infty),E)$, $\Omega_D^E := D([0,\infty),E)$, the spaces of functions from $[0,\infty)$ into E which are continuous, right continuous with left limits, respectively. For $s \geq 0$, $\omega \in \Omega_C^E$, let $X(s,\omega) := \omega(s)$ and let

$$\mathcal{F}_t := \bigcap_{\varepsilon>0} \sigma\{X(s)\colon 0 \leq s \leq t+\varepsilon\}.$$

For a $\mathcal{F}_t$-stopping time τ, let $\mathcal{F}_\tau := \{A\colon \text{for } s \in [0,\infty),\ A\cap\{\tau\leq s\} \in \mathcal{F}_s\}$.

An <u>E-valued diffusion process</u> is given by a family of probability measures $\{P_\mu\colon \mu \in E\}$ on Ω_C^E such that

(2.1) *(Measurable)* the mapping $\mu \to P_\mu$ is measurable from E into $\mathcal{P}(\Omega_C^E)$, the space of probability measures on Ω_C^E furnished with the topology of weak convergence,

(2.2) $P_\mu(\{\omega : X(0,\omega) = \mu\}) = 1$,

(2.3) *(Strong Markov Property)* for $\mu \in E$, $t \geq 0$, $f \in L_\infty(E)$, and all a.s. finite F_t-stopping times τ, $E_\mu(\, f(X(\tau+t)) \,|F_\tau) = T(t)\, f(X(\tau))$, μ-a.s., where $T(t)\, f(\mu) = \int f(X(t,\omega))\, P_\mu(d\omega)$.

Given a subset $A \subset C_b(E) \times L_\infty(E)$, a measurable E-valued stochastic process $\{Y(t) : t \geq 0\}$ defined on a probability space (Ω, G, P) with filtration G_t is said to be a <u>solution to the martingale problem</u> for A if for every $(f,g) \in A$,

(2.4) $f(Y(t)) - \int_0^t g(Y(s))ds$ is a G_t-martingale.

A family of probability laws $\{P_\mu : \mu \in E\}$ is said to be a $P(\Omega_D^E)[P(\Omega_C^E)]$-<u>solution to the initial value martingale problem for A</u> if for every $(f,g) \in A$,

(2.5) for each $\mu \in E$, $P_\mu \in P(\Omega_D^E)[P(\Omega_C^E)]$ and $P_\mu(\{\omega : X(0,\omega) = \mu\}) = 1$,

(2.6) for each probability law P_μ the canonical process $\{X(t) : t \geq 0\}$ is a solution to the martingale problem for A.

2.1. <u>THEOREM (Stroock and Varadhan)</u>

Let Φ denote a subset of $L_\infty(E)$ which generates $L_\infty(E)$ under bounded pointwise convergence (e.g. $C_b(E)$). Assume that for each $t \geq 0$ and $F \in \Phi$ there is a function $\Lambda_{F,t} \in L_\infty(E)$ such that for $\mu \in E$,

(2.7) $E_\mu(F(X(t)) = \Lambda_{F,t}(\mu)$

for any $P(\Omega_D^E)[P(\Omega_C^E)]$-solution $\{P_\mu : \mu \in E\}$ of the initial-value problem for A. Then there is at most one $P(\Omega_D^E)[P(\Omega_C^E)]$-solution to the initial-value problem for A and if such a solution exists, then it is measurable and satisfies the strong Markov property (2.3).

Proof. Refer to Stroock and Varadhan ([11]; section 6.2) or Ethier and Kurtz ([3]; section 5.4).

In the finite dimensional case a large class of diffusion processes can be characterized as unique solutions of martingale problems associated with second order elliptic differential operators (refer to Stroock and Varadhan ([11];chapters 6,7)). By analogy a large class of measure-valued diffusion processes can be obtained from martingale problems associated with second order differential operators of the following form: for $F \in D(L)$,

(2.8) $LF(\mu) := A(\mu)(\delta F(\mu)/\delta\mu(x)) + \frac{1}{2} B(\mu)(\delta^2 F(\mu)/\delta\mu(x)\ \delta\mu(y))$

where

(2.9) $\delta F(\mu)/\delta\mu(x)$ is the variational derivative defined by:

$\delta F(\mu)/\delta\mu(x) := \lim_{\varepsilon\downarrow 0} [F(\mu + \varepsilon a_x) - F(\mu)]/\varepsilon$, where $a_x \in M(S)$ consists of a unit atom at $x \in S$;

(2.10) for each $\mu \in E$, $A(\mu)$ is a linear functional defined on $D(A) \subset C(S)$ and $B(\mu)$ is a bilinear functional defined on $D(B) \subset C(S\times S)$,

(2.11) for $F \in D(L)$, $\delta F(\mu)/\delta\mu(.) \in D(A)$ and $\delta^2 F(\mu)/\delta\mu(.)\,\delta\mu(.) \in D(B)$.

The set $D(L) \subset C(E)$, the space of continuous functions on E, must be chosen so that the first and second variational derivatives exist and can be computed and it must be also sufficiently rich so that it generates $L_\infty(E)$ under bounded pointwise convergence. With this motivation we now introduce a family of functions on M(S) which has these properties.

In this development we assume that S is locally compact with one-point compactification S^*. Let S^{*N} denote the N-fold Cartesian product of S^* and

$$\mathcal{C} := \bigcup_{N=0}^{\infty} C(S^{*N}) \text{ (disjoint union), where } C(S^{*0}) := R^1.$$

For $f \in \mathcal{C}$, $N(f) := N$ if $f \in C(S^{*N})$. Similarly we define

$$\mathcal{D} := \bigcup_{N=0}^{\infty} D(S^{*N})$$ where for each N, $D(S^{*N})$ is assumed to be a dense subset of $C(S^{*N})$.

For $f \in \mathcal{C}$, a function on M(S) of the form

(2.12) $F_f(\mu) := \int_{S^*}\cdots\int_{S^*} f(x_1,\ldots,x_{N(f)})\,\mu^{(N(f))}(dx)$, where

$\mu^{(N)}(dx) := \mu(dx_1)\ldots\mu(dx_N)$,

is said to be a <u>monomial</u> on $M(S^*)$. For a compact subset E of $M(S^*)$ we denote by $\Pi_1(E)$ the set of monomials restricted to E. The <u>algebra of polynomials</u> $\Pi(E)$ is defined to be the smallest algebra of functions on E which contains $\Pi_1(E)$. The sets $\Pi_1^{\mathcal{D}}(E)$, $[\Pi^{\mathcal{D}}(E)]$ denote the monomials [polynomials] on E with coefficients $f \in \mathcal{D}$.

2.2. <u>LEMMA</u>

(i) $\Pi^{\mathcal{D}}(E)$ is dense in C(E).

(ii) $\Pi^{\mathcal{D}}(E)$ generates $L_\infty(E)$ under bounded pointwise convergence.

Proof: Let $|||\cdot|||_N$ denote the supremum norm on $C(S^{*N})$. If $f, f_n \in C(S^{*N})$ are such that $|||f - f_n|||_N \to 0$ as $n \to \infty$, then $|F_{f_n}(\mu) - F_f(\mu)| \to 0$ uniformly on E. From this it follows that $\Pi^{D}(E)$ is dense in $\Pi(E)$. If μ and $\nu \in E$ and $\int f(x)\mu(dx) = \int f(x)\nu(dx)$ for all $f \in D(S^*)$, then $\mu = \nu$. Therefore the algebra $\Pi^{\mathcal{D}}(E)$ separates points of E. The Stone Weierstrass theorem then implies that

$\Pi^{D}(E)$ is dense in $C(E)$. Part (ii) follows from (i) since $C(E)$ generates $L_\infty(E)$ under bounded pointwise convergence.

For a function of the form (2.12) the variational derivatives are given by:

$$\delta F_f(\mu)/\delta\mu(x) = \sum_{j=1}^{N(f)} \int_{S^*}\cdots\int_{S^*} f(x_1,\ldots,x_{j-1},x,x_{j+1},\ldots,x_{N(f)})\ \mu^{(N(f)/j)}(dx)$$

$$\delta^2 F_f(\mu)/\delta\mu(x)\delta\mu(y)$$
$$= \sum_{j=1}^{N(f)} \sum_{\substack{k=1\\ j\neq k}}^{N(f)} \int_{S^*}\cdots\int_{S^*} f(x_1,\ldots,x_{j-1},x,x_{j+1},\ldots,x_{k-1},y,x_{k+1},\ldots,x_{N(f)})\mu^{(N(f)/jk)}(dx)$$

where
$$\mu^{(N/j)}(dx) = \prod_{\substack{\ell=1\\ \ell\neq j}}^{N} \mu(dx_\ell); \qquad \mu^{(N/jk)}(dx) = \prod_{\substack{\ell=1\\ \ell\neq j,k}}^{N} \mu(dx_\ell).$$

The operator (2.8) is said to be a <u>differential operator with polynomial coefficients of degree K</u> if acting on a function of the form (2.12), it can be written in the form: for $f \in D$,

(2.13) $$LF_f(\mu) = F_{G_{N(f)}f}(\mu) + \sum_{\ell=-1}^{K} \sum_{j=1}^{N(f)} (F_{A_\ell^{(j)}f}(\mu) - F_f(\mu))$$
$$+ \tfrac{1}{2} \sum_{\ell=-1}^{K} \sum_{j=1}^{N(f)} \sum_{\substack{k=1\\ j\neq k}}^{N(f)} (F_{B_\ell^{(jk)}f}(\mu) - F_f(\mu)) \quad + \quad V(N(f)).F_f(\mu)$$

where $V(n) = c_1 n + c_2 n(n-1)$,

G_N is assumed to be the infinitesimal generator of a strongly continuous semigroup of operators on $C(S^{*N})$ whose domain contains $D(S^{*N})$

A_ℓ is a linear mapping from $D(S^*)$ to $D(S^{*(1+\ell)})$, $\ell = -1,0,1,\ldots,K$, and $A_\ell^{(j)}f$ denotes the action of A_ℓ acting on f as a function of the jth variable,

B_ℓ is a linear mapping from $D(S^{*2})$ to $D(S^{*(1+\ell)})$, $\ell = -1,0,1,\ldots,K$, and $B_\ell^{(jk)}f$ denotes the action of B_ℓ acting on f as a function of the jth and kth variables.

Let Q denote the linear space of functions on D which contains functions of the form:

(2.14) $F(\mu,f) = F_\mu(f) := F_f(\mu)$ for $\mu \in E$, $f \in D$.

Then (2.13) can be rewritten as

(2.15) $LF_f(\mu) = L^{\#}F_\mu(f) + V(N(f))F_\mu(f)$

where for $F_\mu \in Q$,

$$(2.16)\quad L^{\#}F_\mu(f) = F_\mu(G_{N(f)}f) + \sum_{\ell=-1}^{K}\sum_{j=1}^{N(f)} a_\ell[F_\mu(A_\ell^{(j)}f/a_\ell) - F_\mu(f)]$$

$$+ \tfrac{1}{2}\sum_{\ell=-1}^{K}\sum_{j=1}^{N(f)}\sum_{\substack{k=1\\ j\neq k}}^{N(f)} b_\ell[F_\mu(B_\ell^{(jk)}f/b_\ell) - F_\mu(f)]$$

where a_ℓ, b_ℓ are arbitrary positive real numbers and $V(n) = c_1 n + c_2 n(n-1)$.

The application of the duality method grows out of the observation that $L^{\#}$ has the structure of an infinitesimal generator of a D-valued Markov process restricted to functions in Q. The dynamics of the D-valued Markov process $\{Y(t):t \geq 0\}$ (if it exists) with infinitesimal generator $L^{\#}$ involves two basic mechanisms:

(2.17) *(Jump mechanism)* $f \to (A_\ell^{(j)}f)/a_\ell$ with jump rate a_ℓ, $j = 1,2,\ldots,N(f)$,

$f \to (B_\ell^{(jk)}f)/b_\ell$ with jump rate $\tfrac{1}{2}b_\ell$, $j,k = 1,2,\ldots,N(f)$;

(2.18) *(Spatial diffusion semigroup)* between jumps $Y(t)$ evolves in a continuous and deterministic way, namely

$$f \to S_t^{N(f)} f$$

where $\{S_t^N : t \geq 0\}$ represents the semigroup of transformations on $C(S^{*N})$ with infinitesimal generator G_N. We assume that $D(S^{*N})$ is invariant under S_t^N.

If $V(.) = 0$, $L^{\#}$ is the infinitesimal generator of a conservative right continuous D-valued Markov process $\{Y(t):t > 0\}$ and L is the infinitesimal generator of an E-valued diffusion $\{X(t):t \geq 0\}$, then

$$(2.19)\quad E_\mu(F_f(X(t))) = E(F_\mu(Y(t))\,|\,Y(0) = f).$$

Processes X and Y which satisfy the relation (2.19) are said to form a pair of <u>dual processes</u>. The development of the theory of dual processes in a general setting is the objective of the next section.

3. DUALITY RELATIONS FOR SOLUTIONS OF MARTINGALE PROBLEMS

The existence of a process Y dual to a process X can provide a powerful tool to study the latter if it turns out that the dual process is in some sense more tractable. Important applications of dual processes arise in the study of infinite interacting systems (Spitzer [10], Holley and Liggett [4], and Holley and Stroock [6]) as well as finite dimensional diffusions (Holley, Stroock and Williams [5] and Shiga [9]). In addition applications to infinite dimensional diffusions have recently been developed (Shiga [8], Dawson and Hochberg [2]).

Let E_1 and E_2 be metric spaces and $C(E_1 \times E_2)$ denote the space of continuous functions on $E_1 \times E_2$.

Let

(3.1.a) $A_1 := \{(f(.,y),g(.,y)):y \in E_2\}$, where $f,g \in C(E_1\times E_2)$, $\alpha \in C(E_1)$;

(3.1.b) $A_2 := \{(f(x,.),h(x,.)):x \in E_1\}$, where $f,h \in C(E_1\times E_2)$, $\beta \in C(E_2)$.

Let $\{X(t):t \geq 0\}$ be an E_1-valued measurable process which is adapted to a filtration F_t and let τ be a F_t-stopping time. Let $\{Y(t):t \geq 0\}$ be an E_2-valued measurable process which is adapted to a filtration G_t and let σ be a G_t-stopping time. Assume that X and τ are independent of Y and σ. We also assume that for each $T > 0$ there is an integrable random variable Γ_T such that

(3.2.a) $$\sup_{r,s,t\leq T} |\{1 \vee \alpha(X(r\wedge\tau))\}f(X(s\wedge\tau),Y(t\wedge\sigma))| \leq \Gamma_T , \quad (x\wedge y) := \min(x,y),$$

(3.2.b) $$\sup_{r,s,t\leq T} |\{1 \vee \beta(Y(r\wedge\sigma))\}f(X(s\wedge\tau),Y(t\wedge\sigma))| \leq \Gamma_T , \quad (x\vee y) := \max(x,y),$$

(3.2.c) $$\sup_{r,s,t\leq T} |\{1 \vee \alpha(X(r\wedge\tau))\}g(X(s\wedge\tau),Y(t\wedge\sigma))| \leq \Gamma_T ,$$

(3.2.d) $$\sup_{r,s,t\leq T} |\{1 \vee \beta(Y(r\wedge\sigma))\}h(X(s\wedge\tau),Y(t\wedge\sigma))| \leq \Gamma_T ,$$

and there is a constant c_T such that

(3.2.e) $$\sup_{t\leq T} \left|\int_0^{t\wedge\tau} \alpha(X(u))du\right| + \sup_{t\leq T} \left|\int_0^{t\wedge\sigma} \beta(Y(u))du\right| \leq c_T .$$

3.1. <u>THEOREM</u>

Assume that for each $y \in E_2$,

(3.3.a) $f(X(t\wedge\tau),y) - \int_0^{t\wedge\tau} g(X(s),y)ds$ is a F_t-martingale,

and for each $x \in E_1$,

(3.3.b) $f(x,Y(t\wedge\sigma)) - \int_0^{t\wedge\sigma} h(x,Y(s))ds$ is a G_t-martingale.

Then

(3.4) $$E[f(X(t\wedge\tau),Y(0)).\exp(\int_0^{t\wedge\tau} \alpha(X(u))du)] - E[f(X(0),Y(t\wedge\sigma)).\exp(\int_0^{t\wedge\sigma} \beta(Y(u))du)]$$

$$= \int_0^t E[\{\chi_{\{s\leq\tau\}}\{g(X(s),Y((t-s)\wedge\sigma)) + \alpha(X(s))f(X(s\wedge\tau),Y((t-s)\wedge\sigma))\}$$

$$- \chi_{\{(t-s)\leq\sigma\}} \{h(X(s\wedge\tau),Y(t-s)) + \beta(Y(t-s))f(X(s\wedge\tau),Y((t-s)\wedge\sigma))\}\}$$

$$.\exp(\int_0^{s\wedge\tau} \alpha(X(u))du + \int_0^{(t-s)\wedge\sigma} \beta(Y(u))du)]ds ,$$

where $\chi_{\{s\leq\tau\}}$ denotes the indicator function of the event $\{s \leq\tau\}$.

The proof of the theorem is based on the following three lemmas.

3.1.1. LEMMA

Suppose that the function $f(s,t)$ is absolutely continuous in s for each fixed t and absolutely continuous in t for each fixed s. Let $f_1 = \partial f/\partial s$, $f_2 = \partial f/\partial t$, and assume that

$$(3.5)\quad \int_0^T\int_0^T |f_i(s,t)|\,dsdt < \infty,\ i=1,2,\ T>0.$$

Then for almost every t,

$$(3.6)\quad f(t,0) - f(0,t) = \int_0^t [f_1(s,t-s) - f_2(s,t-s)]\,ds .$$

Proof. $\int_0^T\int_0^t (f_1(s,t-s) - f_2(s,t-s))\,dsdt$

$$= \int_0^T\int_0^t f_1(t-s,s)\,dsdt - \int_0^T\int_0^t f_2(s,t-s)\,dsdt$$

$$= \int_0^T\int_s^T f_1(t-s,s)\,dtds - \int_0^T\int_s^T f_2(s,t-s)\,dtds$$

$$= \int_0^T (f(T-s,s) - f(0,s))\,ds - \int_0^T (f(s,T-s) - f(s,0))\,ds$$

$$= \int_0^T (f(s,0) - f(0,s))\,ds .$$

By differentiating with respect to T we obtain

$$\int_0^t (f_1(s,t-s) - f_2(s,t-s))\,ds = f(t,0) - f(0,t),$$

and the proof is complete.

3.1.2. LEMMA

Let $\{\gamma(s): s \geq 0\}$ be a measurable process adapted to a filtration F_t, $M(t)$ be a F_t-martingale, and $\{\alpha(s): s \geq 0\}$ be a measurable process adapted to F_t. If $\phi(t) := \int_0^t \gamma(s)ds + M(t)$, and all expressions are well-defined, then

$$(3.7)\quad E[\phi(t+h).\exp(\int_0^{t+h} \alpha(u)du) - \phi(t).\exp(\int_0^t \alpha(u)du)]$$

$$= E[\int_t^{t+h} \gamma(s)ds.\exp(\int_0^t \alpha(u)du) + (\int_t^{t+h} \alpha(u).\exp(\int_0^u \alpha(r)dr)du).\phi(t+h)] .$$

Proof: $E[\phi(t+h).\exp(\int_0^{t+h} \alpha(u)du) - \phi(t).\exp(\int_0^t \alpha(u)du)]$

$$= E[\int_0^{t+h} \gamma(s)ds.\exp(\int_0^{t+h} \alpha(u)du) + M(t+h).\exp(\int_0^{t+h} \alpha(u)du)$$

$$- \int_0^t \gamma(s)ds.\exp(\int_0^t \alpha(u)du) - M(t).\exp(\int_0^t \alpha(u)du)]$$

$$= E[\int_t^{t+h} \gamma(s)ds.\exp(\int_0^t \alpha(u)du) + (\int_t^{t+h} \alpha(u).\exp(\int_0^u \alpha(r)dr)du).\int_0^{t+h} \gamma(s)ds$$

$$+ M(t+h)(\int_t^{t+h} \alpha(u).\exp(\int_0^u \alpha(r)dr)du) + (M(t+h) - M(t)).\exp(\int_0^t \alpha(u)du)]$$

$$= E[\int_t^{t+h} \gamma(s)ds.\exp(\int_0^t \alpha(u)du) + (\int_t^{t+h} \alpha(u).\exp(\int_0^u \alpha(r)dr)du).\phi(t+h)]$$

and the proof is complete.

3.1.3. LEMMA

Let

$$(3.8)\quad \Phi(s,t) := E[f(X(s\wedge\tau),Y(t\wedge\sigma)).\exp(\int_0^{s\wedge\tau} \alpha(X(u))du).\exp(\int_0^{t\wedge\sigma} \beta(Y(u))du)] .$$

Then

$$(3.9)\quad \Phi(s,t) - \Phi(0,t)$$
$$= E[\int_0^{s\wedge\tau} \{f(X(r\wedge\tau),Y(t\wedge\sigma)).\alpha(X(r\wedge\tau)) + g(X(r\wedge\tau),Y(t\wedge\sigma))\}$$
$$.\exp(\int_0^r \alpha(X(u))du)\exp(\int_0^{t\wedge\sigma} \beta(Y(u))du)\ dr\ ].$$

A similar identity holds for $(\Phi(t,s) - \Phi(t,0))$.

Proof. Let $h > 0$. Using (3.3.a), the independence of X and Y, and Lemma 3.1.2 we obtain:

$$\Phi(s+h,t) - \Phi(s,t)$$
$$= E[f(X((s+h)\wedge\tau),Y(t\wedge\sigma)).\exp(\int_0^{(s+h)\wedge\tau} \alpha(X(u))du).\exp(\int_0^{t\wedge\sigma} \beta(Y(u))du)]$$
$$- E[f(X(s\wedge\tau),Y(t\wedge\sigma)).\exp(\int_0^{s\wedge\tau} \alpha(X(u))du).\exp(\int_0^{t\wedge\sigma} \beta(Y(u))du)]$$

(3.10)

$$= E[f(X((s+h)\wedge\tau),Y(t\wedge\sigma))\{\int_{s\wedge\tau}^{(s+h)\wedge\tau} \alpha(X(r)).\exp(\int_0^{r\wedge\tau} \alpha(X(u))du)dr\}.\exp(\int_0^{t\wedge\sigma} \beta(Y(u))du)]$$
$$+ E[\int_{s\wedge\tau}^{(s+h)\wedge\tau} g(X(r\wedge\tau,Y(t\wedge\sigma))dr.\exp(\int_0^{s\wedge\tau} \alpha(X(u))du).\exp(\int_0^{t\wedge\sigma} \beta(Y(u))du)]$$
$$= E[\{\int_{s\wedge\tau}^{(s+h)\wedge\tau} f(X(r),Y(t\wedge\sigma)).\alpha(X(r)).\exp(\int_0^r \alpha(X(u))du)dr\}.\exp(\int_0^{t\wedge\sigma} \beta(Y(u))du)$$
$$+ \{\int_{s\wedge\tau}^{(s+h)\wedge\tau} \int_r^{(s+h)\wedge\tau} g(X(v),Y(t\wedge\sigma))dv.\alpha(X(r)).\exp(\int_0^r \alpha(X(u))du)dr\}.\exp(\int_0^{t\wedge\sigma} \beta(Y(u))du)$$
$$+ \int_{s\wedge\tau}^{(s+h)\wedge\tau} g(X(r),Y(t\wedge\sigma)).\exp(\int_0^r\alpha(X(u))du)dr\ .\exp(\int_0^{t\wedge\sigma} \beta(Y(u))du)$$
$$+ \int_{s\wedge\tau}^{(s+h)\wedge\tau} g(X(r),Y(t\wedge\sigma)).\{\exp(\int_0^{s\wedge\tau} \alpha(X(u))du)-\exp(\int_0^{r\wedge\tau} \alpha(X(u))du)\}dr\ .$$
$$\exp(\int_0^{t\wedge\sigma} \beta(Y(u))du)\ ]\ .$$

But for $t,(s+h) \leq T$, the absolute values of the second and fourth terms are bounded by

(3.11) $\frac{1}{2}h^2 \, E(\Gamma_T).\exp(c_T)$.

Note that

(3.12) $$\Phi(s,t) - \Phi(0,t) = \sum_{j=1}^{m} (\Phi(s_j,t) - \Phi(s_{j-1},t))$$

where $0 = s_0 < s_1 < \ldots < s_m = s$. If $\max_j (s_j - s_{j-1}) \to 0$, then (3.10), (3.11) and (3.12) yield:

$$\Phi(s,t) - \Phi(0,t) = E[\int_0^{s\wedge\tau} \{f(X(r),Y(t\wedge\sigma)).\alpha(X(r)) + g(X(r),Y(t\wedge\sigma))\}$$
$$. \exp(\int_0^r \alpha(X(u))du).\exp(\int_0^{t\wedge\sigma} \beta(Y(u))du) \, dr]$$

and the proof is complete.

<u>Proof of the Theorem.</u> To complete the proof of the theorem note that by Lemma 3.1.1,

$$\Phi(t,0) - \Phi(0,t) = \int_0^t (\Phi_1(s,t-s) - \Phi_2(s,t-s))ds.$$

From Lemma 3.1.3

$$\Phi_1(s,t) = E[\chi_{\{s<\tau\}}\{f(X(s\wedge\tau),Y(t\wedge\sigma)).\alpha(X(s\wedge\tau)) + g(X(s\wedge\tau),Y(t\wedge\sigma))\}$$
$$.\exp(\int_0^{s\wedge\tau} \alpha(X(u))du).\exp(\int_0^{t\wedge\sigma} \beta(Y(u))du) \,] \, ,$$

$$\Phi_2(t,s) = E[\chi_{\{s<\sigma\}}\{f(X(t\wedge\tau),Y(s\wedge\sigma)).\beta(Y(s\wedge\sigma)) + h(X(t\wedge\tau),Y(s\wedge\sigma))\}$$
$$.\exp(\int_0^{t\wedge\tau} \alpha(X(u))du).\exp(\int_0^{s\wedge\sigma} \beta(Y(u))du)] \, .$$

Therefore,

$$\Phi(t,0) - \Phi(0,t)$$
$$= E[\int_0^t\{\chi_{\{s<\tau\}}\{f(X(s\wedge\tau),Y((t-s)\wedge\sigma).\alpha(X(s\wedge\tau)) + g(X(s\wedge\tau),Y((t-s)\wedge\sigma))\}$$
$$.\exp(\int_0^{s\wedge\tau} \alpha(X(u))du).\exp(\int_0^{(t-s)\wedge\sigma} \beta(Y(u))du)$$
$$- \chi_{\{t-s<\sigma\}}\{f(X(s\wedge\tau),Y(t-s)).\beta(Y(t-s)) + h(X(s\wedge\tau),Y(t-s))\}$$
$$.\exp(\int_0^{s\wedge\tau} \alpha(X(u))du).\exp(\int_0^{(t-s)\wedge\sigma} \beta(Y(u))du)\}ds]$$

and the proof of the theorem is complete.

3.2. <u>COROLLARY</u>. In addition to the assumptions of Theorem 3.1 assume that:

(3.13) $g(x,y) + \alpha(x)f(x,y) = h(x,y) + \beta(y)f(x,y)$. If $a^+ := (a\vee 0)$, then

$$E[f(X(t\wedge\tau),Y(0)).\exp(\int_0^{t\wedge\tau} \alpha(X(u))du)] - E[f(X(0),Y(t\wedge\sigma)).\exp(\int_0^{t\wedge\sigma} \beta(Y(u))du)]$$

$$= E[\{f(X((t-\sigma)^+\wedge\tau),Y(\sigma))\exp(\int_0^{(t-\sigma)^+\wedge\tau}\alpha(X(u))du) - f(X(0),Y(\sigma))\}\exp(\int_0^{\sigma}\beta(Y(u))du)]$$

$$-E[\{f(X(\tau),Y((t-\tau)^+\wedge\sigma))\exp(\int_0^{(t-\tau)^+\wedge\sigma}\beta(Y(u))du) - f(X(\tau),Y(0))\}\exp(\int_0^{\tau}\alpha(X(u))du)]$$

3.3. COROLLARY. Assume that (3.2) and (3.3) are true with $\tau = \sigma = \infty$ and (3.13). Then

$$E[f(X(t),Y(0)).\exp(\int_0^t \alpha(X(u))du)] = E[f(X(0),Y(t)).\exp(\int_0^t \beta(Y(u))du)].$$

3.4. COROLLARY. Let $E_{1,n} \uparrow E_1$, $E_{2,m} \uparrow E_2$; $\tau_n := \inf\{t:X(t) \notin E_{1,n}\}$, and $\sigma_m := \inf\{t:Y(t) \notin E_{2,m}\}$. Assume that τ_n are F_t-stopping times, σ_m are G_t-stopping times and that $X(\tau_n) \in E^+_{1,n}$(closed) $\supset E_{1,n}$, $Y(\sigma_m) \in E^+_{2,m}$(closed) $\supset E_{2,m}$. Let

$$\zeta(n,m) := \sup_{x,y\in E^+_{1_n}} \sup_{u,v\in E^+_{2_m}} \{|f(x,u)|.\exp(\ (\alpha(y) + \beta(v))t)\}.$$

Assume that

(3.14.a) $E[|f(X(t),y)|.\exp(\int_0^t\alpha(X(s))ds)] < \infty$, $E[|f(x,Y(t))|.\exp(\int_0^t \beta(Y(s))ds)] < \infty$,

(3.14.b) for $\varepsilon > 0$, there exists n,m such that $(P(\tau_n \le t) + P(\sigma_m \le t)).\zeta(n,m) < \varepsilon$.

Then,

$$E[f(X(t),y).\exp(\int_0^t \alpha(X(s))ds)] = E[f(x,Y(t)).\exp(\int_0^t \beta(Y(s)ds)] .$$

4. UNIQUENESS RESULTS FOR MARTINGALE PROBLEMS

In this section the duality method is applied to establish the uniqueness of solutions to a class of martingale problems for measure-valued diffusion processes. Since the principal applications involve measure-valued processes on Euclidean spaces and since it makes possible the application of Lemma 2.2, it is assumed that S is a locally compact separable metric space with one point compactification S^*. It should be noted however that results of a similar nature can be obtained when S is a Polish space.

For $\xi > 0$, $E_\xi := \{\mu:\mu(S^*) \le \xi\}$ which is a compact subset of $M(S^*)$ and τ_ξ is the F_t-stopping time defined by $\tau_\xi := \inf\ \{t:X(t) \notin E_\xi\}$.

Let L be a differential operator with polynomial coefficients of degree K as given by (2.13)-(2.16). It is assumed that L satisfies one of the following three sets of hypotheses.

4.1. HYPOTHESES.

(a) *(Conservative Property)* For every $N \ge 1$, $1_{(N)}$ (the function of N variables which assumes the constant value 1) $\in D(S^{*N})$ and $LF_{1_{(N)}} = 0$.

(b) $B_\ell = 0$ if $\ell > 0$.

(c) There is a norm $|||\cdot|||_N$ on $D(S^{*N})$ and constants c, $c_{1,\ell}$, $c_{2,\ell}$ such that for

$f \in D(S^{*N})$,

(i) $|||f|||_N + |||Gf|||_N \leq c.||||f||||_N$,

(ii) $||||B_\ell^{(jk)} f/b_\ell||||_{N(f)+\ell} \leq c_{2,\ell}.||||f||||_{N(f)}$, and

$||||A_\ell^{(j)} f/a_\ell||||_{N(f)+\ell} \leq c_{1,\ell}.||||f||||_{N(f)}$, $\ell = -1,0,1,\ldots,K$; $j,k = 1,\ldots,N(f)$,

where $c_{2,0} \leq 1$ and $c_I := \max_\ell (c_{2,\ell}, c_{1,\ell}) < \infty$.

(d) $V(n)^+ \leq c_2 n$, where c_2 is a constant.

4.2. HYPOTHESES.

(a) *(Conservative property)* (As in 4.1.a.)

(b) Condition 4.1.c is satisfied with $c_{j,\ell} \leq 1$, $j = 1,2$ and $\ell = -1,0,1,\ldots,K$.

(c) There is a constant $c_1 < \infty$ such that $V(n)^+ \leq c_1$.

(d) Consider the Markov jump process on the non-negative integers Z^+ with infinitesimal transitions rates q_{jk} for a transition $j \to k \neq j$, where

$$q_{j,j+\ell} = j.a_\ell + \tfrac{1}{2}j(j-1).b_\ell \ , \quad j \geq 1, \ \ell = -1,1,2,\ldots,K,$$

$$q_{0,\ell} = 0, \quad \ell = 1,\ldots,K.$$

$$q_{jj} = 1 - \sum_{k \neq j} q_{jk} \ .$$

It is assumed that $\{q_{jk}\}$ form the infinitesimal transition matrix for a conservative Markov chain $\{\eta(t):t \geq 0\}$ and $E(\eta(t)) < \infty$.

4.3. HYPOTHESES.

(a) *(Pure Death Dual)* $a_\ell = b_\ell = 0$ if $\ell > 0$.

(b) For each $N \geq 1$, $1_{(N)} \in D(S^{*N})$, $G_N 1_{(N)} = 0$ and $LF_{1_{(N)}} \geq 0$.

(c) Condition 4.1.c (with $c_{2,0} \leq 1$) is satisfied.

(d) The moment problem for $X(t,S^*)$ is well posed for small t; e.g. 4.1.d is satisfied.

A family of probability measures $\{P_\mu:\mu \in M(S^*)\} \subset P\,[\Omega_C^{M(S^*)}]\}$ is a solution to to the initial value martingale problem for $\{(F_f,LF_f):f \in D\}$ if (2.5) is satisfied and for $\mu \in E_\xi$, $f \in D$,

(4.1) $F_f(X(t)) - \int_0^t LF_f(X(s))ds$ is a F_t-martingale with respect to P_μ.

4.4. THEOREM.

Under each of the sets of hypotheses 4.1, 4.2, 4.3, a $P[\Omega_C^{M(S^*)}]$-valued solution $\{P_\mu:\mu \in M(S^*)\}$ of the initial value martingale problem associated with $\{(F_f,LF_f):f \in D\}$ is unique, measurable and satisfies the strong Markov property.

Proof. By Lemma 2.2, $\Pi^D(E_\xi)$ generates $L_\infty(E_\xi)$ under bounded pointwise convergence. Then Theorem 2.1 implies that a $P[\Omega_C^{E_\xi}]$-valued solution to the initial value martingale problem is unique, measurable and satisfies the strong Markov property provided that for each $t \geq 0$ and $f \in D$ there is a function $\Lambda_{f,t} \in L_\infty(M(E_\xi))$ such that for $\mu \in M(E_\xi)$,

$$(4.2) \qquad E_\mu(F_f(X(t)) = \Lambda_{f,t}(\mu) .$$

In fact, it suffices to establish (4.2) for all $0 \leq t \leq t_0$ for some $t_0 > 0$ (independent of μ and f). To verify the latter assertion note that under this condition the argument in the proof of Theorem 2.1 yields the uniqueness and strong Markov property of P_μ on F_{t_0}. Then in the same way the conditional distribution $P_\mu(.|F_{t_0}) = P_{X(t_0)}(.)$ is uniquely determined on $\sigma\{X(s):t_0 \leq s \leq 2t_0\}$ and the proof is completed by induction.

In order to establish (4.2) we first construct a dual process $\{Y(t):t \geq 0\}$ adapted to a filtration G_t which is a right continuous strong Markov process with state space D and distribution $\{P_f:f \in D\}$. The evolution of Y(.) can be informally described as follows. If $N(Y(t_1)) = N$, a set of N independent exponential clocks with means $(1/a_\ell)$ and a set of N(N-1) independent exponential clocks with means $(2/b_\ell)$ are constructed for each $\ell = -1,0,1,\ldots,K$. If T_1 denotes the time at which the first of these clocks rings, then

$$(4.3) \qquad Y(t_1+t) = S_t^{N(Y(t_1))}Y(t_1) \quad \text{for } t_1 \leq t < t_1+T_1 , \text{ and}$$

$$Y(t_1+T_1) = A^{(j)}Y((t_1+T_1)-) \quad \text{or} \quad B^{(jk)}Y((t_1+T_1)-), \text{ depending on which of the clocks rings at time } T_1, \ \ell = -1,0,1,\ldots,K; \ j,k = 1,2,\ldots,N(Y(t_1)).$$

Under Hypotheses 4.1 $N(Y(t))$ is a Markov chain on Z^+ such that the "birth rate" that is, the rate of transitions from $D(S^{*N}) \to D(S^{*N+\ell})$ with $\ell > 0$, is bounded by a linear function of $N(Y(t))$. This implies that $N(Y(t))$ is a conservative Markov chain. In addition a coupling argument can be used to verify that $NB(Y(t))$, the number of "births" in [0,t], is stochastically dominated by K.Z(t) where Z(t) is a Yule process, that is a pure birth process with linear birth rate, on Z^+. Therefore

$$(4.4) \qquad P(NB(Y(t)) = n|N(Y(0))=k) \leq c.n^k.z(t)^n \quad \text{where } z(t) \downarrow 0 \text{ as } t \downarrow 0, \text{ and } c \text{ is a constant.}$$

For $\kappa > 0$, let $\sigma_\kappa := \inf\{t: |||Y(t)|||_{N(Y(t))} + N(Y(t)) > \kappa\}$. Then for $\kappa < \infty$, $\mu \in M(E_\xi)$ and $f \in D$,

$$(4.5) \qquad F_\mu(Y(t\wedge\sigma_\kappa)) - \int_0^{t\wedge\sigma_\kappa} L^{\#}F_\mu(Y(s))ds$$

is a bounded G_t-martingale with respect to P_f and conditions (3.3) are satisfied. Conditions (3.2) are also satisfied for $\tau = \tau_\xi, \sigma = \sigma_\kappa$, so that Theorem 3.1 and Corollary 3.2 can be applied. If $|||Y(0)|||_{N(Y(0))} + N(Y(0)) < \kappa$, $\mu \in E_\xi$, this yields:

$$(4.6) \qquad E_\mu[F(X(t\wedge\tau_\xi),Y(0)] - E[F(X(0),Y(t\wedge\sigma_\kappa)).\exp(\int_0^{t\wedge\sigma_\kappa}V(Y(u))du)]$$

$$= E[(\int_0^{(t-\sigma_\kappa)^+\wedge\tau_\xi} - \int_{\tau_\xi\vee(t-\sigma_\kappa)^+}^{t})(\{L^{\#}F(X(s\wedge\tau_\xi),Y((t-s)\wedge\sigma_\kappa)$$

$$+ V(Y((t-s)\wedge\sigma_\kappa)F(X(s\wedge\tau_\xi),Y((t-s)\wedge\sigma_\kappa))\}\exp(\int_0^{(t-s)\wedge\sigma_\kappa}V(Y(u))du))ds].$$

Under the conservative hypothesis 4.1.a, $E_\mu[(X(t\wedge\tau_\xi,S^*)^n] = E_\mu[(X(0,S^*))^n]$ for all $n \in Z^+$ and $\xi \geq X(0,S^*)$. Therefore $X(t,S^*) = X(0,S^*)$ for all t and $\tau_\xi = \infty$. Under Hypotheses 4.1 it follows from (4.4) that

$$(4.7) \quad E_f[|F_\mu(Y(t))|.\exp(\int_0^t V^+(Y(u))du)] \leq \sum_{n=1}^{\infty} (c_1 n)^{N(f)+1}.\exp(c_2 nt).(z(t))^n.c_I^{2n}$$

$$< \infty \quad \text{for } t \leq t_0, \ c_1 \text{ a constant,}$$

provided that t_0 is chosen so that $\exp(c_2 t_0).z(t_0).c_I^2 < 1$. In addition,

$$(4.8) \quad \lim_{\kappa\to\infty} E_f[\chi_{\{\sigma_{\kappa_\infty} \leq t\}} .\sup_\mu |LF_\mu(Y(\sigma_\kappa))|.\exp(\int_0^{t\wedge\sigma_\kappa} V^+(Y(u))du)]$$

$$\leq \lim_{\kappa\to\infty} \sum_{n=\kappa} (c_1 n)^{N(f)+1}.\exp(c_2 nt).(z(t))^n.c_I^{2n} = 0 .$$

Under Hypotheses (4.2), results analogous to (4.7) and (4.8) follow immediately (with $t_0 = \infty$) since $N(Y(t))$ is conservative and $|||Y(t)||| \leq c.|||Y(0)|||$. Together, (4.6), (4.7) and (4.8) imply that for $t \leq t_0$,

$$(4.9) \quad E_\mu[F_f(X(t))] = E_f[F(\mu,Y(t)).\exp(\int_0^t V(Y(u))du)] := \Lambda_{f,t}(\mu).$$

Since $\Lambda_{f,t}(.) \in L_\infty(E_\xi)$, (4.2) has been established under either Hypotheses (4.1) or (4.2).

In the non-conservative case the state **space** is all of $M(S^*)$ and allowance must be made for the fact that $\Pi^D(M(S^*))$ is not contained in the space of bounded continuous functions on $M(S^*)$. In this case the uniqueness, measurability and strong Markov property of a $P[\Omega_C^{M(S^*)}]$-valued solution to the initial value martingale problem can be obtained from a modification of Theorem 2.1 provided that:

(4.9) for $0 \leq t \leq t_0$, $f \in D$, there is a function $\Lambda_{f,t}$ which is a measurable function on $M(S^*)$, bounded on E_ξ for $0 < \xi < \infty$, and satisfies (4.2), and

(4.10) for $0 \leq t \leq t_0$, the moment problem for $X(t,S^*)$ with distribution determined by $\{P_\mu : \mu \in M(S^*)\}$ has a unique solution.

Under Hypotheses 4.3, $\sigma_\kappa = \infty$ if $|||Y(0)|||_{N(Y(0))} + N(Y(0)) < \kappa$, and conditions (3.2), (3.3) are satisfied with $\sigma_\kappa = \infty$, $\tau = \tau_\xi$, $\xi < \infty$. Furthermore by 4.3.b, $(X(t,S^*))^N$ is a submartingale for each $N \geq 1$. Since $E_\mu(|X(t,S^*)|) < \infty$, $\lim_{\xi\to\infty} P_\mu(\tau_\xi \leq t) = 0$. Therefore,

$$(4.11) \quad \lim_{\xi\to\infty} E_\mu(|X(t\wedge\tau_\xi,S^*)|^N.\chi_{\{\tau_\xi \leq t\}})$$

$$\leq \lim_{\xi\to\infty} E_\mu(|X(t,S^*)|^N.\chi_{\{\tau_\xi \leq t\}}) = 0 .$$

Then, for any $f \in D$, $\lim_{\xi\to\infty} E_\mu[F_f(X(t\wedge\tau_\xi))] = E_\mu[F_f(X(t))]$. Then (4.6) and (4.11) yield:

$$(4.12) \quad E_\mu[F_f(X(t))] = E_f[F(\mu,Y(t)).\exp(\int_0^t V(Y(u))du))] := \Lambda_{f,t}(\mu).$$

Since $\Lambda_{f,t}(.)$ is a measurable function which is bounded on the sets E_ξ, it remains to verify that the moment problem for $X(t,S^*)$ has a unique solution. But condition 4.1.d implies that

(4.13) $m_N := E_\mu((X(t,S^*)^N) \leq c_0 \cdot \exp(c_0 N(t+1))$ where c_0 is a constant,

which in turn yields the Carleman condition

$$\sum_{j=1}^{\infty} (m_{2j})^{-(1/2j)} = \infty.$$

This completes the proof under Hypotheses 4.3.

4.5. EXAMPLE.

Fleming and Viot introduced a continuous state version of the Ohta Kimura stepwise mutation model for describing allelic frequencies within a population undergoing mutation, random genetic drift and selection. This model is formulated as a probability-measure-valued diffusion process on R^* defined by a differential operator with polynomial coefficients of degree 2. The differential operator is given by

$$(4.14)\quad LF_f(\mu) = F_{G_{N(f)}f}(\mu) + \sum_{j=1}^{N(f)} (F_{M^{(j)}f}(\mu) - F_f(\mu)) + \tfrac{1}{4}\sum_{j=1}^{N(f)}\sum_{\substack{k=1\\ j\neq k}}^{N(f)} (F_{2B^{(jk)}f}(\mu) - F_f(\mu))$$
$$+ (N(f) - \tfrac{1}{4}N(f)(N(f)-1))F_f(\mu)\ ,$$

where G_N denotes the Laplace operator in N dimensions, $m(.,.) \in D(R^{*2})$,

$$M^{(j)}f(x_1,\ldots,x_N) := (m(x_j,x_{N+1}) - m(x_{N+1},x_{N+2}))f(x_1,\ldots,x_N),\quad \text{and}$$

$$B^{(jk)}f(x_1,\ldots,x_N) := \begin{cases} f(x_1,\ldots,x_j,\ldots,x_{k-1},x_j,\ldots,x_N) & \text{if } j<k,\\ f(x_1,\ldots,x_{k-1},x_j,\ldots,x_j,\ldots,x_N) & \text{if } j>k,\end{cases}$$

and $D(R^{*N})$ are the functions whose derivatives of order two or less belong to $C(R^{*N})$.

The differential operator L given by (4.14) satisfies Hypotheses 4.1 and therefore Theorem 4.4 yields the uniqueness, measurability and strong Markov property of a solution to the associated initial value problem (if it exists). (The existence of a solution can be established by other methods.) The duality method has also been used to study limit theorems for this model in the selectively neutral case (Dawson and Hochberg [2]).

4.6. EXAMPLE.

In this example we consider branching diffusion together with turbulent transport in R^3 which is an extension of the model of P.L. Chow [1] for molecular diffusion with turbulent transport. The differential operator associated with this model is given by:

$$(4.15)\quad LF_f(\mu) = F_{G_{N(f)}f}(\mu) + \sum_{j=1}^{N(f)}\sum_{\substack{k=1\\ j\neq k}}^{N(f)} (F_{B^{(jk)}f}(\mu) - F_f(\mu)) + N(f)(N(f)-1).F_f(\mu),$$

where $B^{(jk)}$ is as in Example 4.5 . For $\alpha,\beta = 1,2,3,$ let $\rho_{\alpha\beta}(x)$ be smooth functions on R^3 with

$$\sum_{j=1}^{N} \sum_{k=1}^{N} \sum_{\alpha,\beta=1}^{3} \rho_{\alpha\beta}(x_j - x_k)\xi_{j\alpha}\xi_{k\beta} \geq 0$$

for any set of real numbers $\{\xi_{j\alpha}\}$. If $x_j \in R^3 = (x_{j,1}, x_{j,2}, x_{j,3})$, then

$$G_N f(x_1,\ldots,x_N) := \sum_{\substack{j=1 \\ j\neq k}}^{N} \sum_{k=1}^{N} \sum_{\alpha,\beta=1}^{3} \rho_{\alpha\beta}(x_j - x_k).\partial^2/\partial x_{j,\alpha}\partial x_{k,\beta}(f(x_1,\ldots,x_N))$$
$$+ \kappa \sum_{j=1}^{N} \sum_{\alpha=1}^{3} \partial^2/\partial x_{j,\alpha}^2 (f(x_1,\ldots,x_N)) .$$

For sufficiently large κ the operators G_N are uniformly elliptic and generate strongly continuous semigroups on $C_0((R^3)^N)$. Then the differential operator L given by (4.15) satisfies Hypotheses 4.3 and Theorem 4.4 can be applied to establish the uniqueness, measurability and strong Markov property of a solution to the associated initial value martingale problem.

REFERENCES.

1. P.L. Chow, Function-space differential equations associated with a stochastic partial differential equation, Indiana Univ. Math. J. 25(1976), 609-627.

2. D.A. Dawson and K.J. Hochberg, Wandering random measures in the Fleming-Viot model, Annals of Probability, to appear.

3. S.N. Ethier and T. C. Kurtz, Markov Processes: Characterization and Convergence, to appear.

4. R. Holley and T. Liggett, Ergodic theorems for weakly interacting systems and the voter model, Annals of Probability 3(1975), 643-663.

5. R. Holley, D. Stroock and D. Williams, Applications of dual processes to diffusion theory, Proc. Symp. Pure Math. 31(1977), 23-36, American Mathematical Society.

6. R. Holley and D. Stroock, Dual processes and their applications to infinite interacting systems, Advances in Mathematics 32(1979), 149-174.

7. E. Pardoux, Stochastic partial differential equations and filtering of diffusion processes, Stochastics 3 (1979), 127-167.

8. T. Shiga, An interacting system in population genetics, J. Math. Kyoto Univ. 20 (1980), 213-242.

9. T. Shiga, Diffusion processes in population genetics, J. Math. Kyoto Univ. 21 (1981), 133-151.

10. F. Spitzer, Interaction of Markov processes, Advances in Mathematics 5(1970), 246-290.

11. D.W. Stroock and S.R.S. Varadhan, Multidimensional diffusion processes, Grundlehren der mathematischen Wissenschaften 233(1979), Springer-Verlag, Berlin, Heidelberg, New York.

OPTIMAL STOPPING OF CONTROLLED MARKOV PROCESSES

Nicole EL KAROUI

with Jean-Pierre LEPELTIER and Bernard MARCHAL

I-INTRODUCTION

We are concerned with optimal stopping of Markov Processes, controlled by change of absolutely continuous probability measures. We use the results on optimal stopping depending on a parameter, in order to construct by iteration the value-function of this optimization problem.

We then show that the separation principle is satisfied: we can first choose the optimal stopping rule, and then study a classical problem of instantaneous control, that we solve under appropiate assumptions.

II - THE MODEL

In free evolution, <u>the state of the system</u> is an E-valued Markov Process, $\mathcal{X} = (\Omega, \underline{\underline{F}}_t , X_t , P_\mu ; \mu \in \mathcal{P}^+(E))$, where E is an lusinian space. We suppose that $\mathcal{X}$ is a "good" Markov Process, <u>Feller Process</u> if E is compact, or more generally that $\mathcal{X}$ is a <u>right Process</u>, roughly speacking that $\mathcal{X}$ is a Feller Process (with branching points) for an adequate topology on E.

For example, $\mathcal{X}$ is a diffusion process on R^n, a diffusion with jumps, a killing diffusion, a reflected diffusion on a convex domain in R^n,.....

<u>A admissible control</u> is a adapted process u_t, taking values in a compact set U.

Let $\mathcal{D}$ the family of <u>admissible controls</u>, and $\mathcal{T}$ the family of <u>stopping times</u>.

(2.1) We suppose that $\mathcal{D}$ is closed under <u>concatenation</u> :

for every u,v in $\mathcal{D}$ and every stopping time S,

$u \overset{S}{\circ} v = u \, 1_{]0,S]} + v \, 1_{]S,\infty]}$ belongs to $\mathcal{D}$.

(2.2) Under the policy $u \in \mathcal{D}$, the law of X. is the probability

$$P^u_\mu = L^u_\infty \cdot P_\mu$$

where:- $L^u_\cdot$ is a uniformly integrable martingale > 0 .

- the family $\{ L^u_\cdot \, ; \, u \in \mathcal{D} \}$ is compatible :

if $u, v \in \mathcal{D}$ and $u_t = v_t \quad \forall\, t \in [S,T[\quad , S,T \in \mathcal{C}$

$$\frac{L_t^u}{L_S^u} = \frac{L_t^v}{L_S^v} \quad \forall\, t \in [S,T[\qquad \text{a.s.}$$

Particularly, $\forall\, t \geq S \quad \dfrac{L_t^{u \overset{S}{\circ} v}}{L_S^{u \overset{S}{\circ} v}}$ does not depend on u a.s.

(2.3) If a is a constant policy, $\mathcal{X}^a = (\Omega, \underline{\underline{F}}_t, X_t, P^a_\mu ; \mu \in \mathcal{P}^+(E))$ is a Markov Process, Feller or right as $\mathcal{X}$.

Remark: Usually, $L^u_\cdot$ is the exponential martingale associated with a family of stochastic integrals.

The reward process is $C_t^u + \exp - H_t^u \; g(X_t)$

Here, $C_t^u = \int_0^t e^{-H_s^u} c(X_s,u_s)\, ds$, and $H_t^u = \int_0^t h(X_s,u_s)\, ds$

We suppose that $c(x,a)$, $h(x,a)$, and $g(x)$ are bounded functions. $g(x)$ is the terminal pay-off.

Our control problem is to find an optimal policy $\hat{u} \in \mathcal{D}$, and a stopping time $\hat{T}$ wich maximizes the reward functionnal:

$$E_\mu^u (C_T^u + \exp - H_T^u \; g(X_T))$$

We write $\quad J_x(u,T,g) = E_x^u (C_T^u + \exp - H_T^u \; g(X_T))$

We use the "martingale approach", remarkably explained by M.H.A. DAVIS in [1] and R.ELLIOTT in [2] , and the formulation of "Bellman's principle of optimality" as a supermartingale inequality for the process $C_S^u + \exp- H_S^u \; W(S)$, where $W(S)$ is the value-process :

$$W(S) = P_\mu - \operatorname{essup}_{T \geq S, v \in \mathcal{D}} E^v \left[C_T^v - C_S^v + \exp-(H_T^v - H_S^v)\; g(X_T) / \underline{\underline{F}}_S \right]$$

More precisely we have:

THEOREM 1 : PRINCIPLE OF OPTIMALITY

Let μ be a fixed initial law

a) $J(u,S) = C_S^u + \exp- H_S^u \; W(S)$ is a P^u_μ- $\mathcal{C}$-supermartingale;

i.e. S and $T \in \mathcal{C}$, $S \leq T \quad E^u [J(u,T) / \underline{\underline{F}}_S] \leq J(u,S) \quad P^u_\mu$ .a.s.

b) A policy $(\hat{u}, \hat{T})$ is optimal iff :

- $W(\hat{T}) = g(X_{\hat{T}}) \quad P^{\hat{u}}_\mu$ a.s.
- $C^{\hat{u}}_{S \wedge \hat{T}} + \exp - H^{\hat{u}}_{S \wedge \hat{T}} \; W(S \wedge \hat{T})$ is a $P^{\hat{u}}_\mu$ - $\mathcal{C}$-martingale

The proof is similar to the proof of relations 5.3 and 5.11 de [1] because we assume (2.1) and (2.2).

Our control problem is Markovian. We at first show that the value-process is Markovian, i.e. $W(S) = \hat{w}g(X_S)$ P^u_μ .a.s.
Here $\hat{w}g(x) = \sup_{T\geq 0,\ u\in\mathcal{D}} E^u_x\left[C^u_T + \exp -H^u_T\ g(X_T)\right]$ is called the $\mathcal{D}$-value function.
However the family $\mathcal{D}$ is too complicated to study directly the properties of $\hat{w}g(x)$. Thus we work with the set $\mathcal{D}_e$ of control admissible which are step functions along stopping times, i.e.,

$$\mathcal{D}_e = \{u\in\mathcal{D};\ u = \Sigma_{1\leq n}\ u_n\ 1_{]T_n,T_{n+1}]}\ \}$$

(T_n) is a increasing sequence of stopping times, such that $\sup_n T_n = +\infty$ and u_n is an $\underline{\underline{F}}_{T_n}$-random variable.

REMARK: $\mathcal{D}_e$ is the "smallest set", closed under concatenation which contains the constant controls.

We construct the value-function $wg(x)$ on the set $\mathcal{D}_e$ of step controls by impulse control technics and verify that the dynamic programming principle is satisfied for $\mathcal{D}_e$. From this we easily deduce the separation principle for control problem.
Under appropriate assumptions on the continuity of $a\to L^a$, $c(x,a)$, $h(x,a)$ and $t\to g(X_t)$, we show that the $\mathcal{D}_e$-value-function is equal to $\hat{w}g(x)$.

III - CONSTRUCTION OF THE $\mathcal{D}_e$ - VALUE FUNCTION

Let $u\in\mathcal{D}_e$ be a step control $u = \Sigma_{1\leq i}\ u_i\ 1_{]T_i,T_{i+1}]}$
Between T_i and T_{i+1}, the evolution of $\mathcal{X}$ is governed by the probability $P^{[u_i(\omega)]}$, i.e., if $A\in\underline{\underline{F}}_{T_{i+1}}$ $P^u_\mu(A/\underline{\underline{F}}_{T_i}) = P^{u_i(.)}_\mu(A/\underline{\underline{F}}_{T_i})$ a.s.by (2.3).
Each u belonging to $\mathcal{D}_e$ can be seen as a policy of impulse control with impulse times (T_n). We use the methods of impluse control, [3] , [4] but without impulse cost, to solve our control problem.

To obtain wg by approximation, let

$\mathcal{D}_n$ = { n-step control of $\mathcal{D}_e$, i.e, $T_n = +\infty$ }

$w_n g(x) = \sup\{J_x(u,T,g);\ T\in\mathcal{T}$ and $u\in\mathcal{D}_n\}$ and

$$Rg(x) = w_1 g(x) = \sup\{\ J_x(a,T,g);\ T\in\mathcal{T} \text{ and } a\in U\ \}$$

The functions $w_n g$ and the operator Rg have the following properties:

PROPOSITION 2:

a) For each n, $w_n g = R^n g$, where R^n is the n_th power of R .

b) The sequence $w_n g$ increases to wg .

c) For $u\in\mathcal{D}_e$ and $S\in\mathcal{T}$, if $u^S_t = u_{t\wedge S}$

$$J_\mu(u,S,wg) = \sup\{\ J_\mu(v,T,g);\ v^S = u^S\ ,\ v\in\mathcal{D}_e\ ,\ T\geq S\ \}$$

PROOF: The principal difficulty is to interchange the supremum and integration signs for the operator R .

Let $R^a g(x) = \sup \{ J_x(a,T,g) ; T \in \mathscr{C} \}$ be the value-function of the optimal stopping problem for the parametrized $\mathfrak{X}^a$ process , and $Rg(x) = M(R^{\cdot}g)(x)$ where $Mf(x) = \sup_a f(x,a)$

From the theory of analytic function, we obtain the following result:

LEMMA 3 : Let f(x,a) be an analytic function on E×U compact.

a) $Mf(x) = \sup_a f(x,a)$ is analytic

b) For each probability measure μ on E,

$$\mu(Mf) = \sup_{u(x)} \int \mu(dx)\, f(x,u(x)),$$ where u(x) is a Borel-function .

c) Let $(\Omega, \underline{\underline{G}}, Q)$ be a complete probability space and Z a $\underline{\underline{G}}$-measurable, E-valued variable:

$$Mf(Z) = Q\text{-esssup} \{ f(Z,V) ; V\ \underline{\underline{G}}\text{-measurable, U-valued variable} \}$$

This lemma applied to the optimal stopping problem leads to following result:

(3.1) If $\mu \in \mathscr{P}^+(E)$, $u \in \mathfrak{D}_e$ and S is a stopping time then:

$$J_\mu(u,S,Rg) = \sup \{ J_\mu(u^S_\circ \delta ,T,g) ; \delta\ \underline{\underline{F}}_S\text{-measurable random variable}\}$$

where $(u^S_\circ \delta)_t = u_t 1_{\{ t \leq S\}} + \delta\ 1_{\{t > S\}}$

By iteration, it is clear that:

(3.2) $J_\mu(u,S,R^n g) = \sup \{J_\mu(v,T,g) ; T \geq S \text{ and } v^S = u^S , v \in \mathfrak{D}_n(u,S) \}$

where $v \in \mathfrak{D}_n(u,S)$ iff v has not more than n impulsions after S.

It remains to identify the limit $w_\infty g$ of the non-decreasing sequence $w_n g$. We write, for $v \in \mathfrak{D}_e$,

$$J_\mu(v,T,g) = J_\mu(v, T \wedge T_n, g) + E^v_\mu[1_{\{T > T_n\}}(C^v_T - C^v_{T_n} + e^{-H^v_T} g(X_T) - e^{-H^v_{T_n}} g(X_{T_n}))]$$

g and c(x,a) are bounded; therefore, by Lebesgue theorem,

$J_\mu(v,T,g) - J_\mu(v,T \wedge T_n,g)$ converges to 0.

On the other hand, the policy v^{T_n} belongs to $\mathfrak{D}_{n+1}$,and

$$J_\mu(v,T \wedge T_n,g) = J_\mu(v^{T_n}, T \wedge T_n, g) \leq \mu(w_n g) \leq \mu(w_\infty g)$$

We deduce that: $\sup \{ J_\mu(v,T,g) ; v \in \mathfrak{D}_e \} = \mu(w_\infty g)$ and that $w_\infty g = wg$ and also by passage to the limit in (3.2) that c) is true.

Thus the principle of optimality and the principle of dynamic programming hold:

THEOREM 4: For each initial law μ

a) $\exp -H^u_t\, wg(X_t)$ is the $\mathfrak{D}_e$ value-process and the theorem 1 holds for $\mathfrak{D} = \mathfrak{D}_e$

b) For all stopping times S,

(4.1) $wg(x) = \sup_{T,u\in\mathcal{D}_e} E_x^u\left[1_{\{T<S\}} (C_T^u + e^{-H_T^u} g(X_T)) + 1_{\{T\geq S\}} (e^{-H_S^u} wg(X_S) + C_S^u)\right]$

c) Suppose $g \geq 0$, and let $D^\lambda = \inf \{t \geq 0;\ \lambda\, wg(X_t) \leq g(X_t)\}$ $\lambda\in [0,1[$

(4.2.) Then $wg(x) = \sup \{J_x(u, D^\lambda, wg);\ u\in\mathcal{D}_e\}$

wg is the value-function of instantaneous control problem with play-off wg at time D^λ.

PROOF: From Proposition 2 c), it is clear that $\exp{-H_t^u}\, wg(X_t)$ is the value-process of this optimality problem on $\mathcal{D}_e$, and that:

$\sup_{T,u} E_x^u \left[1_{\{T<S\}} (C_T^u + \exp{-H_T^u}\, g(X_T)) + 1_{\{T\geq S\}} (C_S^u + \exp{-H_S^u}\, wg(X_S))\right] =$

$\sup_{T,u} \sup_{V^S=u^S, U\geq S} E_x^u \left[1_{\{T<S\}} (C_T^u + e^{-H_T^u} g(X_T)) + 1_{\{T\geq S\}} (C_U^V + \exp{-H_U^V}\, g(X_U))\right]$

$\geq wg(x)$,for we can put $v = u$ and $U = T\vee S$

On the other hand,the left hand side is always bounded, since $g \leq wg$, by $\sup_{T,u} E_x^u \left[C_{T\wedge S}^u + \exp{-H_{T\wedge S}^u}\ wg(X_{T\wedge S})\right] \leq wg(x)$ by Proposition 2 c) .

Equality (4.1) follows .

Let $S = D^\lambda$; $g(X_T) \leq \lambda wg(X_T)$ a.s. on $\{T<S\}$ and by (4.1)

$wg(x) \leq \sup_{T,u} E_x^u \left[1_{\{T<S\}} (C_T^u + \exp{-H_T^u}\, \lambda wg(X_T)) + 1_{\{T\geq S\}} (C_S^u + \exp{-H_S^u}\, wg(X_S))\right]$

$\leq \sup_{T,u}\left[\lambda E_x^u \left[C_T^u + \exp{-H_T^u}\, wg(X_T)\right] + (1-\lambda)\, E_x^u \left[1_{\{T\geq S\}} (C_S^u + \exp{-H_S^u}\, wg(X_S))\right]\right]$

$\leq \lambda\, wg(x) + (1-\lambda) \sup_u E_x^u \left[C_S^u + \exp{-H_S^u}\, wg(X_S)\right]$

Then $wg(x) \leq \sup_u E_x^u \left[C_S^u + \exp{-H_S^u}\, wg(X_S)\right]$

By Proposition 2c), the reverse inequality holds and (4.2) is proved.

REMARK:We extend this construction and results in [5] .

III EXISTENCE OF OPTIMAL CONTROLS

To ensure existence of an optimal stopping time, and optimal policy we make the following assumptions :

(H1) g is non-negative and bounded, and $t \to g(X_t)$ is right-continuous and regular P a.s. , or equivalently

$\lim_n E_\mu^u(g(X_{T_n})) = E_\mu^u(g(X_T))$ $\forall u\in\mathcal{D}_e$

for an increasing (or decreasing) sequence of stopping times (T_n) with $\lim_n T_n = T$

(H2) There exists $\alpha > 0$ such that $h(x,a) \geq \alpha > 0$

(H3) The family $\{ L_T^u ;\ u\in\mathcal{D}_e \}$ is uniformly integrable for $T\in\mathcal{C}$

Under (H1), there exists an ε-optimal stopping time .

More precisely we have:

THEOREM 5: Under the assumption (H1)

a) For all $\mu\varepsilon\mathcal{P}^{+}(E)$, $t \to wg(X_t)$ is right continuous P_μ a.s.

b) The stopping time $\hat{D}^{\varepsilon} = \inf\{t\geq 0 ; wg(X_t) \leq g(X_t) + \varepsilon\}$ is ε-optimal.

c) The stopping time $D = \inf\{ t \geq 0 \; g(X_t) = wg(X_t)\}$ is optimal, and

(5.1.) $\quad wg(x) = \sup_u \; E_x^u [C_D^u + \exp -H_D^u \; g(X_D)]$

if the assumptions (H1), (H2), (H3) are satisfied.

PROOF:

a) Let $\overset{+}{w}g(x) = \lim_{t\to o} E_x[wg(X_t)]$. This limit exists because wg(X.) is a strong semimartingale.

Under (H1), $wg \geq \overset{+}{w}g \geq g$

Using Theorem 4a) $C_t^u + \exp -H_t^u \, \overset{+}{w}g(X_t)$ is a right continuous supermartingale, and :

$$\begin{aligned} wg(x) &= \sup\{J_x(u,T,g) \; ; \; u\varepsilon\mathcal{D}_e , T \geq 0\} \\ &= \sup\{J_x(u,T,\overset{+}{w}g); \; u\varepsilon\mathcal{D}_e, T\geq 0\} = \overset{+}{w}g(x) \end{aligned}$$

b) Using Theorem 4, we prove that, if $\tilde{C}_t^u = C_t^u + \|g\|\int_o^t e^{-H_s^u} dH_s^u$ and $\tilde{g} = g + \|g\|$, $w\tilde{g}(x) = wg(x) + \|g\|$.

Let $\tilde{D}^{\lambda}$ be the stopping time $\inf\{t; \lambda(wg + \|g\|)(X_t) \leq g(X_t) + \|g\|\}$

$\hat{D}^{\varepsilon}$ the stopping time $\inf\{t; \; wg(X_t) \leq g(X_t) + \varepsilon\}$

Because wg is bounded, $\hat{D}^{\varepsilon} \leq \tilde{D}^{\lambda}$ for $\varepsilon \geq (1-\lambda)(\|g\| + \|wg\|)$ and

$$wg(x) = \sup_{u\varepsilon\mathcal{D}_e} E_x^u [C_{\hat{D}^\varepsilon}^u + \exp -H_{\hat{D}^\varepsilon}^u wg(X_{\hat{D}^\varepsilon})]$$

$$(5.2) \qquad = \sup_{u\varepsilon\mathcal{D}_e} E_x^u [C_{\tilde{D}^\lambda}^u + \exp -H_{\tilde{D}^\lambda}^u \, wg(X_{\tilde{D}^\lambda})]$$

g(X.) and wg(X.) are right continuous. The inequality $wg(X_{\hat{D}^\varepsilon}) \leq g(X_{\hat{D}^\varepsilon}) + \varepsilon$ is true, and $\hat{D}^{\varepsilon}$ is ε-optimal.

c) In order to show (5.1), we are going to prove that $J_x(u,D^\lambda,g)$ converges uniformly to $J_x(u,D^*,g)$ if $\lambda\to 1$, where $D^* = \lim_{\lambda\to 1} D^\lambda \leq D$.

Under (H2) and (H3),

$$\begin{aligned} |J_x(u,D^\lambda ,g) - J_x(u,D^* ,g)| &\leq K \, E_x^u\Big[\int_{D^\lambda}^{D^*} e^{-\alpha s} ds + e^{-\alpha D^*}|g(X_{D^*}) - g(X_{D^\lambda})|\Big] \\ &\leq K \sup_u E_x[L_{D^*}^u 1_{\{L_{D^*}^u \geq c\}}] + Kc \, E_x\Big[\int_{D^\lambda}^{D^*} e^{-\alpha s} ds + e^{-\alpha D^*}|g(X_{D^\lambda}) - g(X_{D^*})|\Big] \end{aligned}$$

The right hand have 0 as limit.

We take the limit in (5.2).

$$wg(x) = \sup_{u\varepsilon\mathcal{D}_e} E_x^u [C_{D^*}^u + \exp -H_D^u \, g(X_{D^*})]$$

D^* is optimal, and by Theorem 1, $D^* \geq D$. Then $D = D^*$, and (5.1) is proved.

Notice that wg is the value-function associated to instantaneous control problem, with terminal time D.
In order to prouve the existence an optimal policy, we can apply the classical results, ([6] p.218).
We suppose, for simplicity, that $L^u_\cdot$ is a exponential continuous martingale associated to :

$$N_t^u = \int_0^t \phi(X_s, u_s)\, dM_s, \quad \text{where } M \text{ is a continuous martingale}$$

which is a functionnal additive of Markov Process
Under (H3), and if $a \to \phi(x,a)$, $c(x,a)$, and $h(x,a)$ are continuous, there exists a borel function $u(x)$, that $u(X_{t-})$ is a optimal policy.
([6] Théorem 3.59.) .

REFERENCES

[1] M.H.A. DAVIS: Martingale methods in stochastic control
Stochastic Control Theory and Stochastic Differential Systems. Bonn 1979.
Lect.Notes in Cont. and Inf. Springer n°16

[2] R. ELLIOTT The martingale calculus and applications
Stochastic Control Theory and Stochastic Differential Systems. Bonn.1979

[3] M. ROBIN Contrôle impulsionnel des Processus de Markov
Thèse de Doctorat Paris IX 1978.

[4] J.P. LEPELTIER, B. MARCHAL Theorie générale du Contrôle impulsionnel
à paraitre dans Stochastics

[5] N. EL KAROUI, J.P.LEPELTIER, B.MARCHAL Construction du semi-groupede Nisio associé au contrôle de Processus de Markov.
à paraitre. 1982.

[6] N. EL KAROUI Les aspects probabilistes du Contrôle stochastique
Ecole d'Eté de Probabilités de Saint-Flour 1979
Lect.Notes in Math. SpringerVerlag n° 876.

Nicole EL KAROUI
Ecole Normale Supérieure
5, rue Boucicaut
92260-Fontenay-aux-Roses
FRANCE

TWO PARAMETER FILTERING EQUATIONS FOR JUMP PROCESS SEMIMARTINGALES

By

ATA AL-HUSSAINI, UNIVERSITY OF ALBERTA, CANADA

and

ROBERT J. ELLIOTT, UNIVERSITY OF HULL, ENGLAND

1. INTRODUCTION

A random event which occurs in the positive quadrant is considered. The signal process is a two parameter semimartingale associated with this event. The observation process is a second random event in the positive quadrant, and the signal influences the observation because the distribution function of the observation is a function of the signal. This description of the filtering problem is based on the reference probability model of Zakai [7] . Using the calculus of two parameter semimartingales, ([2], [3]), horizontal, vertical and diagonal filtering equations are obtained. Intuitively the problem can be considered as searching for a random point in the quadrant using noisy observations.

Analogous formulae, for the filtering of a continuous two parameter semimartingale associated with the Brownian sheet, were obtained by Wong [6] and Korezlioglu, Mazziotto and Szpirglass [4] . Filtering equations using two parameter Poisson process observations of a such a continuous semimartingale were obtained under certain symmetry hypotheses by Mazziotto and Szpirglas in [5] . In the notation of section 3 below the symmetry hypotheses of [5] would, for example, include the assumptions that $H = \tilde{H}$, and that $R = 0$. Such hypotheses would considerably simplify our calculations and give rise to results which superficially resemble those of [5] . For the fundamental situation associated with the two parameter jump process we derive filtering formulae without such hypotheses. The calculations are explicit and our results new.

Professor Al-Hussaini wishes to thank the Department of Pure Mathematics of the University of Hull for its hospitality in September and October 1981 when work on this paper was commenced.

2. SIGNAL AND OBSERVATION PROCESS

THE SIGNAL PROCESS

The signal process is a semimartingale related to an event that occurs at a random 'time' (S_1,S_2) in the positive quadrant $[0,\infty]^2$. We can, therefore, take the underlying probability space to be $\Omega_1 = [0,\infty]^2$ with a probability measure P_1.

Consider the processes

$$p_i^* = p_i^*(t_i,\omega) = I_{t_i \geq S_i} , \quad i = 1,2 .$$

The σ - field $F_{t_i}^{i*}$ on the ith factor of Ω_1 is the completion of $\sigma\{I_{s_i \geq S_i} : s_i \leq t_i\}$, and $F^*_{t_1,t_2}$ is the product $F_{t_1}^{1*} \times F_{t_2}^{2*}$. We shall assume the usual condition (F4) is satisfied, that is that $F^*_{t_1}$ and $F^*_{t_2}$ are conditionally independent given $F^*_{t_1 t_2}$. In [2] it is proved that in the present situation this is equivalent to the components S_1 and S_2 of the event being independent.

Write F^* for $F^*_{\infty\infty}$,

and $\quad F_{u_i}^{i*} = P_1[S_i > u_i] , \quad i = 1,2 .$

Then

$$\tilde{p}_i^* = \tilde{p}_i^*(t_i,\omega) = - \int_{]0,t_i \wedge S_i]} (F_{u_i -}^{i*})^{-1} dF_{u_i}^{i*}$$

and $\quad q_i^* = p_i^* - \tilde{p}_i^* , \quad i = 1,2 .$

NOTATION 2.1.

For suitable integrands $\theta(u_1,u_2) \in L^1(\Omega_1,F^*,P_1)$ we shall write, for example,

$$\theta.p_1^*\tilde{p}_2^* = \int_0^{t_2} \int_0^{t_1} \theta(u_1,u_2) dp_{u_1}^* d\tilde{p}_{u_2}^* . \quad \text{(See [1] and [3].)}$$

DEFINITION 2.2.

The signal process is then the semimartingale

$$X_{t_1 t_2} = X_{00} + \theta^1.p_1^*p_2^* + \theta^2.\tilde{p}_1^*p_2^* + \theta^3.p_1^*\tilde{p}^*_2 + \theta^4.\tilde{p}_1^*\tilde{p}_2^* ,$$

where X_{00} is a constant and $\theta^i \in L^1(\Omega_1,F^*,P_1)$, $i = 1,2,3,4$.

REMARKS 2.3.

Here we have assumed that X is constant on the axes. Terms of the form

$\psi.p_i^*$ and $\psi'.\widetilde{p}_i^*$ could be introduced into X. However, such one dimensional stochastic integrals do not give rise to more interesting results, but only more complex calculations.

For the same reason we shall suppose that $F_{u_i}^{i*}$ is continuous for $i = 1,2$.

THE OBSERVATION PROCESS.

The process X cannot be observed directly, but a second random event which occurs at a 'time' (T_1,T_2) in $[0,\infty]^2$ can be observed. We shall describe below how the distribution of (T_1,T_2) is a function of X.

The underlying probability space for (T_1,T_2) can be taken as $\Omega_2 = [0,\infty]^2$, with a probability measure P_2. We assume that under the original measures T_1 and T_2 are independent and, furthermore, (T_1,T_2) is independent of (S_1,S_2). Therefore, the signal and observation processes can be considered on the product space $\Omega = \Omega_1 \times \Omega_2$ with the product measure $P = P_1 \times P_2$.

Write

$$F_{u_i}^i = P_2[T_i > u_i] \, , \; i = 1,2,$$

and again assume, to simplify the calculations, that $F_{u_i}^i$ is continuous. On each factor of Ω_2 $\mathcal{F}_{t_i}^i$ is the completion of $\sigma\{I_{s_i \geq T_i} : s_i \leq t_i\}$, $\mathcal{F}_{t_1t_2}$ is the product $\mathcal{F}_{t_1}^1 \times \mathcal{F}_{t_2}^2$ and $\mathcal{F} = \mathcal{F}_{\infty\infty}$. Associated with (T_1,T_3) we have the processes, (to which Notation 2.1 will also apply):

$$p_i = p_i(t_i,\omega) = I_{t_i \geq T_i} \, ,$$

$$\widetilde{p}_i = \widetilde{p}_i(t_i,\omega) = - \int_{]0,t_i \wedge T_i]} (F_{u_i-}^i)^{-1} dF_{u_i}^i \, ,$$

and $q_i = p_i - \widetilde{p}_i$, $i = 1,2$.

3. THE REFERENCE PROBABILITY MODEL.

Suppose Q_2 is a probability measure equivalent to P_2 on $(\Omega_2,\mathcal{F})$. There is then a Radon-Nikodym derivative

$$\frac{dQ_2}{dP_2} = L \in L^1(\Omega_2, F) .$$

We can consider the two parameter martingale $L_{t_1t_2} = E[L|F_{t_1t_2}]$. Because Q_2 is equivalent to P_2, $L > 0$ a.s. and $L_{t_1t_2} > 0$ a.s. for each $(t_1, t_2) \in [0,\infty]^2$. We shall assume in the situations discussed below that P_2 and Q_2 have the same marginals, so that

$$L_{t_1 0} = L_{0 t_2} = L_{00} = 1 \text{ for all } t_1, t_2 .$$

Again, all our calculations go through without this hypothesis, but are then even more complicated. Applying the martingale representation result of [1] to $\{L_{t_1t_2} - 1\}$ we have that there is a function $g \in L^1(\Omega_2, F)$ such that

$$L_{t_1t_2} = 1 + \int_0^{t_1} \int_0^{t_2} g(u_1,u_2) dq_{u_2} dq_{u_1} .$$

THE FILTERING MODEL

With this representation result, and the reference probability model for filtering of Zakai [7] in mind, we suppose there is a function $g : R \to R$ such that for each sample path X of the signal process $g(X_{u_1u_2}) \in L^1(\Omega_2, F)$. Furthermore, if

$$L_{t_1t_2}(X) = L_{t_1t_2} = 1 + \int_0^{t_1} \int_0^{t_2} g(X_{u_1u_2}) dq_{u_2} dq_{u_1} ,$$

then we assume that $L_{t_1t_2} > 0$ a.s. and the distribution of the observation event (T_1, T_2), given the sample path X , is described by an equivalent probability measure $Q_2 = Q_2(X)$, where

$$\frac{dQ_2(X)}{dP_2} = L_{\infty\infty}(X) > 0 \quad a.s.$$

Note that under $Q_2(X)$ the components T_1 and T_2 are not, in general, independent.

On the space $(\Omega, F^* \times F)$ the measure $P = P_1 \times P_2$ is replaced by the equivalent probability measure $Q = P_1 \times Q_2$.

NOTATION 3.1.

Write

$$X_{t_1 t_2} = X_{00} + \alpha . p_1^* + \beta . \tilde{p}_1^* \qquad (3.1)$$

where

$$\alpha = \theta^1 . p_2^* + \theta^3 . \tilde{p}_2^* \text{ and } \beta = \theta^2 . p_2^* + \theta^4 . \tilde{p}_2^* .$$

Also

$$X_{t_1 t_2} = X_{00} + \bar{\alpha} . p_2^* + \bar{\beta} . \tilde{p}_2^* \qquad (3.2)$$

where

$$\bar{\alpha} = \theta^1 . p_1^* + \theta^2 . \tilde{p}_1^* \text{ and } \bar{\beta} = \theta^3 . p_1^* + \theta^4 . \tilde{p}_1^* .$$

Define

$$H(t_1, t_2) = L^{-1}_{t_1 -, t_2} \int_0^{t_2} g(X_{t_1 s_2}) dq_{s_2}$$

$$\tilde{H}(t_1, t_2) = L^{-1}_{t_1, t_2 -} \int_0^{t_1} g(X_{s_1 t_2}) dq_{s_1}$$

so that

$$L_{t_1 t_2}(X) = L_{t_1 t_2} = 1 + \int_0^{t_1} L_{s_1 - t_2} H(s_1, t_2) dq_{s_1} \qquad (3.3)$$

$$= 1 + \int_0^{t_2} L_{t_1 s_2 -} \tilde{H}(t_1, s_2) dq_{s_2} . \qquad (3.4)$$

For any random variables

$$U, V, W \in L^1(\Omega, F^A_{t_1 t_2} \times F_{t_1 t_2})$$

write

$$\bar{U} = E_P[U | F_{t_1 t_2}]$$

$$\hat{U} = E_Q[U | F_{t_1 t_2}] = (\bar{L}_{t_1 t_2})^{-1} (\overline{L_{t_1 t_2} U}) .$$

Conditional covariances are denoted by:

$$K(U, V) = E_Q[(U - \hat{U})(V - \hat{V}) | F_{t_1 t_2}]$$

$$= \widehat{UV} - \hat{U}\hat{V}$$

and

$$K(U, V, W) = E_Q[(U - \hat{U})(V - \hat{V})(W - \hat{W}) | F_{t_1 t_2}] .$$

4. HORIZONTAL AND VERTICAL FILTERS.

Keeping t_2 fixed and using the Ito product rule for one parameter semimartingales we have from (3.1) and (3.3):

$$(LX)(t_1,t_2) = X_{00} + \int_0^{t_1} L_{s_1-,t_2}(\alpha(s_1,t_2)dp^*_{s_1} + \beta(s_1,t_2)d\tilde{p}^*_{s_1})$$

$$+ \int_0^{t_1} X_{s_1-,t_2} L_{s_1-,t_2} H(s_1,t_2)dq_{s_1} \quad .$$

Therefore,

$$\overline{(LX)}(t_1,t_2) = \bar{X}_{00} - \int_0^{t_1} \bar{L}_{s_1-,t_2}(\hat{\alpha}(s_1,t_2) + \hat{\beta}(s_1,t_2))dF^{1*}_{s_1}$$

$$+ \int_0^{t_1} \bar{L}_{s_1-,t_2} \widehat{XH}(s_1-,t_2)\, dq_{s_1} \qquad (4.1)$$

where

$$\hat{\alpha}(s_1,t_2) = (\bar{L}_{s_1-,t_2})^{-1}E_P[\alpha(s_1,t_2)L_{s_1-,t_2}|S_1 = s_1 \text{ and } F_{t_1t_2}]$$

$$\hat{\beta}(s_1,t_2) = (\bar{L}_{s_1-,t_2})^{-1}E_P[\beta(s_1,t_2)L_{s_1-,t_2}|S_1 = s_1 \text{ and } F_{t_1t_2}]$$

and

$$\widehat{XH}(s_1-,t_2) = (\bar{L}_{s_1-,t_2})^{-1}(\overline{X_{s_1-,t_2}L_{s_1-,t_2}H(s_1,t_2)}).$$

From (3.3) we have that

$$\bar{L}_{t_1t_2} = 1 + \int_0^{t_1} \bar{L}_{s_1-,t_2} \hat{H}(s_1,t_2)dq_{s_1}$$

where

$$\hat{H}(s_1,t_2) = (\bar{L}_{s_1-,t_2})^{-1}(\overline{L_{s_1-,t_2}H(s_1,t_2)}).$$

Therefore, keeping t_2 fixed,

$$(\bar{L}_{t_1,t_2})^{-1} = 1 - \int_0^{t_1} (\bar{L}_{s_1-,t_2})^{-1} \hat{H}(s_1,t_2)dq_{s_1}$$

$$+ I_{t_1\geq T_1}(\bar{L}_{T_1,t_2})^{-1} \hat{H}(T_1,t_2)^2 \qquad (4.2).$$

Forming the product of (4.1) and (4.2) we have, after some algebra, the following expression:

$$\hat{X}_{t_1t_2} = \frac{\overline{LX}(t_1,t_2)}{\bar{L}_{t_1t_2}}$$

$$= \hat{X}_{00} - \int_0^{t_1} (\hat{\alpha}(s_1,t_2) + \hat{\beta}(s_1,t_2))dF^{1*}_{s_1}$$

$$+ \int_0^{t_1} \frac{(\widehat{XH} - \hat{X}\hat{H})}{1 + \hat{H}_{s_1t_2}}(s_1-,t_2)(dq_{s_1} - \hat{H}(s_1,t_2)d\tilde{p}_{s_1}).$$

REMARKS 4.1.

It is proved in [3] that the process

$$\nu^1_{t_1t_2} = q_{t_1} - \int_0^{t_1} \hat{H}(s_1,t_2)d\tilde{p}_{s_1}$$

is a 1-martingale for the filtration $\{F_{t_1t_2}\}$ under the measure Q . Similarly the process

$$\nu^2_{t_1t_2} = q_{t_2} - \int_0^{t_2} \hat{\hat{H}}(t_1,s_2)d\tilde{p}_{s_2}$$

is a 2-martingale for the filtration $\{F_{t_1t_3}\}$ under the measure Q .

DEFINITION 4.2.

$\{\nu^1_{t_1t_2}\}$ is the horizontal innovations process.

$\{\nu^2_{t_1t_2}\}$ is the vertical innovations process.

The horizontal filtering equation derived above can then be written in the following form:

THEOREM 4.3.

Horizontal Filtering Equation :

$$\hat{X}_{t_1t_2} = \hat{X}_{00} - \int_0^{t_1}(\hat{\alpha}(s_1,t_2) + \hat{\beta}(s_1,t_2))dF^{1*}_{s_1} + \int_0^{t_1} \frac{K(X,H)}{1 + \hat{H}}(s_1-,t_2)d\nu^1_{s_1t_2} .$$

Similarly we can establish the following formula:

THEOREM 4.4.

Vertical Filtering Equation :

$$\hat{X}_{t_1t_2} = \hat{X}_{00} - \int_0^{t_1} (\hat{\bar{\alpha}}(t_1,s_2) + \hat{\bar{\beta}}(t_1,s_2))dF^{2*}_{s_2} + \int_0^{t_2} \frac{K(X,\tilde{H})}{1+\hat{\tilde{H}}}(t_1,s_2-)d\nu^2_{t_1,s_2} .$$

Here

$$\hat{\bar{\alpha}}(t_1,s_2) = (\bar{L}_{t_1,s_2-})^{-1}E_P[\bar{\alpha}(t_1,s_2)L_{t_1,s_2-}|S_2 = s_2 \text{ and } F_{t_1,s_2}]$$

$$\hat{\bar{\beta}}(t_1,s_2) = (\bar{L}_{t_1,s_2-})^{-1}E_P[\bar{\beta}(t_1,s_2)L_{t_1,s_2-}|S_2 = s_2 \text{ and } F_{t_1,s_2}] .$$

5. THE DIAGONAL FILTERING EQUATION

NOTATION 5.1

If $Y_{t_1t_2}$, $(t_1,t_2) \in [0,\infty]^2$, is a process write

$\Delta^1 Y_{t_1t_2}$ for the process $Y_{t_1,t_2} - Y_{t_1-,t_2}$

$\Delta^2 Y_{t_1t_2}$ for the process $Y_{t_1t_2} - Y_{t_1,t_2-}$

and $\Delta Y_{t_1t_2}$ for the process $Y_{t_1,t_2} + Y_{t_1-,t_2-} - Y_{t_1-,t_2} - Y_{t_1,t_2-}$.

Write:

$$\gamma(s_1,s_2) = (L_{s_1-,s_2-})^{-1}g(X_{s_1,s_2})$$

$$\hat{\gamma}(s_1,s_2) = (\bar{L}_{s_1-,s_2-})^{-1}(\overline{L_{s_1-,s_2-}\gamma(s_1,s_2)})$$

$$R(s_1,s_2) = \gamma(s_1,s_2) - H(s_1,s_2-)\tilde{H}(s_1-,s_2)$$

$$\begin{aligned}\hat{R}(s_1,s_2) &= (\bar{L}_{s_1-,s_2-})^{-1}(\overline{L_{s_1-,s_2-}R(s_1,s_2)}) \\ &= \hat{\gamma}(s_1,s_2) - \widehat{H\tilde{H}}(s_1,s_2) ,\end{aligned}$$

where

$$\widehat{H\tilde{H}}(s_1,s_2) = L_{s_1-s_2-}^{-1}(\overline{L_{s_1-,s_2-}H(s_1,s_2-)\tilde{H}(s_1-,s_2)}) ,$$

and

$$\begin{aligned}\hat{R}_K(s_1,s_2) &= \hat{\gamma}(s_1,s_2) - \hat{H}(s_1,s_2-)\hat{\tilde{H}}(s_1-,s_2) \\ &= \hat{R} + K(H,\tilde{H}) .\end{aligned}$$

Define

$$F = K(R,X) + K(X,H,\tilde{H})$$

and $$G = (1+\hat{H})K(X,\tilde{H}) + (1+\hat{\tilde{H}})K(X,H).$$

Keeping t_1 fixed, and using the differentiation formula, we have the following results. (See equation (5.3) of [2]).

LEMMA 5.2.

$$\hat{H}(t_1,t_2) = (\bar{L}_{t_1-,t_2})^{-1}(\overline{L_{t_1-,t_2}H(t_1,t_2)})$$

$$= (\bar{L}_{t_1-,t_2})^{-1}\int_o^{t_2}\bar{g}(t_1,s_2)dq_{s_2}$$

$$= \int_o^{t_2}\hat{R}_K(t_1,s_2)dq_{s_2} - \int_o^{t_2}\frac{\hat{R}_K(t_1,s_2)\hat{\tilde{H}}(t_1-,s_2)}{1+\hat{\tilde{H}}(t_1-,s_2)}\,dp_{s_2}$$

$$= \frac{\hat{R}_K}{1+\hat{\tilde{H}}}\cdot p_2 - \hat{R}_K\cdot\tilde{p}_2$$

Similarly

$$\hat{\tilde{H}}(t_1,t_2) = \frac{\hat{R}_K}{1+\hat{H}}\cdot p_1 - \hat{R}_K\cdot\tilde{p}_1 \quad .$$

Using Theorem 3.3 of [3] we have the following two parameter expression for $(\bar{L})^{-1}$:

LEMMA 5.3.

$$(\bar{L}_{t_1t_2})^{-1} = 1 + (\Delta\bar{L}^{-1}_{s_1s_2})\cdot p_1p_2$$

$$+ \Delta^1\Big(\frac{\hat{\tilde{H}}(s_1,s_2)}{\bar{L}_{s_1,s_2-}}\Big)\cdot p_1\tilde{p}_2 + \Delta^2\Big(\frac{\hat{H}(s_1,s_2)}{\bar{L}_{s_1-,s_2}}\Big)\cdot\tilde{p}_1p_2$$

$$+ ((\bar{L}_{s_1-,s_2-})^{-1}(\hat{H}(s_1,s_2-)\hat{\tilde{H}}(s_1-,s_2) - \hat{R}_K(s_1,s_2))\cdot\tilde{p}_1\tilde{p}_2 \quad .$$

REMARKS 5.4.

To determine the two parameter filtering formula for $\hat{X} = (\bar{L})^{-1}(\overline{LX})$ one might try to discuss this product using the above expression for $(\bar{L})^{-1}$. However, the calculations rapidly become very complex. Instead it is better to substitute the vertical filtering formula in the horizontal formula.

For

$i = 1,2,3,4$ write

$$\hat{\theta}^{i}_{s_1,s_2} = (\bar{L}_{s_1-,s_2-})^{-1} E_P\left[L_{s_1-,s_2-}\, \theta^{i} \,|\, S_1 = s_1, S_2 = s_2 \text{ and } F_{s_1,s_2}\right]$$

$$= (\bar{L}_{s_1-,s_2-})^{-1}\{\theta^{i}_{s_1 s_2}(1 + \int_0^{s_1-}\int_0^{s_2-} g(X_{oo} - \int_0^{u_1}\int_0^{u_2} \theta^4 d\tilde{p}_2 d\tilde{p}_1) dq_{u_1} dq_{u_2})\} .$$

and

$$\Pi = \Pi(s_1,s_2) = (1 + \hat{H}(s_1,s_2-))(1 + \hat{\tilde{H}}(s_1-,s_2)) .$$

Then the following expressions are obtained :

$$\hat{\alpha}(s_1,t_2) = - \int_0^{t_2} (\hat{\theta}^1_{s_1 s_2} + \hat{\theta}^2_{s_1,s_2}) dF^{2*}_{s_2} + \int_0^{t_2} \frac{K(\alpha,H)}{1 + \hat{\tilde{H}}}(s_1,s_2-)\, dv^2_{s_1,s_2},$$

with an analogous formula for $\hat{\beta}(s_1,t_2)$,

$$\frac{1}{1 + \hat{H}} = 1 + \frac{\hat{R}_K}{(1 + \hat{H})^2} \cdot \tilde{p}_2 - \frac{\hat{R}_K}{(1 + \hat{H})(\Pi + \hat{R}_K)} \cdot p_2 ,$$

$$\hat{X}\hat{H} = - \int_0^{t_2} \hat{H}(\hat{\alpha} + \hat{\beta}) dF^{2*}_{s_2} + \frac{\hat{R}_K K(X,\tilde{H})}{(1 + \hat{\tilde{H}})^2} \cdot p_2$$

$$+ \frac{(\hat{H}K(X,\tilde{H}) + \hat{X}\hat{R}_K)}{(1 + \hat{\tilde{H}})} \cdot v^2 ,$$

$$\widehat{XH} = - \int_0^{t_2} (\widehat{H\alpha} + \widehat{H\beta}) dF^{2*}_{s_2} + \frac{(\widehat{\alpha X} - \widehat{XH}\hat{\tilde{H}})}{1 + \hat{\tilde{H}}} \cdot v^2 .$$

Using the differentiation rule and substituting in the formula of Theorem 4.3 we obtain the following expression:

THEOREM 5.5

Diagonal Filtering Equation:

$$\hat{X}_{t_1t_2} = \hat{X}_{00} + \int_0^{t_1}\int_0^{t_2} (\hat{\theta}^1 + \hat{\theta}^2 + \hat{\theta}^3 + \hat{\theta}^4) dF^{2*} dF^{1*}$$

$$- \int_0^{t_1}\int_0^{t_2} \frac{K(H, \bar{\alpha} + \bar{\beta})}{1 + \hat{H}} dF^{2*} d\nu^1 - \int_0^{t_1}\int_0^{t_2} \frac{K(\tilde{H}, \alpha + \beta)}{1 + \hat{\tilde{H}}} d\nu^2 dF^{1*}$$

$$+ \int_0^{t_1}\int_0^{t_2} \hat{R}_K \left\{ \frac{K(X,H)}{(1 + \hat{H})^2} - \frac{(F + G)}{(1 + \hat{H})(\Pi + \hat{R}_K)} \right\} d\tilde{p}_2 \, d\nu^1$$

$$+ \int_0^{t_1}\int_0^{t_2} \hat{R}_K \left\{ \frac{K(X,\tilde{H})}{(1 + \hat{\tilde{H}})^2} - \frac{(F + G)}{(1 + \hat{\tilde{H}})(\Pi + \hat{R}_K)} \right\} d\nu^2 d\tilde{p}_1$$

$$+ \int_0^{t_1}\int_0^{t_2} \frac{(\Pi F - \hat{R}_K G)}{(\Pi + \hat{R}_K)} \cdot (d\nu^2 d\nu^1 - \hat{R}_K d\tilde{p}_2 d\tilde{p}_1) .$$

REMARKS 5.6

From Theorem 4.5 of [3], for $\theta \in L^1(\Omega, F)$

$\theta . \nu^2 \nu^1 - (\widehat{\theta R_K}) . \tilde{p}_2 \tilde{p}_1$ is weak martingale for the filtration $\{F_{t_1, t_2}\}$ under the measure Q. Therefore, the process $\nu^2 \nu^1 - \hat{R}_K . \tilde{p}_2 \tilde{p}_1$ can be thought of as the diagonal innnovations process. Note that when $R = 0$, $\hat{R}_K = K(H, \tilde{H})$, $\Pi + \hat{R}_K = E_Q[(1 + H)(1 + \tilde{H}) | F_{t_1 t_2}]$ and our formulae resemble those of [5] .

REFERENCES

1. AL-HUSSAINI, A and ELLIOTT, R. J. Weak martingales associated with a two parameter jump process. Lecture Notes in Control and Information Sciences 16, 253-263. Springer-Verlag. Berlin, Heidelberg, New York. 1979.

2. AL-HUSSAINI, A. and ELLIOTT, R. J. Martingales, potentials and exponentials associated with a two parameter jump process. Stochastics 5(1981).

3. AL-HUSSAINI, A. and ELLIOTT, R. J. Stochastic calculus for a two parameter jump process. Lecture Notes in Mathematics 863, 233-244, Springer-Verlag, Berlin, Heidelberg, New York, 1981.

4. KOREZLIOGLU, H, MAZZIOTTO, G. and SZPIRGLAS, J. Nonlinear filtering equations for two paramter semimartingales. Pre-print. Centre National D'Etudes des Telecommunications.

5. MAZZIOTTO, G. and SZPIRGLAS, J. Equations du Filtrage pour un Processus

de Poisson Mélange à Deux Indices. Stochastics 4 (1980), 89-119.

6. WONG, E. Recursive Filtering for Two Dimensional Random Fields. IEEE Trans. in Information Theory. 21 (1975).

7. ZAKAI, M. On the Optimal Filtering of Diffusion Processes. Zeits. für Wahrs. 11(1969), 230-249.

SPACE-TIME MIXING IN A BRANCHING MODEL

Klaus Fleischmann
Academy of Sciences of the GDR
Institute of Mathematics
DDR-1080 Berlin, Mohrenstrasse 39

Summary

For some stationary spatially homogeneous measure-valued branching processes we have asymptotic independence in the space-time diagram.

1. Introduction

Mostly branching processes describe mathematically the time evolution of finite populations. Under usual assumptions these processes die out or explode. Thus, processes stationary in time can only be expected if infinite populations are considered. A good way to illustrate such an infinite population is to think of an unbounded space, say a Euclidean space $R^d =: G$, in which the population is situated. More precisely, we shall use locally finite measures μ defined on G, i.e. measures μ for which $\mu(B)$ is finite for all bounded Borel subsets of G. In this way $\mu(B)$ describes the mass (e.g. number of particles) of the considered population in the region B.

In a natural way we have for the measures μ space translations

$$T_x \mu = \int \mu(dy)\, \delta_{x+y} \qquad (x \in G).$$

Let $(\mu_t)_{-\infty < t < +\infty}$ be a stochastic process with locally finite measures as its states.

Definition 1.1. (μ_t) is called space-time mixing of all orders if for all finite sequences $t_0, \dots, t_k$ of time points and $x_0, \dots, x_k \in G$ the vector

$$[T_{x_0} \mu_{t_0}, \dots, T_{x_k} \mu_{t_k}]$$

will be asymptotically independent in distribution as $[x_i, t_i] - [x_j, t_j] \longrightarrow \infty$ for all $i \neq j$.

Roughly spoken, space-time mixing means that in the space-time diagram the behaviour in finitely many regions will be asymptotically independent if all distances between the regions tend to infinity.

Our purpose is to formulate the space-time mixing of all orders for a relatively general stationary branching process.

Strong formulations of the model, the results and their proofs are contained in [1].

2. Model Assumptions

On principle we assume that, for all x in G, a unit mass δ_x at position x generates a cluster χ_x, i.e. a random finite measure χ_x defined on G. Moreover we presuppose a spatial homogeneity, i.e. χ_x coincides in distribution with $T_x\chi_0$ for all x where $\chi_0 =: \chi$ is the cluster generated by a unit mass in the origin of R^d. Furthermore we shall use the basic assumption in branching theory, namely that different masses generate independent progenies. Consequently, in addition, we have necessarily to require that χ is an infinitely divisible random measure, i.e. for all natural numbers k it can be represented as a sum of k independent identically distributed random measures. Finally, we assume that χ is critical, i.e. its expected total mass $\mathbb{E}\,\chi(G)$ equals 1.

Using these assumptions, we can define a branching process $(\Phi_t)_{t\geq 0}$ in the following way. At time 0 we start with the Lebesgue measure ℓ on G, i.e. $\Phi_0 := \ell$. Then all 'small parts $d\,\delta_x$' of the initial population Φ_0 are clustered independently as well as spatially homogeneously, and the superposition will give us the first generation Φ_1. (A strong formulation for this intuitive explanation can be given in terms of Lévy-Khinchin representations, cf.[5] or [3].) Given Φ_1 we continue in an analogous manner getting Φ_2 etc.

The following convergence lemma holds (cf. Hermann [3]).

Lemma 2.1. Φ_t converges in distribution to some limit population Φ_∞ which is cluster-invariant (in distribution) and infinitely divisible, and its intensity measure $\mathbb{E}\,\Phi_\infty$ equals ℓ or the zero measure o.

For the interesting problem of separating the extinction case $\Phi_\infty = o$ we refer to [4] or, in the point process case, to [7], section 12.6.

An open problem is the question whether in some particular 'non-stable' cases $\Phi_\infty = o$ there exist cluster-invariant populations (which would have infinite asymptotic densities).

3. The Result

From now on we restrict our considerations to the 'stable' case $\mathbb{E}\,\Phi_\infty = \ell$.

Let $(\Psi_t)_{t \in Z}$ be a stationary branching process with clusters χ and $\Psi_t = \Phi_\infty$ (in distribution) where Z is the set of all integers. Without any additional assumptions we have

Theorem 3.1. (Ψ_t) is space-time mixing of all orders. (Here we mention that the theorem remains valid if we only presuppose that G is a locally compact second countable Hausdorff Abelian topological group.)

Analogously to the point process case it can be expected that Φ_∞ is always tail trivial, i.e. it has short range correlation (cf. [2]).

An open problem is to find criteria for still higher mixing properties in time, as e.g. Bernoulli property (except the very particular case of stochastic translations $\chi = \delta_\xi$ for some random position ξ, cf. [9]).

4. Two Abstract Criterions

Starting from the stationary branching process (Ψ_t) let us consider the random measure

$$\nu := \sum_{t \in Z} \Psi_t \times \delta_t$$

defined on the group $R^d \times Z$ for which we have the space-time translations

$$T_{[x,s]}\,\nu := \sum_{t \in Z} T_x \Psi_t \times \delta_{t+s} \qquad (x \in R^d, s \in Z).$$

ν is infinitely divisible, too. This leads us to the following abstraction.

Let A be a complete separable metric space, M be the set of all locally finite measures μ defined on A and P be an infinitely divisible distribution defined on M. The Lévy measure (canonical measure) of P is denoted by $\widetilde{P}$ (cf. [6] or [3]).

Let G be a locally compact second countable Hausdorff Abelian topological group and $g \curvearrowright S_g$ be a homomorphism of G into the group of all one-to-one transformations of A onto itself. The mapping $[a,g] \curvearrowright S_g a$ is assumed to be continuous and $\{ S_g a : a \in B, g \in K \}$

should be bounded for all bounded subsets B of A and compact subsets K of G.

The last property ensures that the set-up

$$T_g\mu := \mu(S_g a \in \cdot)$$

transfers the continuous G-flow (S_g) in A into a continuous G-flow (T_g) in M. In this way the group G acts continuously in A and M. New we assume in addition that P is invariant with respect to (T_g).

P is called mixing of order k > 0 if

$$P(T_{g_0} Y_0 \cap \dots \cap T_{g_k} Y_k) - P(Y_0) \cdots P(Y_k)$$

tends to zero as $g_i - g_j \longrightarrow \infty$ $(i \neq j)$ for all $g_0, \dots, g_k \in G$ and measurable subsets $Y_0, \dots, Y_k$ of M.

In generalization of results of Nawrotzki [8] (and earlier papers on point processes cited there) we have the following two theorems.

Theorem 4.1. P is mixing of order k if and only if

$$\tilde{P}(\mu(B) > \varepsilon, T_g\mu(B) > \varepsilon) \xrightarrow[g \to \infty]{} 0$$

for all bounded Borel subsets B of A and all positive numbers ε.

Let us mention here that the condition in the theorem does not depend on the order k. In other words, for infinitely divisible random measures the properties mixing and mixing of all orders coincide.

P is called ergodic if all measurable subsets of M which are almost-invariant with respect to (T_g) are trivial with respect to P.

Let (γ_n) be a sequence of distributions defined on G such that the variational distance $\| \sigma_1 * \gamma_n - \sigma_2 * \gamma_n \|$ tends to zero as $n \longrightarrow \infty$ for all distributions σ_1, σ_2 defined on G which are absolutely continuous with respect to a Haar measure ℓ on G. Such (γ_n) is called weakly asymptotically uniformly distributed.

Theorem 4.2. P is ergodic if and only if

$$\int \gamma_n(dg)\, \tilde{P}(\mu(B) > \varepsilon, T_g\mu(B) > \varepsilon) \xrightarrow[n \to \infty]{} 0$$

for all bounded measurable $B \subset A$ and $\varepsilon > 0$.

5. To the Proofs

Let F be the set of all bounded non-negative functions f on A with bounded support. The Laplace functional $\hat{P}$ of P is defined by

$$\hat{P}(f) := \int P(d\mu)\, e^{-\mu f} \qquad (f \in F)$$

where $\mu f := \int \mu(dx) f(x)$.

Theorem 4.1 is based on some elementary inequalities:

Lemma 5.1. Let $f_0, \ldots, f_k \in F$ and $\varepsilon > 0$. Then

$$0 \leq \hat{P}\left(\sum_i f_i\right) - \prod_i \hat{P}(f_i) \leq \text{const} \sum_i \int_{\mu f_i > \varepsilon} \tilde{P}(d\mu)\, \mu f_i$$

$$+ \text{const} \sum_{i<j} \tilde{P}(\mu f_i > \varepsilon,\ \mu f_j > \varepsilon)$$

where the constants only depend on k. On the other side

$$\hat{P}(f_1 + f_2(T_g \cdot)) - \hat{P}(f_1)\hat{P}(f_2) \geq \text{const}^+\ \tilde{P}(\mu f_1 > \varepsilon,\ T_g\mu f_2 > \varepsilon)$$

where the positive constant does not depend on g.

For theorem 4.2 additionally an abstract statistical ergodic theorem is used.

In order to prove space-time mixing of (ψ_t) first of all time ergodicity is shown. Then we derive space ergodicity of Φ_∞. Space-time mixing then follows by using theorems 4.1 and 4.2, several tools in random measure theory and properties of sequences of convolution powers of distributions.

References

[1] K. Fleischmann: Mixing properties of infinitely divisible random measures. Carleton Mathematical Lecture Note, Ottawa (in preparation)

[2] K. Fleischmann, K. Hermann and K. Matthes: Kritische Verzweigungsprozesse mit allgemeinem Phasenraum, VIII., Math. Nachr. (submitted)

[3] K. Hermann: Critical measure-valued branching processes in discrete time with arbitrary phase space, I., Math. Nachr. 103, 63-107 (1981)

[4] K. Hermann: Kritische maßwertige Verzweigungsprozesse in diskreter Zeit. Dissertation. Akad. Wiss. DDR 1981

[5] M. Jiřina: Branching processes with measure-valued states. Trans. 3rd Prague Conf. Inf. Theory Stat. Dec. Functions Random Proc. 333-357 (1964)

[6] O. Kallenberg: Random Measures. Akademie-Verlag Berlin 1975; Academic Press London 1976

[7] K. Matthes, J. Kerstan and J. Mecke: Infinitely Divisible Point Processes. Wiley & Sons, Chichester 1978

[8] K. Nawrotzki: Mischungseigenschaften stationärer unbegrenzt teilbarer zufälliger Maße. Math.Nachr. 38, 97-114 (1968)

[9] T. Shiga and Y. Takahashi: Ergodic properties of the equilibrium process associated with infinitely many Markovian particles. Publ. Res. Inst. Math. Sci., Kyoto Univ. 9, 505-516 (1974)

LOGARITHMIC TRANSFORMATIONS AND STOCHASTIC CONTROL

Wendell H. Fleming[1]

Lefschetz Center for Dynamical Systems
Division of Applied Mathematics
Brown University
Providence, Rhode Island 02912

1. Introduction. We are concerned with a class of problems described in a somewhat imprecise way as follows. Consider a linear operator of the form $L + V(x)$, where L is the generator of a Markov process x_t and the "potential" $V(x)$ is some real-valued function on the state space Σ of x_t. We are interested in probabilistic representations for solutions $\phi(s,x)$ to the backward equation

$$\frac{d\phi}{ds} + L\phi + V(x)\phi = 0, \quad s \leq T, \tag{1.1}$$

with data $\phi(T,x) = \Phi(x)$ at a final time T. It is well known that, under suitable assumptions,

$$\phi(s,x) = E_{sx}\ \{\Phi(x_T)\exp\int_s^T V(x_t)dt\} \tag{1.2}$$

gives such a representation. For instance, if $x_t = x_s + w_t - w_s$, with w_t a brownian motion, then (1.2) is just the Feynman-Kac formula. We seek a different kind of probabilistic representation for $I = -\log\phi$, if $\phi(s,x)$ is a positive solution to (1.1). In this representation the generator L is replaced by another generator $L^{\underline{u}}$ of a Markov process ξ_t (possibly time inhomogeneous.) The operator $L^{\underline{u}}$ is chosen to solve an optimal stochastic control problem of the following kind. The logarithmic transformation $I = -\log\phi$ changes (1.1) into the nonlinear equation

$$\frac{dI}{ds} + H(I) - V(x) = 0, \text{ where} \tag{1.3}$$

$$H(I) = -e^{I}L(e^{-I}). \tag{1.4}$$

The function H is concave. For a fairly wide class of Markov processes, we wish to write (1.3) as the dynamic programming equation associated with a suitable optimal stochastic control problem for Markov processes. The stochastic control problem is specified by giving: (a) a suitable control space U; for each constant control $u \in U$, the generator L^u of a Markov process; and (c) a cost function $k(x,u)$ associated with constant control u and state x. See [6, Chap. VI]. It is re-

---

[1]This research was supported in part by the National Science Foundation under contract MCS 79-03554 and in part by the Air Force Office of Scientific Research under contract AF-AFOSR 81-0116.

quired that

(1.5) $$H(I)(x) = \min_{u \in U} [L^u I(x) + k(x,w)], \quad x \in \Sigma .$$

Then (1.3) becomes a dynamic programming equation:

(1.6) $$\frac{dI}{ds} + \min_{u \in U} [L^u I + k(x,u) - V(x)] = 0.$$

Time and state dependent controls $\underline{u}(s,x)$, in feedback form, with values in the control space U are allowed. The stochastic control problem is to find a feedback $\underline{u}$ minimizing

(1.7) $$J(s,x;\underline{u}) = E_{sx}\int_s^T [k(\xi_t,u_t) - V(\xi_t)]dt + \Psi(\xi_T) \quad ,$$

where ξ_t is the (controlled) Markov process with generator $L^{\underline{u}}$, $\xi_s = x$, and

$$u_t = \underline{u}(t,\xi_t), \quad \Psi = -\log \Phi .$$

The Verification Theorem of optimal stochastic control theory [6, p.159] asserts that if I is a "well behaved" solution to (1.3) with $I(T,x) = \Psi(x)$ and if certain other technical conditions hold, then

(1.8) $$I(s,x) = \min_{\underline{u}} J(s,x;\underline{u}) .$$

Moreover, an optimal feedback control $\underline{u}(s,x)$ is found by minimizing $L^u I(s,x) + k(x,u)$ over the control space U.

In this paper we take $\Sigma \subset R^n$, a subset of n-dimensional euclidean space. In §2 we review the case when x_t is a diffusion process on R^n. For nondegenerate diffusions, an appropriate stochastic control problem is immediately suggested by the form of equation (1.3). In §3 we consider jump Markov processes x_t, and associated stochastic control problems. The choice of an appropriate control problem is less immediate for jump processes than for diffusions. In his Ph.D. thesis S-J Sheu [11] uses a different control formulation, valid for a wide class of generators L(§4). The optimal control in his sense leads to the change of probability measures described in (4.5). In §5 we give a formal derivation indicating why stochastic control methods can be used to obtain asymptotic estimates for exit probabilities for a family x_t^ε of nearly deterministic jump processes. The results are not new (see [1][12]); the interest is in the stochastic control method. Rigorous proofs are given in [11] using such methods.

In §6 we consider briefly the Donsker-Varadhan formula for the dominant eigenvalue λ_1 of $L+V$, from a control viewpoint. For nondegenerate diffusions the stochastic control representation obtained for λ_1 is the same as Holland's [9].

2. Diffusion processes. Let x_t be a diffusion in n-dimensional R^n, with generator

$$Lf = \frac{1}{2} \operatorname{tr} a(x) f_{xx} + b(x) \cdot f_x \tag{2.1}$$

$$\operatorname{tr} a(x) f_{xx} = \sum_{i,j=1}^{n} a_{ij}(x) \frac{\partial^2 f}{\partial x_i \partial x_j},$$

and with f_x the gradient. In this case,

$$H(I) = \frac{1}{2} \operatorname{tr} a(x) I_{xx} + b(x) \cdot I_x - \frac{1}{2} I_x' a(x) I_x \tag{2.2}$$

We may take $U = R^n$, $u = (u_1,\text{---},u_n)$,

$$L^u I = \frac{1}{2} \operatorname{tr} a(x) I_{xx} + u \cdot I_x \tag{2.3}$$

$$k(x,u) = \frac{1}{2}(b(x)-u)' a^{-1}(x)(b(x)-u). \tag{2.4}$$

For a feedback control $\underline{u}$, the drift coefficient $b(x)$ in (2.1) is changed to drift coefficient $\underline{u}(s,x)$ in the operator $L^{\underline{u}}$.

The stochastic control representation (1.8) was used in [3] to give a stochastic control proof of results of Ventsel-Freidlin type for some large deviations problems for nearly deterministic diffusions. In those results $a(x)$ is replaced by $\varepsilon a(x)$, ε small. In [4] the logarithmic transformation was used to obtain stochastic representations for positive solutions to the heat equation with a potential term, and to obtain the "classical mechanical limit." In [5] [10] the same logarithmic transformation was applied to solutions to the pathwise equation of nonlinear filtering. Large deviations results for the nonlinear filter problem are obtained by Hijab [8] elsewhere in this volume.

In [7] Hernandez-Lerma obtained similar results for certain degenerate diffusions, for which the matrix $(a_{ij}(x))$, $i,j=1,\text{---},m < n$ is positive definite and $a_{ij}(x)=0$ if $i > m$ or $j > m$.

3. Jump processes. To motivate our choice of stochastic control problem, let us begin with a simple special case in which the process x_t jumps only by a fixed increment y (as for example for a Poisson process.) In this case the generator L takes the form

$$Lf(x) = a(x)[f(x+y)-f(x)].$$

From (1.4)

$$H(I)(x) = a(x)(1 - \exp [I(x) - I(x+y)]).$$

The dual function to the convex function e^r is $u - u \log u$ $(u > 0)$:

$$(3.1) \qquad e^r = \max_{u>0} \; [u - u \log u + ur] \; .$$

The max occurs when $\log u = r$. Let

$$(3.2) \qquad L^u I(x) = ua(x)[I(x+y)-I(x)], \quad u > 0$$

$$(3.3) \qquad k(x,u) = a(x)(u \log u - u+1).$$

By taking $r = I(x)-I(x+y)$ in (3.1) and changing signs (to replace max by min), we get the required form (1.5) for $H(I)$. In this special case the control u is scalar, with $u > 0$. A constant control u changes the jumping rate from $a(x)$ to $ua(x)$. A feedback control $\underline{u}(s,x)$ changes the rate at time s and state x from $a(x)$ to $\underline{u}(s,x)a(x)$. If $I(s,x) = -\log \phi(s,x)$ as in §1, then the optimal feedback control is $\underline{u}^*(s,x) = \phi(s,x)^{-1}\phi(s,x+y)$.

Let us now consider a jump process x_t with generator of the form

$$(3.4) \qquad Lf(x) = a(x) \int_{R^n} [f(x+y) - f(x)] \, \pi(x,dy).$$

Here $f \in B(R^n)$, the space of bounded Borel measurable functions on R^n. We assume that $a \in B(R^n)$ and that $\pi(x,\cdot)$ is a probability measure with $\pi(\cdot,A)$ Borel measurable for each Borel set A and $\pi(x,\{0\}) = 0$. Additional conditions on a and π need to be imposed later. Motivated by the special case above, we control the jumping distribution, replacing $a(x)\pi(x,dy)$ by $a(x)\underline{u}(s,x;y)\pi(dy)$. To formalize this idea, we introduce the control space

$$(3.5) \qquad U = \{u(\cdot)\colon u,u^{-1} \in B(R^n), \; u(y) > 0 \text{ for all } y \in R^n \}.$$

Suitable $L^{u(\cdot)}$ and $k(x,u(\cdot))$ are obtained by integrating (3.2), (3.3) with respect to $\pi(x,dy)$:

$$(3.6) \qquad L^{u(\cdot)}I(x) = a(x) \int_{R^n} [I(x+y)-I(x)]u(y)\pi(x,dy)$$

$$(3.7) \qquad k(x,u(\cdot)) = a(x) \int_{R^n} [u(y)\log u(y)-u(y)+1]\pi(x,dy).$$

We get as in equation (1.5)

$$(3.8) \qquad H(I)(x) = \min_{u(\cdot)\in U} \; [L^{u(\cdot)}I(x)+k(x,u(\cdot))]$$

If $\phi(s,x)$ is a positive solution to (1.1) and $I = -\log \phi$, then the optimal feedback control is

$$\underline{u}^*(s,x;\cdot) = \frac{\phi(s,x+\cdot)}{\phi(s,x)} \quad . \tag{3.9}$$

As outlined in the next section, it is sometimes more convenient to consider instead a related control problem. In particular, the formulation in §4 is the one used in [11] to give control method proofs of the results on the exit problem mentioned in §5.

4. The Sheu formulation. In [11] another kind of control problem is considered. Let L be a bounded linear operator on $C(\Sigma)$, the space of continuous bounded functions on Σ, such that L obeys a positive maximum principle. (In particular, L may be of the form (3.4) above.) For $w = w(\cdot)$ a positive function with $w, w^{-1} \in C(\Sigma)$, define the operator $\tilde{L}^w$ by

$$\tilde{L}^w f = \frac{1}{w}\,[L(wf) - fLw]. \tag{4.1}$$

In addition, define $K^w(x)$ by

$$K^w = \tilde{L}^w(\log w) - \frac{1}{w}L(w). \tag{4.2}$$

For unbounded L, additional restrictions on w are needed in order that $\tilde{L}^w$ and K^w be well defined.

From the duality (3.1) between e^r and $u \log u - u$, it is not difficult to show [11] that for $I \in C(\Sigma)$

$$H(I) = \min_w\, [\tilde{L}^w I + K^w] \ . \tag{4.3}$$

The minimum is attained for $w = \exp(-I)$. For L the generator of a jump process, the two formulations are related by $\tilde{L}^w = L^{\underline{u}}$, where $\underline{u}$ is the (stationary) feedback control defined by

$$\underline{u}(x;y) = \frac{w(x+y)}{w(x)} \quad . \tag{4.4}$$

Moreover, $K^w(x) = k(x,\underline{u}(x;\cdot))$.

In Sheu's formulation, the control problem is to choose $w_t(\cdot)$ for $s \le t \le T$ to minimize

$$\mathcal{J}(s,x;w) = E_{sx}\left\{\int_s^T [K^{w_t}(\xi_t) - V(\xi_t)]dt + \Psi(\xi_T)\right\},$$

where ξ_t is a Markov process with generator $\tilde{L}^{w_t}$ and with $\xi_s = x$. Here we assume that L is the generator of a Markov process x_t which implies in particular $L1 = 1$.

Suppose that ϕ is a positive solution to (1.1), with $\phi(s,\cdot)$, $\phi(s,\cdot)^{-1} \in C(\Sigma)$ and with $V \in C(\Sigma)$. We can use (4.3) together with the Verification Theorem in stochastic control to conclude that $I(s,x) \leq \mathcal{J}(s,x;w)$ with equality when $w_t^* = \phi(t,\cdot)$. Thus the control $w_t^* = \phi(t,\cdot)$ is optimal in this sense. For jump processes this agrees with (3.9), according to (4.4).

The change of generator from L to $\tilde{L} = \tilde{L}^{w_t^*}$ corresponds to a change of probability measure, from P to $\tilde{P}$, as follows:

$$\tilde{E}_{sx} f(\xi_t) = \frac{E_{sx}[f(x_t)\phi(x_T)]}{E_{sx}\phi(x_T)}, \quad s \leq t \leq T, \quad f \in C(\Sigma). \tag{4.5}$$

This is seen from the following argument. The denominator of the right side is $\phi(s,x)$. Let

$$\psi(s,x) = E_{sx}[f(x_t)\phi(x_T)] = E_{sx}[f(x_t)\phi(t,x_t)] .$$

Since ϕ and ψ both satisfy (1.1) with $V = 0$, the quotient $v = \psi\phi^{-1}$ satisfies

$$\frac{\partial v}{\partial s} = -\left[\frac{L\psi}{\phi} - \frac{\psi L\phi}{\phi^2}\right] = -\frac{1}{\phi}\left[L(v\phi) - v\, L\phi\right],$$

$$\frac{\partial v}{\partial s} + \tilde{L}v = 0, \quad s \leq t, \tag{4.6}$$

with $v(t,x) = f(x)$ as required.

The author wishes to thank M. Day for a helpful suggestion related to (4.5).

5. Asymptotic estimates for exit probabilities.

Let x_t^ε be a family of Markov processes, $s \leq t \leq T$, depending on a small parameter $\varepsilon > 0$, such that x_t^ε tends (in a suitable sense) to a deterministic limit x_t^0 as $\varepsilon \to 0$. Let ϕ^ε denote the probability that x^ε belongs to a set Γ of trajectories which does not include trajectories "near" x^0. Typically ϕ^ε is exponentially small. Its asymptotic rate of decay to 0 can be found from the theory of large deviations [1][12][13]. In the exponent a constant I^0 appears, which is the minimum of a certain action functional over a set of smooth paths.

In many instances these asymptotic estimates can also be obtained by introducing a stochastic control problem of the kind indicated in previous sections, for each $\varepsilon > 0$ [3] [11]. With this method a (stochastic) optimization problem appears for each

$\varepsilon > 0$, not just in the limit as $\varepsilon \to 0$.

Let us consider the special case when ϕ^ε is an exit probability:

$$\phi^\varepsilon(s,x) = P_{sx}(\tau^\varepsilon \le T),$$

where τ^ε is the exit time of x_t^ε from a bounded, open set $D \subset R^n$, and where $x_t^0 \in D$ for $s \le t \le T$. We consider nearly deterministic jump processes, as follows. Nearly deterministic diffusions were considered in [3] [7]. Following Vent'cel [12] let us rescale the jump process in §3, replacing y by εy and $a(x)$ by $\varepsilon^{-1}a(x)$ to obtain the generator for x_t^ε:

$$L_\varepsilon f(x) = \varepsilon^{-1}a(x) \int_{R^n} [f(x+\varepsilon y)-f(x)]\, \pi(x,dy). \tag{5.1}$$

Fix $x_s^\varepsilon = x$. For $s \le t \le T$, the path x^ε tends in probability as $\varepsilon \to 0$ (D-metric) to x^0, where x_t^0 satisfies

$$\frac{dx_t^0}{dt} = a(x_t^0) \int_{R^n} y\pi(x_t^0,dy), \quad s \le t \le T, \tag{5.2}$$

with $x_s^0 = x$. The exit probability $\phi^\varepsilon(s,x)$ is a positive solution to

$$\frac{\partial\phi^\varepsilon}{\partial s} + L_\varepsilon\phi^\varepsilon = 0 \tag{5.3}$$

in $(-\infty,T) \times D$. The logarithmic transformation $I^\varepsilon = -\varepsilon \log \phi^\varepsilon$ changes (5.3) into

$$\frac{\partial I^\varepsilon}{\partial s} + \varepsilon\, H_\varepsilon(\varepsilon^{-1}I^\varepsilon) = 0, \tag{5.4}$$

where $H_\varepsilon(I) = -e^{I}L_\varepsilon(e^{-I})$. Then

$$\varepsilon H_\varepsilon(\varepsilon^{-1}I) = a(x) \int_{R^n} (1 - \exp[\frac{I(x)-I(x+\varepsilon y)}{\varepsilon}])\pi(x,dy) \tag{5.5}$$

For $I(x)$ such that I, I_x are continuous, bounded

$$\lim_{\varepsilon\to 0} \varepsilon\, H_\varepsilon\, (\varepsilon^{-1}I) = H_0(x,I_x),$$

with I_x the gradient and

$$H_0(x,p) = a(x) \int_{R^n} (1 - e^{-p \cdot y})\pi(x,dy). \tag{5.6}$$

This suggests (but certainly does not prove) that I^ε tends to a limit I^0 as $\varepsilon \to 0$, where I^0 satisfies (perhaps in some generalized sense)

$$\text{(5.7)} \qquad \frac{\partial I^0}{\partial s} + H(x, I_x^0) = 0.$$

Now (5.7) is the dynamic programming equation for the deterministic control problem with control space U as in §3, with running cost $k(\xi_t, u_t(\cdot))$, and with dynamics

$$\text{(5.8)} \qquad \frac{d\xi_t}{dt} = b(\xi_t, u_t(\cdot)),$$

$$b(x, u(\cdot)) = a(x) \int_{R^n} y\, u(y) \pi(x, dy).$$

Sheu [11] proved that indeed $I^\varepsilon \to I^0$ as $\varepsilon \to 0$ under the following hypotheses:

(i) $a(\cdot)$ is bounded, positive, and Lipschitz;

(ii) $\pi(x,dy) = g(x,y)\pi_1(dy)$ with π_1 a probability measure, $\pi_1(\{0\}) = 0$, $g(\cdot, y)$ uniformly Lipschitz, and $0 < c_1 < g(x,y) \leq c_2$;

(iii) $\int_{R^n} \exp(\alpha |y|^2) \pi_1(dy) < \infty$ for some $\alpha > 0$;

(iv) the convex hull of the support of π_1 contains a neighborhood of 0.

Condition (iv) insures that $H_0(x,p)$ is the dual of the usual "action integrand" $A(\xi, \dot{\xi})$ in large deviation theory, where for $\xi, \dot{\xi} \in R^n$

$$\text{(5.9)} \qquad A(\xi, \dot{\xi}) = \min_{u(\cdot)} \{k(\xi, u(\cdot)) : \dot{\xi} = b(\xi, u(\cdot))\}.$$

Then

$$\text{(5.10)} \qquad I^0(s,x) = \min_{\xi_\cdot} \int_s^\theta A(\xi_t, \dot{\xi}_t)\,dt, \quad x \in D.$$

The minimum is taken among C^1 paths $\xi_\cdot$ with $\xi_s = x$ such that ξ_t first reaches ∂D at time $\theta \leq T$. The requirement in (5.10) that ξ_t exit from D by time T is suggested by the boundary condition $I^\varepsilon(T,x) = +\infty$ for $x \in D$. This corresponds in the limit as $\varepsilon \to 0$ to an infinite penalty for failure to reach ∂D by time T.

In both [3] and [11] the stochastic control method used to show that $I^\varepsilon \to I^0$ depends on comparison arguments involving an optimal stochastic control process when $\varepsilon > 0$ and an optimal ξ^0 in (5.10) when $\varepsilon = 0$.

6. <u>The dominant eigenvalue</u>. In [2] Donsker and Varadhan gave a variational formula [(6.4)below] for the dominant eigenvalue λ_1 of $L + V$. Another derivation of this formula is given in [11], using the family of operators $\tilde{L}^w$ mentioned in §4.

When L is the generator of a nondegenerate diffusion process, Holland [9] expressed λ_1 as the minimum average cost per unit time in a stochastic control problem. Let us first indicate <u>formally</u> how this idea extends to more general generators L. Then we impose strong restrictions on L, and give a short derivation of (6.4).

Assume that L+V has a positive eigenfunction ϕ_1 corresponding to $\lambda_1:(L+V)\phi_1=\lambda_1\phi_1$. Let $I_1 = -\log\phi_1$. Then

(6.1) $$-H(I_1) + V = \lambda_1 .$$

Assuming that there is a stochastic control representation (1.5) for H(I), equation (6.1) becomes

(6.2) $$\min_{u\in U} [L^u I_1(x) + k(x,u)] - V(x) = -\lambda_1.$$

Equation (6.2) is the dynamic programming equation for the following average cost per unit time control problem. We admit stationary controls $\underline{u}(\cdot)$ such that the controlled process with generator $L^{\underline{u}}$ has an equilibrium distribution μ . The criterion to be minimized is

(6.3) $$J(\mu,\underline{u}) = \int_\Sigma [k(x,\underline{u}(x)) - V(x)]d\mu(x).$$

(If there is a unique equilibrium distribution $\mu = \mu^{\underline{u}}$ then reference to μ on the left side of (6.3) is unnecessary.) The principle of optimality states that $-\lambda_1 \leq J(\mu,\underline{u})$ with equality provided $\underline{u}^*(x)$ gives the minimum over $u \in U$ of $L^u I_1(x) + k(x,u)$.

Let us now assume that Σ is compact, that the generator L is bounded on $C(\Sigma)$ and $V \in C(\Sigma)$. As in [2] for any probability measure μ on Σ let

$$\mathscr{I}(\mu) = \sup_I \int_\Sigma H(I)d\mu = -\inf_{\phi>0}\int_\Sigma \frac{L\phi}{\phi}\,d\mu \;,$$

where $I, \phi \in C(\Sigma)$. The Donsker-Varadhan formula is

(6.4) $$\lambda_1 = \sup_\mu \left[\int_\Sigma V d\mu - \mathscr{I}(\mu)\right].$$

Let

$$P(I,\mu) = \int_\Sigma [-H(I) + V]d\mu .$$

The function P is convex in I and linear in μ . Formula (6.4) will follow if we can find I_1, μ_1 with the saddle point property:

(6.5) $$P(I_1,\mu) \leq \lambda_1 \leq P(I,\mu_1) \text{ for all } I, \mu.$$

(This idea was known to Donsker and Varadhan a long time ago, and figures in their proof [2] of (6.4).) If there is a positive eigenfunction ϕ_1, then we take $I_1=-\log\phi_1$.

From (6.1) we have in fact $P(I_1,\mu) = \lambda_1$ for all probability measures μ on Σ. To get the right hand inequality, choose $\underline{u}^*$ as above and assume that $L^{\underline{u}^*}$ is bounded on $C(\Sigma)$. The corresponding Markov process ξ_t^* has an equilibrium distribution μ_1, and

$$\int_\Sigma (L^{\underline{u}^*} I)\,d\mu_1 = 0, \text{ for all } I \in C(\Sigma). \tag{6.6}$$

(If $L^{\underline{u}^*}$ is unbounded we need to assume the existence of μ_1, and to restrict I to the domain of $L^{\underline{u}^*}$). By taking $u = \underline{u}^*(x)$ in (1.5) we have for $I \in C(\Sigma)$

$$L^{\underline{u}^*} I + k(x,\underline{u}^*) - V \geq H(I) - V.$$

By integrating both sides with respect to μ_1,

$$-\lambda_1 = J(\mu_1,\underline{u}^*) \geq -P(I,\mu_1),\ \lambda_1 \leq P(I,\mu_1),$$

as required.

In order to derive (6.4) in this way we had to impose unnecessarily restrictive hypotheses. In particular, we assumed that λ_1 is a dominant eigenvalue in the strict sense that $(L + V)\phi_1 = \lambda_1\phi_1$, with $\phi_1 > 0$. Actually, (6.4) holds if L is the generator of a strongly continuous, nonnegative semigroup T_t on $C(\Sigma)$, such that $T_t 1 = 1$, L has domain dense in $C(\Sigma)$, and L satisfies the maximum principle [2]. With such assumptions λ_1 is a dominant eigenvalue in the sense that the spectrum of $L + V$ is contained in $\{z: \text{Re } z \leq \lambda_1\}$ and $\lambda_1 - (L + V)$ does not have an inverse.

REFERENCES

[1] R. Azencott, Springer Lecture Notes in Math. No. 774, 1978.

[2] M. D. Donsker and S. R. S. Varadhan, "On a variational formula for principal eigenvalue for operators with a maximum principle, Proc. Nat. Acad. Sci. USA 72(1975) 780-783.

[3] W. H. Fleming, Exit probabilities and optimal stochastic control, Applied Math. Optimiz. 4 (1978) 329-346.

[4] W. H. Fleming, Stochastic calculus of variations and mechanics, to appear in J. Optimiz. Th. Appl.

[5] W. H. Fleming and S. K. Mitter, Optimal control and nonlinear filtering for nondegenerate diffusion processes, to appear in Stochastics.

[6] W. H. Fleming and R. W. Rishel, Deterministic and Stochastic Optimal Control, Springer-Verlag, 1975.

[7] O. Hernandez-Lerma, Exit probabilities for a class of perturbed degenerate systems, SIAM J. on Control and Optimiz. 19 (1981) 39-51.

References (cont.)

[8] O. Hijab, Asymptotic nonlinear filtering and large deviations, this volume.

[9] C. J. Holland, A minimum principle for the principal eigenvalue of second order linear elliptic equations with natural boundary conditions, Communications Pure Appl. Math. 31(1978) 509-519.

[10] E. Pardoux, The solution of the nonlinear filter equation as a likelihood function, Proc. 20th Conf. on Decision and Control, Dec. 1981.

[11] S. J. Sheu, PhD Thesis, Brown University, 1982.

[12] A. D. Ventsel, Rough limit theorems on large deviations for Markov stochastic processes, Theory of Probability and its Appl. 21(1976) 227-242, 499-512.

[13] A. D. Ventsel and M. I. Freidlin, On small random perturbations of dynamical systems, Russian Math. Surveys, 25(1970) 1-55.

GENERALIZED GAUSSIAN RANDOM SOLUTIONS OF CERTAIN EVOLUTION EQUATIONS

Luis G. Gorostiza*

Centro de Investigación y de Estudios Avanzados, IPN
and
Instituto de Investigación en Matemáticas Aplicadas y Sistemas, UNAM
México

Certain generalized Gaussian processes which arise as high density limits of supercritical branching random fields (see [1], [4]) possess interesting properties. In this note we prove some of these properties. We remark the fact that the processes obey deterministic evolution equations with generalized random initial conditions.

Let $\mathcal{S}(R^d)$ denote the Schwartz space of infinitely differentiable rapidly decreasing real functions on R^d topologized by the norms

$$\|\phi\|_p = \max_{0\leq|k|\leq p} \sup_x \Pi_{j=1}^d (1+|x_j|)^p |D^k\phi(x)|, \quad \phi \in \mathcal{S}(R^d), \quad p = 0,1,\ldots,$$

where $x = (x_1,\ldots,x_d)$, $k = (k_1,\ldots,k_d)$, $|k| = k_1+\ldots+k_d$, $D^k = \partial^k/\partial x_1^{k_1}\ldots\partial x_d^{k_d}$. Let $\mathcal{S}'(R^d)$ denote the topological dual of $\mathcal{S}(R^d)$, $\langle\cdot,\cdot\rangle$ the canonical bilinear form on $\mathcal{S}'(R^d)\times\mathcal{S}(R^d)$, and $\|\cdot\|_{-p}$ the operator norm on the dual of the $\|\cdot\|_p$-completion of $\mathcal{S}(R^d)$. The Schwartz spaces $\mathcal{S}(R^d\times R^+)$ and $\mathcal{S}'(R^d\times R^+)$ are similarly defined. The standard Gaussian white noise on R^d will be written W; it is the $\mathcal{S}'(R^d)$-valued random variable whose characteristic functional is $E\exp\{i\langle W,\phi\rangle\} = \exp\{-\frac{1}{2}\int_{R^d}\phi^2(x)dx\}$. (See [3]).

Let A be the infinitesimal generator, $\{T_t, t\geq 0\}$ the semigroup and $p_t(x,dy)$ the transition probability of a time-homogeneous Markov process $\{X_t, t\geq 0\}$ whose state space is all of R^d. We assume that

$$T_t : \mathcal{S}(R^d) \to L^2(R^d),$$
$$T : \mathcal{S}(R^d\times R^+) \to L^2(R^d)$$
$$\phi(x,t) \to \int_0^\infty T_t\phi(\cdot,t)(x)dt$$

and

$$A:\mathcal{S}(R^d) \to \mathcal{S}(R^d)$$

and that these mappings are continuous.

We are interested in random $\mathcal{S}'(R^d)$-valued solutions of the deterministic evolution equation $\partial f/\partial t = A^*f$, where A^* is the adjoint of A, with initial condition $f_0 = W$.

Let $\{M_t, t\geq 0\}$ denote the centered $\mathcal{S}'(R^d)$-valued Gaussian process with covariance

* Research supported in part by CONACYT grant PCCBNA 005167.

functional

$$\mathrm{Cov}(\langle M_s,\phi\rangle, \langle M_t,\psi\rangle) = \int_{R^d}\int_{R^d} \phi(y)\psi(z)K(s,dy;t,dz), \quad \phi,\psi \in S(R^d), \quad s,t\geq 0,$$

where

$$K(s,dy;t,dz) = \int_{R^d} p_s(x,dy)p_t(x,dz)dx, \quad s,t\geq 0.$$

The existence of this process follows from its representation as a stochastic integral

$$\langle M_t,\phi\rangle = \int_{R^d} T_t\phi(x)W(dx), \quad \phi\in S(R^d), \quad t\geq 0;$$

indeed, the integral is well defined as a consequence of the first assumption on T_t, and by the properties of such integrals it determines $\{M_t\}$ as a centered Gaussian random field with covariance functional

$$\mathrm{Cov}(\langle M_s,\phi\rangle, \langle M_t,\psi\rangle) = \int_{R^d} T_s\phi(x)T_t\psi(x)dx, \quad \phi,\psi\in S(R^d), \quad s,t\geq 0$$

(see [3], [7]), which coincides with the previous expression. We note that $M_o = W$ and $K(0,dy;0,dz)$ is the unit mass at the origin of $R^d\times R^d$.

THEOREM. <u>The process $\{M_t,t\geq 0\}$ is Markovian, it possesses a continuous version in the topology of $S'(R^d)$ (i.e. for each T>0 there is an integer p>0 such that $\{M_t\}$ is $\|\cdot\|_{-p}$-continuous on [0,T] almost-surely), and viewed as a space-time process it satisfies the evolution equation</u>

$$\partial M/\partial t = A^*M, \quad t\geq 0,$$
$$M_o = W.$$

REMARKS.

1. All the randomness of $\{M_t\}$ comes from the initial condition $M_o = W$.
2. We recall that a Gaussian random field is "classical" (or continuous) if and only if its covariance measure has a density with respect to Lebesgue measure (covariance kernel) and it is continuous (see [3]). In our case, for $K(s,dy;t,dz)$ to have a density $k(s,y;t,z)$ it is necessary that $p_t(x,dy)$ have a density $p_t(x,y)$, and the covariance kernel is

$$k(s,y;t,z) = \int_{R^d} p_s(x,y)p_t(x,z)dx, \quad y,z\in R^d.$$

Hence for t>0, M_t is classical if and only if there is a density $p_t(x,y)$ and $(y,z)\to k(t,y;t,z)$ is continuous.

Independently of the assumptions on T_t and A, a covariance measure K of the type above need not have a density. For example, if $\{X_t\}$ is a pure-jump Markov process with holding time parameter $\lambda(x)$ and jump distribution $P(dy;x)$, i.e.

$$A\phi(x) = \lambda(x)\int_{R^d} (\phi(y)-\phi(x))P(dy;x),$$

then $p_t(x,dy)$ has an atom at x.

For time-homogeneous diffusion operators A the situation concerning the existence of a continuous covariance kernel k does not seem simple at first sight, among other reasons because there are different definitions of diffusion. Under reasonable conditions on $a_{ij}(\cdot)$ and $b_i(\cdot)$ there corresponds to

$$A = \tfrac{1}{2}\Sigma_{i,j=1}^d a_{ij}(\cdot)\partial^2/\partial x_i\partial x_j + \Sigma_{i=1}^d b_i(\cdot)\partial/\partial x_i$$

a transition density (see [6]), but the continuity of $(y,z) \to k(t,y;t,z)$ is another question. We leave it open.

3. For Brownian motion things are nice, as they should be. In this case $A = \tfrac{1}{2}\Delta$ and $p_t(x,y) = \exp\{-\|x-y\|^2/2t\}(2\pi t)^{-d/2}$. The assumption on A is clearly met. We now verify the assumptions on T_t. For $\phi \in S(R^d)$,

$$\|T_t\phi\|_{L^2}^2 = \int_{R^d} (T_t\phi(x))^2dx = \int_{R^d} (E_x\phi(X_t))^2dx \le \int_{R^d} E_x\phi^2(X_t)dx$$

$$= \int_{R^d}\int_{R^d} \phi^2(y)e^{-\|y-x\|^2/2t}(2\pi t)^{-d/2}dydx$$

$$\le \sup_y(\phi(y)\,\Pi_{j=1}^d(1+|y_j|)^2)^2\int_{R^d}\int_{R^d} e^{-\|y-x\|^2/2t}(2\pi t)^{-d/2}\Pi_{j=1}^d(1+|y_j|)^{-4}dydx$$

$$\le \|\phi\|_2^2\int_{R^d}\Pi_{j=1}^d(1+|y_j|)^{-4}dy.$$

For $\phi \in S(R^d\times R^+)$,

$$\|T\phi\|_{L^2}^2 = \int_{R^d}\left(\int_0^\infty T_t\phi(\cdot,t)(x)dt\right)^2dx = \int_{R^d}\left(\int_0^\infty\int_{R^d}\phi(y,t)e^{-\|y-x\|^2/2t}(2\pi t)^{-d/2}dydt\right)^2dx$$

$$= \int_0^\infty\int_0^\infty\int_{R^d}\int_{R^d}\phi(y,t)\phi(z,s)\int_{R^d} e^{-\|y-x\|^2/2t}(2\pi t)^{-d/2}e^{-\|z-x\|^2/2s}(2\pi s)^{-d/2}dxdydzdtds$$

$$= \int_0^\infty\int_0^\infty\int_{R^d}\int_{R^d}\phi(y,t)\phi(z,s)e^{-\|y-z\|^2/2(s+t)}(2\pi(s+t))^{-d/2}dydzdtds$$

$$\leq \sup_{x,t\geq 0} (\phi(x,t)\Pi_{j=1}^{d}(1+|x_j|)^2(1+t)^2)^2\int_0^\infty\int_0^\infty\int_{R^d}\int_{R^d} e^{-\|y-z\|^2/2(s+t)}(2\pi(s+t))^{-d/2}$$

$$\cdot \Pi_{j=1}^{d}(1+|y_j|)^{-2}\Pi_{j=1}^{d}(1+|z_j|)^{-2}(1+t)^{-2}(1+s)^{-2}dydzdtds$$

$$\leq \|\phi\|_2^2\int_{R^d}\Pi_{j=1}^{d}(1+|z_j|)^{-2}dz\left(\int_0^\infty(1+t)^{-2}dt\right)^2,$$

since $\Pi_{j=1}^{d}(1+|y_j|)^{-2}\leq 1$.

Therefore the Gauss-Markov process $\{M_t\}$ in this case has covariance kernel

$$k(s,y;t,z) = e^{-\|y-z\|^2/2(s+t)}(2\pi(s+t))^{-d/2}, \quad s,t\geq 0$$

(because $p_t(x,y) = p_t(y,x)$ and the Chapman-Kolmogorov equation), hence it is classical for $t>0$, and it satisfies the heat equation

$$\partial M/\partial t = \tfrac{1}{2}\Delta M, \quad t\geq 0$$
$$M_0 - W.$$

In additon the process is self-similar, i.e. for any constant $\alpha>0$ the distribution of $\{M_t\}$ is invariant under the transformation $\langle\alpha^{-d/2}M_{\alpha^2 t},\phi(\alpha^{-1}\cdot)\rangle$, as can be seen from the covariance kernel.

PROOF. For the Markov property of $\{M_t\}$ it suffices to show that given $t_0<t$ and $\phi\in S(R^d)$ there is a $\tilde{\phi}\in S(R^d)$ such that

$$E\{(\langle M_t,\phi\rangle - \langle M_{t_0},\tilde{\phi}\rangle)\langle M_s,\psi\rangle\} = 0 \text{ for all } s\leq t_0 \text{ and } \psi\in S(R^d),$$

and this can be done using the covariance of $\{M_t\}$ and the Chapman-Kolmogorov equation for $p_t(x,dy)$, with $\tilde{\phi} = T_{t-t_0}\phi$.

It follows from the covariance of $\{M_t\}$ (or more easily from the stochastic integral representation) that

$$E(\langle M_t,\phi\rangle - \langle M_s,\phi\rangle)^2 = \int_{R^d}(T_t\phi(x)-T_s\phi(x))^2dx, \quad \phi\in S(R^d).$$

Then, using the semigroup property of T_t, the relation $\int_0^t T_s A\phi ds = T_t\phi-\phi$ and the Schwarz inequality, we have for $s<t$

$$E(\langle M_t,\phi\rangle - \langle M_s,\phi\rangle)^2 = \int_{R^d} (T_s(T_{t-s}\phi(x)-\phi(x)))^2 dx$$

$$= \int_{R^d}\left(\int_0^{t-s} T_{s+r}A\phi(x)dr\right)^2 dx \le (t-s)\int_{R^d}\int_0^{t-s} (T_{s+r}A\phi(x))^2 drdx$$

$$= (t-s)\int_0^{t-s}\int_{R^d} (T_{s+r}A\phi(x))^2 dxdr \le K(t-s)^2,$$

where K is a constant (depending on each bounded interval [0,T] such that $0\le s<t<T$). The Dudley-Fernique theorem ([2], Theorem 7.1) then implies that $\{\langle M_t,\phi\rangle\}$ is sample-continuous for each $\phi\in S(R^d)$, and the stated norm-continuity of $\{M_t\}$ follows by Mitoma's extension [5].

To obtain the evolution equation we view $\{M_t\}$ as a space-time field. By the second assumption on T_t (definition of T),

$$\langle M,\phi\rangle = \int_0^\infty \langle M_t,\phi(\cdot,t)\rangle dt = \int_{R^d}\left(\int_0^\infty T_t\phi(\cdot,t)(x)dt\right)W(dx), \quad \phi\in S(R^d\times R^+)$$

determines $\{M_t, t\ge 0\}$ as a generalized space-time Gaussian random field M. Let

$$\tilde{W}_t = \partial M_t/\partial t - A^*M_t,$$

meaning $\tilde{W} = \partial M/\partial t - A^*M$ as a space-time field; hence for $\phi\in S(R^d\times R^+)$,

$$\langle \tilde{W},\phi\rangle = \int_0^\infty \langle \tilde{W}_t,\phi(\cdot,t)\rangle dt,$$

and

$$\langle \tilde{W},\phi\rangle = \langle \partial M/\partial t - A^*M,\phi\rangle = -\langle M,\partial\phi/\partial t + A\phi\rangle = -\int_0^\infty \langle M_t,(\partial\phi/\partial t + A\phi)(\cdot,t)\rangle dt.$$

Since A is continuous, $\tilde{W}$ is a space-time Gaussian random field. Therefore to verify the evolution equation we need only show that $\tilde{W}$ has the right covariance. We have

$$\mathrm{Cov}(\langle \tilde{W},\phi\rangle, \langle \tilde{W},\psi\rangle) = \int_0^\infty\int_0^\infty \mathrm{Cov}(\langle \tilde{W}_s,\phi(\cdot,s)\rangle, \langle \tilde{W}_t,\psi(\cdot,t)\rangle)dsdt,$$

and on the other hand

$$\mathrm{Cov}(\langle \tilde{W},\phi\rangle, \langle \tilde{W},\psi\rangle) = \int_0^\infty\int_0^\infty \mathrm{Cov}(\langle M_s,(\partial\phi/\partial t + A\phi)(\cdot,s)\rangle, \langle M_t,(\partial\psi/\partial t + A\psi)(\cdot,t)\rangle)dsdt$$

$$= \int_0^\infty\int_0^\infty\int_{R^d} T_s(\partial\phi/\partial t + A\phi)(x,s)T_t(\partial\psi/\partial t + A\psi)(x,t)dxdsdt,$$

hence

$$\int_0^\infty\int_0^\infty \mathrm{Cov}(\langle \tilde{W}_s,\phi(\cdot,s)\rangle,\langle \tilde{W}_t,\psi(\cdot,t)\rangle)dsdt$$

$$= \int_{R^d} E_x\left(\int_0^\infty (\partial\phi/\partial t+A\phi)(X_s,s)ds\right) E_x\left(\int_0^\infty (\partial\psi/\partial t+A\psi)(X_t,t)dt\right)dx,$$

where E_x is the expectation under the distribution P_x of $\{X_t\}$ starting from $x \in R^d$. But

$$\phi(X_t,t) - \int_0^t (\partial\phi/\partial t+A\phi)(X_s,s)ds, \quad t \geq 0$$

is a P_x-martingale commencing from $\phi(x,0)$, and since martingales have constant expectation and $\phi(\cdot,\infty) = 0$, then

$$E_x\int_0^\infty (\partial\phi/\partial t+A\phi)(X_s,s)ds = -\phi(x,0).$$

Similarly for ψ. Substituting above we obtain

$$\int_0^\infty\int_0^\infty \mathrm{Cov}(\langle \tilde{W}_s,\phi(\cdot,s)\rangle,\langle \tilde{W}_t,\psi(\cdot,t)\rangle)dsdt = \int_{R^d} \phi(x,0)\psi(x,0)dx,$$

and therefore

$$\mathrm{Cov}(\langle \tilde{W}_s,\phi(\cdot,s)\rangle,\langle \tilde{W}_t,\psi(\cdot,t)\rangle) = \delta_2(s,t)\int_{R^d} \phi(x,0)\psi(x,0)dx,$$

where δ_2 is the Dirac delta function on R^2 centered at the origin. This means that

$$\tilde{W}_t = \begin{cases} W, & t = 0 \\ 0, & t > 0 \end{cases}$$

and hence $\{M_t\}$ satisfies the evolution equation.

REFERENCES

1. Dawson, D. and Gorostiza, L.G. Limit theorems for supercritical branching random fields. In preparation.

2. Dudley, R.M. (1967). The sizes of compact subsets of Hilbert space and continuity of Gaussian processes, J. Functional Analysis, Vol. 1, 290-330.

3. Gelfand, I.M. and Vilenkin, N.J. (1966). Generalized Functions, Vol. 4, Academic Press, New York.

4. Gorostiza, L.G. (1981). Limites gaussiennes pour les champs aléatoires ramifiés supercritiques, in "Aspects Statistiques et Aspects Physiques des Processus Gaussiens", Colloque International du C.N.R.S., Saint-Flour, France, 1980. Editions du C.N.R.S. No. 307, 385-398.

5. Mitoma, I. (1981). On the norm-continuity of S'valued Gaussian processes, Nagoya Math. J., Vol. 82, 209-220.

6. Stroock, D.W. and Varadhan, S.R.S. (1979). Multidimensional Diffusion Processes, Springer-Verlag, Berlin.

7. Wong, E. and Zakai, M. (1974). Martingales and stochastic integrals for processes with a multi-dimensional parameter. Z. Wahrscheinlichkeitstheorie verw. Gebiete, Vol. 29, 109-122.

EXTREMAL CONTROLS FOR COMPLETELY OBSERVABLE DIFFUSIONS*

U.G. Haussmann
Department of Mathematics
University of British Columbia

1. Introduction. We consider the basic control problem:

$$(1.1) \qquad \inf\{J[u]: u \in \mathcal{U}\}$$

where

$$(1.2) \qquad J[u] = E\left\{\int_0^T \ell(t,x(t),u(t))dt + c(x(T))\right\}$$

and $\mathcal{U}$ is some class of admissible controls, and where the state x satisfies the equation

$$(1.3) \qquad dx = f(t,x(t),u(t))dt + \sigma(t,x(t))dw \ , \ x(0) = x_0 \ .$$

Here x and w are processes defined on some probability space which may depend on u, and w is a standard Brownian motion. More details will be given in the next section where several classes of controls will be considered. We begin with the following notation and assumptions (more will follow). $\mathcal{C}_t$ is the Banach space of continuous functions, $[0,t] \to \mathbb{R}_n$, $\mathcal{C}_T = \mathcal{C}$, $\mathcal{G}$ is the Borel σ-algebra on $\mathcal{C}$ and $\{\mathcal{G}_t\}$ is the canonical filtration on $\mathcal{C}$. Write $\|x\|_t = \sup\{|x(s)|: 0 \le s \le t\}$. If $x: [0,T] \times \Omega \to \mathbb{R}_n$ is a continuous process, i.e. x is a $\mathcal{C}$-valued random variable, then $\mathcal{F}_t^x = x^{-1} \circ \mathcal{G}_t$ is the σ-algebra generated by the past of x . On $[0,T]$ we use the Borel σ-algebra. U is a Borel set in some Euclidean space with Borel subsets $\mathcal{B}_U$. $\mathbb{R}_n$ is n-dimensional Euclidean space and $\mathbb{R}_{n\times m}$ is the space of $n \times m$ matrices.

(1.4) (i) $f: [0,T] \times \mathbb{R}_n \times U \to \mathbb{R}_n$ is Borel measurable and continuous in u uniformly in t for each x .

(ii) $\sigma: [0,T] \times \mathbb{R}_n \to \mathbb{R}_{n\times m}$ is Borel measurable.

(iii) $|f(t,x,u)| + |\sigma(t,x)| \le K_1(1 + |x|)$,
$|f(t,x,u) - f(t,y,u)| + |\sigma(t,x) - \sigma(t,y)| \le K_2|x-y|$,
for some $K_1 < \infty$, $K_2 < \infty$.

(1.5) (i) $\ell: [0,T] \times \mathbb{R}_n \times U \to \mathbb{R}$ is Borel measurable, and continuous in (x,u) for each t, continuous in x uniformly in u for each t.

(ii) $c: \mathbb{R}_n \to \mathbb{R}$ is continuous.

(iii) $|\ell(t,x,u)| \le K_3(1 + |x|^q)$, $|c(x)| \le K_3(1 + |x|^q)$ for some $q < \infty$.

We show in section three that a control which satisfies the necessary condition of the maximum principle, i.e. an extremal control, is in fact 'optimal' provided that certain convexity hypotheses are satisfied. This result is proved along the lines of the corresponding deterministic case, [9]. Here 'optimal' means optimal within the class of controls which are adapted processes on a given probability space with filtration and Brownian motion. However usually the 'non-anticipative' controls are adapted processes defined on some probability space with filtration and Brownian motion which may change with u . In section two we see that $\inf J[u]$ is the same over all the usual control classes, so that any extremal control is optimal under the convexity hypothesis. In section four we apply this result to show that the extremal controls computed for several examples in [5] are in fact optimal. It is also useful in establishing ε-optimality for certain problems, c.f. section three and [6] .

One final note. We shall always assume that the probability spaces are chosen so that the solution x of (1.3) is not just continuous w.p.1, but in fact all trajectories are continuous.

2. Control Classes. Various sets of controls have been introduced in the literature. We shall consider several of these here and observe that they all lead to the same infimum in (1.1).

Definition 2.1: $\mathcal{U}_L$, the set of control laws, consists of the $\{\mathcal{G}_t\}$ progressively measurable functions $u:[0,T]\times\mathcal{C}\to U$ for which there is a probability space $(\Omega,\mathcal{F},P)$ with filtration, which may depend on u and which carries a continuous adapted process x and a Brownian motion w such that $(x(\cdot),u(\cdot,x(\cdot)),w(\cdot))$ satisfy (1.3) w.p.1.

Definition 2.2: $\mathcal{U}_A$, the set of adapted controls (adapted to x), consists of all separable, measurable processes u defined on a probability space $(\Omega,\mathcal{F},P)$, which may depend on u and which carries two continuous processes x , w such that $(w_t,\mathcal{F}_t^x)$ is a standard Brownian motion, u is $\{\mathcal{F}_t^x\}$ progressively measurable, U-valued, and $(x(\cdot),u(\cdot),w(\cdot))$ satisfy (1.3) w.p.1.

Definition 2.3: $\mathcal{U}_N$, the set of non-anticipative controls (non-anticipative with respect to w) , consists of all separable, progressively measurable, and U-valued processes u defined on π , a probability space with filtration and Brownian motion, i.e. $\pi=(\Omega,\mathcal{F},P,\{\mathcal{F}_t\},w)$, which may depend on u .

Definition 2.4: $\mathcal{U}_M$, the set of Markovian control laws, consists of all Borel measurable functions $u:[0,T]\times\mathbb{R}_n\to U$ for which there is a probability space $(\Omega,\mathcal{F},P)$ with filtration, which may depend on u and which carries two continuous adapted processes x , w such that w is a Brownian motion and $(x(\cdot),u(\cdot,x(\cdot)),w(\cdot))$

satisfy (1.3).

Definition 2.5: For fixed $\pi = (\Omega, F, P, \{F_t\}, w)$ write $\mathcal{U}_N(\pi)$ for the corresponding controls in $\mathcal{U}_N$. Similarly for $\mathcal{U}_M(\pi)$, $\mathcal{U}_L(\pi)$.

Remark 2.6. If $u \in \mathcal{U}_L$, then according to [2], [10] there exists a Brownian motion $(\bar{w}_t, F_t^x)$ such that

$$dx = f(t,x(t),\ u(t,x(\cdot)))dt + \sigma(t,x(t))d\bar{w}\ ,$$

so that $u(\cdot,x(\cdot)) \in \mathcal{U}_A$, i.e. $\mathcal{U}_L \hookrightarrow \mathcal{U}_A$.

For $u \in \mathcal{U}_A$ we can take $F_t = F_t^x$, i.e. $u \in \mathcal{U}_N$, so

$$\mathcal{U}_M \hookrightarrow \mathcal{U}_L \hookrightarrow \mathcal{U}_A \subset \mathcal{U}_N\ .$$

We also have for each π

$$\mathcal{U}_M(\pi) \hookrightarrow \mathcal{U}_L(\pi) \hookrightarrow \mathcal{U}_N(\pi)\ .$$

Remark 2.7. As we are assuming (1.4), then for $u \in \mathcal{U}_N$ or $u \in \mathcal{U}_N(\pi)$, (1.3) has a unique, continuous adapted solution x^u. Gronwall's inequality, Burkholder's inequality and (1.4),(1.5) imply that there are constants K_q, K_4 such that

$$E\,\|x^u\|_t^q \le K_q E|x(0)|^q\ ,\ \text{any}\ \ q < \infty\ ,$$

$$E\,\|x^u - y^u\|_T \le K_4 E|x(0) - y(0)|$$

for any $u \in \mathcal{U}_N$. Here y is the solution of (1.3) with initial condition $y(0)$. Now (1.5) implies

$$J[u] \le E\int_0^T |\ell(t,x^u,u(t))|dt + |c(x^u(T))|$$

$$\le (T+1)K_3(1 + K_q|x_0|^q)\ .$$

Remark 2.8. From the above we have

$$\inf\{J[u]:\ u \in \mathcal{U}_N(\pi)\} \le \inf\{J[u]:\ u \in \mathcal{U}_L(\pi)\}\ .$$

In [7], chapter 3, theorem 1.7, it is shown in fact that equality holds.

We also have

(2.9) $$\inf\{J[u]: u \in \mathcal{U}_M\} = \inf\{J[u]: u \in \mathcal{U}_A\}$$

provided U is compact and either

(2.10) σ is bounded, continuous, invertible, and f, ℓ, c are bounded, c.f. [1], theorem IV-6; or

(2.11) (i) for each (t,u) σ, f, c, ℓ are continuously differentiable in x with uniformly bounded derivatives in the case of σ, f and with $(c_x \equiv \partial c/\partial x)$ $|c_x(x)| + |\ell_x(t,x,u)| \le K(1+|x|^q)$,

(ii) $$f(t,x,u) = \phi(t,x) + \begin{pmatrix} 0 \\ g(t,x,u) \end{pmatrix} , \quad \sigma(t,x) = \begin{pmatrix} 0 \\ \sigma_2(t,x) \end{pmatrix}$$

with ϕ Borel, $g(t,x,u) \in \mathbb{R}_m$ $\sigma_2(t,x) \in \mathbb{R}_{m\times m}$, g, σ, ℓ continuous, σ_2 bounded, invertible with bounded inverse, and with ϕ , continuously differentiable in x with bounded derivative,

(iii) $$dx = \phi(t,x)dt + \sigma(t,x)dw$$

has a transition density $p(s,x,t,y)$ satisfying

$$\int_{s'}^{T} \int_{\mathbb{R}_n} p(s,x,t,y)^{\mu} dy\, dt < \infty$$

for some $\mu > 1$ and all $s' > s$, c.f. [4] , corollary 3.9, where it is shown that (2.9) holds with $\mathcal{U}_A$ replaced by $\mathcal{U}_N$.

We point out that if for each $u \in \mathcal{U}_L$ the solution of (1.3) is unique in law (true under (2.11)) then for any π

$$\inf\{J[u]: u \in \mathcal{U}_L\} = \inf\{J[u]: u \in \mathcal{U}_L(\pi)\}$$

and consequently

$$\inf\{J[u]: u \in \mathcal{U}_N\} = \inf\{J[u]: u \in \mathcal{U}_N(\pi)\} .$$

Finally we remark that if U is compact then the equality of the inf over $\mathcal{U}_N(\pi)$ and $\mathcal{U}_L(\pi)$ can be established even if f, σ, ℓ, c depend on the past of x rather than just $x(t)$. Moreover inf J over the predictable controls also assumes this common value.

3. Extremal Controls. We now define the extremal controls as those which satisfy the condition of the maximum principle. We assume (1.4), (1.5), (2.11)(i) .

Definition 3.1. If $(x(t),u(t),w(t))$ satisfy (1.3), on $(\Omega,F,P,\{F_t\})$, then p, the adjoint process of x, is given by $p(t)=E\{\bar{p}(t)|F_t\}$ and

(3.2) $$\bar{p}(t)=-c_x(x(T))\Phi(T,t)-\int_t^T \ell_x(s,x(s),u(s))\Phi(s,t)ds$$

where the ith-column of $\Phi(s,t)$ is the unique solution of

(3.3) $$dy=f_x(s,x(s),u(s))yds+\sum_{k=1}^{m}\sigma_x^k(s,x(s))y\,dw_k \quad y(t)=e_i ,$$

with σ^k being the kth-column of σ, w_k being the kth component of w, and e_i being the ith column of the $n\times n$ identity I. Note that $p(t)$ is a row vector.

If $u\in U_A$ or U_L then $p(t)=E\{\bar{p}(t)|F_t^x\}$ and if $u\in U_M$ then

$$p(t)=E\{\bar{p}(t)|F_t^x\}=E\{\bar{p}(t)|F[x(t)]\}$$

where $F[x(t)]$ is the σ-algebra generated by $x(t)$. In this case we can write $p(t)=\tilde{p}(t,x(t))$ for a measurable function $\tilde{p}$. We shall, from now on, use p and $\tilde{p}$ interchangeably, and in fact drop the $\tilde{}$.

Definition 3.4. $\hat{u}$ is extremal if for each $\mu\in U$, t not in some null set,

(3.4) $$p(t)f(t,\hat{x}(t),\hat{u}(t))-\ell(t,\hat{x}(t),\hat{u}(t))\geq p(t)f(t,\hat{x}(t),\mu)-\ell(t,\hat{x}(t),\mu) \text{ w.p.1},$$

where p is the adjoint process for $\hat{x}$, and $\hat{x}$ is the solution of (1.3) corresponding to $\hat{u}$.

The following results show that necessarily an optimal control is extremal in certain cases, i.e. a Pontryagin type maximum principle holds.

Theorem 3.5. Assume (1.4), (1.5), (2.11)(i). If $\hat{u}$ is optimal in $U_N(\pi)$ for some π, and if $\hat{u}$ is $\{F_{t-}\}$ adapted, then $\hat{u}$ is extremal.

Proof: This follows from [8], with perturbed controls $u_{t-\delta}(s)=E\{\hat{u}(t)|F_{t-\delta}\}$.

Corollary 3.6. Assume U is compact and (1.4), (1.5), (2.11)(i). If $\hat{u}$ is optimal in U_A, then $\hat{u}$ is extremal.

Proof: For the fixed π corresponding to $(\hat{x},\hat{u})$ we have $\hat{u}\in U_N(\pi)$. Note $F_t\equiv F_t^x$. Now remarks 2.6 and 2.8 imply that $\hat{u}$ is optimal in $U_N(\pi)$. Since $\hat{x}$ has continuous sample paths, then $F_{t-}^x=F_t^x$, hence $\hat{u}$ is F_{t-}^x adapted. By the theorem $\hat{u}$ is extremal.

Corollary 3.7. *Assume* U *is compact and* (2.11). *If* $\hat{u}$ *is optimal in* $\mathcal{U}_M$, *then it is extremal*.

Proof. This follows either from [7], [8] or from remark 2.8 and the above. In fact (2.9) implies that $\hat{u}$ is optimal in $\mathcal{U}_A$ so the previous corollary gives the result.

Remark 3.8. We know that under the hypotheses of corollary 3.7, there exists $\hat{u}$ optimal in $\mathcal{U}_M$, c.f. [8] .

We are now interested in the converse question: if u is extremal is it optimal?

We define the Hamiltonian

$$H(t,x,p,u) = p\,f(t,x,u) - \ell(t,x,u)$$

for $p \in \mathbb{R}_n$ (row vector), $x \in \mathbb{R}_n$, $u \in U$, $t \in [0,T]$; however the following function corresponds more closely to the Hamiltonian in the calculus of variations:

$$H^*(t,x,p) = \sup_{u \in U} H(t,x,p,u) \ .$$

If $\hat{u}$ is extremal then $H(t,\hat{x}(t),p(t),\hat{u}(t)) = H^*(t,\hat{x}(t),p(t))$ a.e.

Theorem 3.9. *Assume* (1.4), (1.5), (2.11)(i) . *If* (i), $\hat{u} \in \mathcal{U}_N(\pi)$ *is extremal where* $(\hat{x},\hat{u},w)$ *is a solution of* (1.3), *if* (ii),

$$\sigma^k(t,x) = D^k(t)x + e^k(t) \ , \ k = 1,\ldots,m \ ,$$

where $D^k:[0,T] \to \mathbb{R}_{n\times n}$, $e^k:[0,T] \to \mathbb{R}_n$ *are bounded and measurable, if* (iii), $c(\cdot)$ *is convex, and if* (iv), $A \subset \mathbb{R}_n$ *is an open convex set such that for each* t w.p.1 $H^*(t,x,p(t))$ *is concave in* x *for* x *in* A , *then for any* $\beta \in (0,1)$ *and some constant* K_0 *depending only on* β *and the bounds and growth constants of* (1.4), (1.5), (2.11)(i) ,

$$J[\hat{u}] \le \inf\{J[u]: u \in \mathcal{U}_N(\pi)\} + K_0[\Pr\{\inf\{\tau(u): u \in \mathcal{U}_N(\pi)\} < T\}]^\beta$$

where $\tau(u)$ *is the first exit time of* x^u *from* A .

Proof: Write $\Phi(t)$ for $\Phi(t,0)$ and $\Psi(t) = \Phi(t)^{-1}$. Then

$$d\Psi = -\Psi[f_x(t,\hat{x}(t),\hat{u}(t)) - \sum_{k=1}^m D^k(t)D^k(t)]dt - \Psi \sum_{k=1}^m D^k(t)dw_k \ , \ \Psi(0) = I \ .$$

From the pathwise uniqueness of solutions of (3.3), it follows that

$\Phi(s,t) = \Phi(s)\Psi(t)$.

From (3.2) we obtain

$$\bar{p}(t) = \bar{p}(0)\Psi(t) + \int_0^t \ell_x(s,\hat{x}(s),\hat{u}(s))\Phi(s)ds\Psi(t) \ .$$

If (x,u,w) is any solution of (1.3) with $u \in \mathcal{U}_N(\pi)$, then

$$d(\Psi(t)x(t)) = -\Psi(t)[f_x(t,\hat{x}(t),\hat{u}(t))x(t) - f(t,x(t),u(t)) + \sum_k D^k(t)e^k(t)]dt + \Psi(t)\sum_k e^k(t)dw_k \ ,$$

and

$$\begin{aligned}\bar{p}(T)\,x(T) - \bar{p}(0)x(0) = \int_0^T & -\bar{p}(t)[f_x(t,\hat{x}(t),\hat{u}(t))x(t) \\ & - f(t,x(t),u(t)) + \sum_k D^k(t)e^k(t)] + \ell_x(t,\hat{x}(t),\hat{u}(t))x(t)dt \\ & + \bar{p}(0)\sum_k \int_0^T \Psi(t)e^k(t)dw_k \\ & + \sum_k \int_0^T\int_0^t \ell_x(s,\hat{x}(s),\hat{u}(s))ds\Psi(t)e^k(t)dw_k \ ,\end{aligned}$$

so that

$$\begin{aligned}\bar{p}(T)[x(T) - \hat{x}(T)] = \int_0^T & -[\bar{p}(t)f_x(t,\hat{x}(t),\hat{u}(t)) - \ell_x(t,\hat{x}(t),\hat{u}(t))][x(t) - \hat{x}(t)] \\ & + \bar{p}(t)[f(t,x(t),u(t)) - f(t,\hat{x}(t),\hat{u}(t))]dt \ .\end{aligned}$$

But $\bar{p}(T) = -c_x(\hat{x}(T))$ so the convexity of c implies

$$\bar{p}(T)[x(T) - \hat{x}(T)] \geq c(\hat{x}(T)) - c(x(T)) \ .$$

Hence

$$\begin{aligned}&[c(\hat{x}(T)) + \int_0^T \ell(t,\hat{x}(t),\hat{u}(t))dt] - [c(x(T)) + \int_0^T \ell(t,x(t),u(t))dt] \\ &\quad \leq \bar{p}(T)[x(T) - \hat{x}(T)] + \int_0^T \ell(t,\hat{x}(t),\hat{u}(t)) - \ell(t,x(t),u(t))dt \\ &\quad = \int_0^T -H_x(t,\hat{x}(t),\bar{p}(t),\hat{u}(t))[x(t) - \hat{x}(t)] \\ &\qquad + [H(t,x(t),\bar{p}(t),u(t)) - H(t,\hat{x}(t),\bar{p}(t),\hat{u}(t)]dt \ .\end{aligned}$$

Since H and H_x are linear in p, and x,u are F_t adapted, then $\bar{p}$ can be replaced by p when expectations are taken. Moreover, since $H_x(t,\hat{x}(t),p(t),\hat{u}(t))$ is a subgradient (in x) of $H^*(t,\cdot,p(t))$ at $\hat{x}(t)$ and $H^*(t,\cdot,p)$ is concave on A, then

$$J[\hat{u}] - J[u] \le E\, 1_A \int_0^T -H^*(t,x(t),p(t)) + H^*(t,\hat{x}(t),p(t))dt$$

$$+ E(1-1_A)\int_0^T -H_x(t,\hat{x}(t),p(t),\hat{u}(t))[x(t)-\hat{x}(t)]dt$$

$$+ E\int_0^T H(t,x(t),p(t),u(t)) - H(t,\hat{x}(t),p(t),\hat{u}(t))dt$$

$$\le E(1-1_A)\int_0^T -H_x(t,\hat{x}(t),p(t),\hat{u}(t))[x(t)-\hat{x}(t)]$$

$$+ [H(t,x(t),p(t),u(t)) - H(t,\hat{x}(t),p(t),\hat{u}(t))]dt$$

where the last inequality follows by the definition of H^*, and where $1_A(\omega)=1$ if $x^u(t,\omega)\in A$ for almost all $t\in[0,T]$, all $u\in \mathcal{U}_N(\pi)$, and $1_A(\omega)=0$ otherwise. Since c_x, ℓ_x satisfy a growth condition in x and since Φ satisfies a linear equation with bounded coefficients then $E\,\|p\|_T^q<\infty$ for any $q<\infty$. The growth conditions, remark 2.7, and Hölder's inequality now give the result.

Corollary 3.10. <u>Under the assumptions of theorem</u> 3.9, <u>if</u> $A=\mathbb{R}_n$ (i.e. H^* is concave for all x) <u>then</u> $\hat{u}$ <u>is optimal</u>.

It is usually difficult to say when H^* is concave, because for example if $n=1$ we cannot usually say that p has constant sign w.p.1; however we have the following result.

Corollary 3.11. <u>Under the assumption of theorem</u> 3.9 <u>except</u> (iv), <u>if</u>

$$(3.12)\qquad \begin{aligned} f(t,x,u) &= A(t)x + b(t,u)\\ \ell(t,x,u) &= \ell_0(t,x) + \ell_1(t,u)\end{aligned}$$

<u>with</u> $\ell_0(t,\cdot)$ <u>convex on</u> $\mathbb{R}_n$, <u>then</u> $\hat{u}$ <u>is optimal in</u> $\mathcal{U}_N(\pi)$.

Proof: The result follows from the theorem since

$$H^*(t,x,p) = pA(t)x - \ell_0(t,x) + k(t,p)\ ,$$

$$k(t,p) = \sup\{pb(t,u) - \ell_1(t,u) : u\in U\}\ ,$$

so H^* is concave in x.

Observe that if $\hat{u} \in \mathcal{U}_M(\pi)$ (or one of the other classes) is extremal, then it is extremal as an element of $\mathcal{U}_N(\pi)$, thus optimal in $\mathcal{U}_N(\pi)$ and so optimal in $\mathcal{U}_M(\pi)$. If we have law uniqueness then $\hat{u}$ is optimal in $\mathcal{U}_M$.

Remark 3.13: These results of this section can be extended to the case when there are constraints of the form $E\,r(x(T)) = 0$ if r satisfies the same conditions as c. We say that u is extremal if $E\,r(x(T)) = 0$ and if (3.4) holds with $p(t)$ defined as

$$p(t) = E\{\theta\bar{p}(t) + \lambda' r_x(x(T))\Phi(T,t) \,|\, F_t\}$$

for $\theta \geq 0$ and some constant vector λ. If $\theta = 0$ the problem is called abnormal. In the normal case θ is normalized to 1. Theorem 3.5 continues to hold but corollary 3.6 may not since the results of section 2 may fail. On the other hand corollary 3.7 holds because of [3], [4]. Theorem 3.9 holds in the normal case if r is linear, i.e. $r(x) = Rx$ for some matrix R.

Remark 3.14. Suppose (3.12) holds and σ is as in the theorem. In [6] we show how to construct (approximations to) controls u^R which are extremal for solutions of (1.3) when f, σ are altered on $\{|x| > R\}$ to be bounded with bounded x derivative. From remark 2.7 it follows that given $\varepsilon > 0$, we can choose R such that for fixed $\beta \in (0,1)$

$$\Pr\{\sup\{\|x^u\|_T : u \in \mathcal{U}_N\} \geq R\} < (\varepsilon/3K_0)^{1/\beta} .$$

But for R sufficiently large we observe that (for all u) $|J[u] - J_R[u]| < \varepsilon/3$ where $J_R[u]$ is defined by (1.2) with x given by the altered f, σ. Hence

$$J[u^R] \leq \inf\{J[u] : u \in \mathcal{U}_N(\pi)\} + \varepsilon ,$$

i.e. u^R is ε-optimal in $\mathcal{U}_N(\pi)$ if ℓ_0, c are convex in x for each t.

4. Some Examples. In [5], we considered eight simple examples and exhibited extremal controls in each case. Our aim here is to show that seven of these controls are optimal. Sometimes it is necessary to add a convexity hypothesis.

Example 4.1 (Linear Regulator)

(4.2) $f(t,x,u) = A(t)x + B(t)u$, $\sigma(t,x) = \sigma(t)$,

(4.3) $\ell(t,x,u) = x'M(t)x + u'N(t)u$, $c(x) = x'Qx$,

$U = \mathbb{R}_p$.

A,B,σ,M,N are bounded, measurable, and $N(t)>0$, $M(t)\geq 0$, $Q\geq 0$. As indicated in [5] we should guess $\hat{u}(t,x)=K(t)x\in \mathcal{U}_M$ as extremal. Then

$$p(t,x) = -2E_{tx}\{x(T)'Q\Phi(T,t) + \int_t^T x(s)'M(s)\Phi(s,t)ds\}$$

$$d\Phi(s,t) = A(s)\Phi(s,t)ds \ ,$$

$$dx = (A+BK)x\,dt + \sigma dw \ .$$

Hence $E_{tx}\{x(T)\} = \Phi_K(T,t)x$ where

$$d\Phi_K(s,t) = [A(s)+B(s)K(s)]\Phi_K(s,t)ds \ , \ \Phi_K(t,t) = I \ ,$$

so that $\hat{u}$ is extremal if

$$K(t) = -N(t)^{-1}B(t)'P(t)$$

$$P(t) = \Phi(T,t)'Q\Phi_K(T,t) + \int_t^T \Phi(s,t)'M(s)\Phi_K(s,t)ds \ .$$

These last two equations imply that P satisfies the usual Ricatti equation

$$\dot{P} + A'P + PA - PBN^{-1}B'P + M = 0 \ ,$$

$$P(T) = Q \ .$$

Now corollary 3.11 implies that $\hat{u}$ is optimal in $\mathcal{U}_K \equiv \{u\colon [0,T]\times\Omega \to \mathbb{R}_p :$ $|u(t,\omega)| \leq K(1+|x(t)|)$, all $t\}$, specifically it is optimal in $\mathcal{U} = \{u(t,x,\omega) : u(t,x,\omega) = L(t)x + v(t,\omega)$, L , v bounded, measurable$\}$.

<u>Example 4.4</u>. (Predicted Miss) f and σ are still given by (4.2) (with say, B , σ continuous, σ invertible), but $\ell=0$, $c(x)=k(v'x)$ where v is a constant vector and k is convex, even, non-negative and continuously differentiable, $k(y)\leq(1+|y|^q)$. $U=\{u\in\mathbb{R}_p : |u_i|\leq 1\}$. In [5] we show that $\hat{u}(t,x) = -\text{sgn}[B(t)'s(t)s(t)'x]$ is extremal with $s(t)$ given by

$$ds = -A(t)'s\,dt \ , \ s(T) = v \ .$$

Hence $\hat{u}$ is optimal in $\mathcal{U}_M$ or any of the other classes. Examples 3.3, 3.4 and 3.5 of [5] are treated similarly, if we add the hypothesis that $\ell(t,\cdot)$, $k(\cdot)$ are convex.

Example 4.5.

$$dx = B(t)u(t)dt + \sigma(t)dw$$

$$\min E \int_0^T u'N(t)u\, dt \ , \quad |u_i(t)| \le 1 \ ,$$

where $B(t)$, $N(t)$ are bounded, measurable and $N(t) > 0$. We add the constraint

$$E\, x(T) = \alpha \ .$$

Then $p(t) = \lambda'$ (λ is a constant column vector), and in the normal case, i.e. when $\{J[u]: u \in \mathcal{U}\ , \ Ex(T) = \alpha\}$ is more than just a singleton, then $\hat{u}$ is extremal if

$$\hat{u}(t) = \text{sat}\{N(t)^{-1}B(t)'\lambda\}$$

where λ must be computed from

$$\alpha - x_0 = \int_0^T B(t)\hat{u}(t)dt \ .$$

Hence $\hat{u}$ is optimal, c.f. remark 3.13.

Example 3.8 of [5] follows similarly. Example 3.6 of [5] cannot be analysed in this fashion, i.e. we cannot show that the extremal control found in [5] is optimal.

References

[1] J.M. Bismut, Théorie probabiliste du contrôle des diffusions, Mem. Amer. Math. Soc., 4(1976), No. 167.

[2] M.P. Ershov, Representations of Itô processes, Th. Probab. Applic., 17(1972), pp. 165-169.

[3] U.G. Haussmann, General necessary conditions for optimal control of stochastic systems, Math. Programming. Stud., 9(1976), pp. 30-48.

[4] ____________, On the adjoint process for optimal control of diffusion processes, SIAM J. Control and Optimization, 19(1981), pp. 221-243.

[5] ____________, Some examples of optimal stochastic controls, SIAM Review, 23(1981), pp. 292-307.

[6] ____________, On the approximation of optimal stochastic controls, preprint, U.B.C.

[7] N.V. Krylov, Controlled Diffusion Processes, Springer-Verlag, New York, 1980.

[8] H.J. Kushner, Necessary conditions for continuous parameter stochastic optimization problems, SIAM J. Control, 10(1972), pp. 550-562.

[9] A. Seierstad, K. Sydsaeter, Sufficient conditions in optimal control theory, Int. Economic Rev., 18(1977), pp. 367-391.

[10] G. Wong, Representation of martingales, quadratic variation and applications, SIAM J. Control, 9(1971), pp. 621-633.

LÉVY'S STOCHASTIC AREA FORMULA IN HIGHER DIMENSIONS

K. Helmes *and* A. Schwane

Institut für Angewandte Mathematik
Universität Bonn

ABSTRACT

Let $(W_t) = (W_t^1, W_t^2, \cdots, W_t^d)$, $d \geq 2$, *be a d-dimensional standard Brownian motion and let* $A(t)$ *be a bounded measurable function from* $\mathbb{R}^+$ *into the space of* $d \times d$ *skew-symmetric matrices and* $x(t)$ *into* $\mathbb{R}^d$. *We are concerned with a class of stochastic processes* $(L_t^{A,x})$ *a particular example of which is P. Lévy's 'stochastic area'*

$$L_t = \frac{1}{2}\int_0^t (W_s^1 dW_s^2 - W_s^2 dW_s^1).$$

We calculate the joint characteristic function of W_t *and* $L_t^{A,x}$ *and based on this result give a formula for fundamental solutions for the hypoelliptic operators which generate the diffusions* $(W_t, L_t^{A,x})$.

1. INTRODUCTION

In [8] P. Lévy began studying what he called the 'stochastic area' of a 2-dimensional standard Brownian motion, i.e. the area enclosed by the trajectory of the Wiener process and its chord: the stochastic process (L_t). In subsequent papers [9-12] he derived formulae for the characteristic function of L_t and for the characteristic function of L_t given the position of the process (W_t) at time $t>0$; namely, for $\lambda \in \mathbb{R}$, $x \in \mathbb{R}^2$

$$(1) \qquad E\left[exp(i\lambda L_t)\right] = \frac{1}{cosh(\frac{\lambda t}{2})}$$

and

$$(2) \qquad E\left[exp(i\lambda L_t) \mid W_t = x\right] = \frac{t\lambda}{2sinh(\frac{t\lambda}{2})}\; exp\left[\frac{|x|^2}{2t}\left(1 - \frac{t\lambda}{2}coth(\frac{t\lambda}{2})\right)\right],$$

where $|x|$ denotes the Euclidean norm of x, cf. [12, p. 173]. Lévy's proof of Eq. 2 is based on the expansion of (standard) Brownian motions (W_t^i), $i=1,2,\cdots$, in a countable coordinate system, e.g.

$$W_t^i = \frac{t}{\sqrt{\pi}} Y_0^i + \sqrt{\frac{2}{\pi}} \sum_{m=1}^{\infty} \frac{\sin(mt)}{m} Y_m^i, \quad 0 \leq t \leq \pi ,$$

where $Y_0^i, Y_1^i, \cdots$ are independent normally distributed random variables with mean zero and variance 1, see [12] or [1, p. 261]. To obtain the characteristic function of L_t then one simply has to integrate (2) with respect to the distribution of $\rho_t^2 := |W_t|^2$. Actually, Lévy gave two proofs for formula 1 the second one being based on the skew product representation of two dimensional Brownian motion and the formula for the characteristic function of

$$\int_0^t (W_s^1)^2 ds .$$

Recently, M. Yor simplified Lévy's proof employing a result on Bessel processes and an elementary result used by D. Williams for the ' first time' to give an elegant proof of a stopped Brownian motion formula originally due to H.M. Taylor, [15, 16]. However, Yor's proof still rests on a non-elementary result in the theory of stochastic differential equations, a criterion due to T. Yamada and S. Watanabe on Itô-uniqueness for non-linear stochastic equations. A fine account of all these ideas can be found in [6, pp. 384-391].

In this note we shall show that Eq. 2 is an immediate consequence of the Cameron-Martin-Girsanov theorem combined with the solution formula for linear stochastic differential equations. This kind of proof does not only work in the 2-dimensional set-up considered by Lévy but extends naturally to dimensions $d \geq 2$. Thus we shall consider stochastic processes $(L_t^{A,x})$ which are specified by a d-dimensional Wiener process $(W_t)_{t \geq 0}$, $W_0 = 0$, a deterministic function $\mathbf{x}(t)$ from $\mathbb{R}^+$ into $\mathbb{R}^d$ and $A(t)$ into the space of skew-symmetric matrices, viz.

$$(3) \qquad L_t^{A,x} = \int_0^t \langle A(s)[W(s)+x(s)], dW(s) \rangle, \quad t \geq 0,$$

where the stochastic integral in (3) is defined in the sense of K. Itô and $\langle \cdot, \cdot \rangle$ denotes the Euclidean scalar product in $\mathbb{R}^d$.

As a special case take $x(t) \equiv 0$ and $A(t) \equiv J$,

$$J = \begin{pmatrix} 0 & -1 \\ +1 & 0 \end{pmatrix} ,$$

then the process $(L_t^{J,0}/2)$ is the same as Lévy's stochastic area (L_t).

We shall study the stochastic process $(W_t, L_t^{A,x})$ and, for fixed $t>0$, calculate its characteristic function. The proof of the result is inspired by work of R. Liptser and A. Shiryayev [13, vol. 2, p. 12] on filtering of random processes.

As has been noticed by B. Gaveau [4] formulae like (1) and (2) yield estimates as well as explicit expressions for fundamental solutions for the generators of the diffusions $(W_t, L_t^{A,x})$, a class of hypoelliptic operators which naturally arises in some problems in analysis and geometry. So, in particular, when we choose $x(t) \equiv 0$, $A(t) \equiv A$, A skew-symmetric, we thus give new proofs of some of the results, e.g. Theorem 4.2.1 and 4.3.1, obtained in [4] using the expansion of Brownian motion described above. By the method used in [2] we can also give a formula for the fundamental solution of the standard sub-Laplacian of any simply connected nilpotent Lie group of order 2. The expression found is further exploited in the special case of 'generalised Heisenberg groups', a class of nilpotent groups introduced by A. Kaplan [7]. By an elegant method Kaplan has shown that the standard sub-Laplacian of these Lie groups admit fundamental solutions analogous to that known for the Heisenberg group.

The following notation will be adopted throughout:

$|y|^2 := \langle y,y\rangle := \sum y_i^2$, $y \in \mathbb{C}^d$.

$H_z(t)$, $z \in C$, $t \geq 0$, denotes the matrix solution to the differential equation

$$\dot{H}_z(t) = zA(t)H_z(t), \quad H_z(0) = \mathrm{Id}.$$

$\Gamma_\Lambda(t)$, $\Lambda \in \mathbb{R}$, $t \in [0,T]$, $T>0$ fixed, denotes the unique symmetric non-positive definite matrix defined by the Riccati equation

$$\dot{\Gamma}_\Lambda(t) = \Lambda^2 A(t)A^*(t) - \Gamma_\Lambda^2(t), \quad 0 \leq t \leq T, \quad \Gamma_\Lambda(T) = 0.$$

* indicates transposition of real- and complex-valued matrices.

$\Phi(t)$ denotes the unique solution to the matrix differential equation

$$\dot{\Phi}(t) = \Gamma_\Lambda(t)\Phi(t), \quad \Phi(0) = \mathrm{Id}.$$

$U_t^{(z)}$ is defined as

$$U_t^{(z)} = z \int_0^t H_z^{-1}(s)A(s)x(s)\,ds.$$

2. THE JOINT CHARACTERISTIC FUNCTION

The main result of this note is the following formula for the joint characteristic function of the two random variables $X_t := W_t + x(t)$ and $L_t^{A,x}$, $t>0$. The proof which is rather lenghty and will thus appear elsewhere is based on Girsanov's measure transformation technique and analytic continuation of the function

$$z \to E\left[exp\{z \int_0^T <A(s)X_s, dW_s> + i<\gamma, X_T>\}\right],$$

defined on $(-c,+c)$, $c>0$ 'small', to the domain $\{z\in \mathbb{C} |\ |Re(z)| \leq c/2\}$.

Theorem 1. Let $A:[0,T] \to S_d$, S_d the space of $d\times d$ skew-symmetric matrices, and $x:[0,T] \to \mathbb{R}^d$, $T>0$ fixed, be bounded measurable functions. Assume that for every $t\in [0,T]$, $z\in C$ the matrices $H_z(t)$ and $A^2(t)$ commute, i.e.

(4) $$H_z(t)A^2(t) = A^2(t)H_z(t).$$

Then for $\Lambda\in \mathbb{R}$ and $\gamma\in \mathbb{R}^d$

(5) $$\begin{aligned} F(\Lambda,\gamma) :&= E\left[exp\{i\Lambda L_T^{A,x} + i<\gamma, X_T>\}\right] \\ &= exp\Big[i<\gamma,[x(T) + H_{i\Lambda}(T)U_T^{(i\Lambda)}]> - \\ &\quad - \frac{\Lambda^2}{2}\int_0^T |A(s)x(s) + A(s)H_{i\Lambda}(s)U_s^{(i\Lambda)}|^2 ds + \\ &\quad + \frac{1}{2}\int_0^T |i\Phi^*(s)^{-1}\Phi^*(T)H_{i\Lambda}^*(T)\gamma + \Lambda^2\Phi^*(s)^{-1} \cdot \\ &\qquad \cdot \int_s^T \Phi^*(r)[H_{i\Lambda}^*(r)A^2(r)x(r)U_r^{(i\Lambda)}]dr|^2 ds + \\ &\quad + \frac{1}{2}\int_0^T Sp(\Gamma_\Lambda(s))ds\Big]. \end{aligned}$$

Remark 1. Condition (4) is satisfied if, for instance, $A(t) \equiv A$, $A\in S_d$, or $A(t)$ is skew-symmetric <u>and</u> orthogonal for all $t\in [0,T]$; more generally, $A(t)\in S_d$ and $A^2(t) = a(t)Id$, $a(t)$ real-valued bounded measurable function. Note that for $d = 2$ the matrices J and J^* are the only ones which are skew-symmetric and orthogonal; in dimension 3 there are none but for $d = 4$, for instance, any matrix A having the representation

$$A = \begin{pmatrix} cJ & -B^* \\ B & \pm cJ \end{pmatrix},$$

where $B = \sqrt{1-c^2}\,O$, $|c| < 1$ and O is an orthogonal matrix such that

$\det(O) = \pm 1$ (-1 if +c is chosen and vice versa), possesses both properties.

Formula (5) can be exploited in some special cases. Since $U_t^{(z)} \equiv 0$ if $x(t) \equiv 0$ the next result follows immediately from (5).

Corollary 1. *If* $x(t) \equiv 0$, *then*

$$F(\Lambda,\gamma) = exp\left[\frac{1}{2}\int_0^T Sp(\Gamma_\Lambda(s))ds - \frac{1}{2}\int_0^T |\Phi^*(s)^{-1}\Phi^*(T)H^*_{i\Lambda}(T)\gamma|^2 ds\right]$$

and

$$F(\Lambda,0) = exp\left[\frac{1}{2}\int_0^T Sp(\Gamma_\Lambda(s))ds\right].$$

Corollary 2. *If* $x(t) \equiv 0$ *and* $A(t) \equiv A$, *then*

$$F(\Lambda,\gamma) = \prod_{k=1}^{[d/2]} \frac{1}{cosh(\Lambda t a_k)} exp\left[-\frac{1}{2}[(O\gamma)^2_{2k-1}+(O\gamma)^2_{2k}]\frac{tanh(\Lambda t a_k)}{\Lambda}\right] \tag{6}$$

where O *is an orthogonal matrix such that* $O'AO$ *is a skew-symmetric matrix formed from* $[d/2]$ *diagonal blocks*

$$\begin{pmatrix} 0 & -a_k \\ a_k & 0 \end{pmatrix}, \qquad 1 \le k \le [d/2],$$

and the numbers a_k *are simple algebraic functions of the entries of the matrix* A.

Proof. Since A is constant we know explicitly $H_{i\Lambda}(t)$, $\Gamma_\Lambda(t)$ and $\Phi(t)$ up to multiplication with the orthogonal matrix O. Then by some lenghty but easy computation formula (6) is derived from (5).

Corollary 3. *If* $x(t) \equiv x$, $A(t) \equiv A$ *then*

$$F(\Lambda,\gamma) = \prod_{k=1}^{[d/2]} \frac{1}{cosh(\Lambda t a_k)} exp\left[-\frac{1}{2}\{(O[\gamma+\Lambda Ax])^2_{2k-1}+(O[\gamma+\Lambda Ax])^2_{2k}\}\cdot \cdot\frac{tanh(\Lambda t a_k)}{\Lambda}\right]\cdot exp(i\langle\gamma,x\rangle).$$

Proof. The assertion follows from formula (6) if there we substitute γ by $(\gamma + \Lambda Ax)$. The formula can also be derived directly from (5) taking into account the formulae for $H_{i\Lambda}(t)$, $\Gamma_\Lambda(t)$ and $\Phi(t)$.

Remark 2. Lévy's formula (1) follows from (6) since for this example $a_1 = 1$. Formula (2) is also derived from (6) by taking conditional expectation on the left hand side of Eq. 6 first and doing Fourier transformation on both sides after; multiplication by $(2\pi t)\,exp(-|x|^2/2t)$ then yields the assertion.

3. FUNDAMENTAL SOLUTIONS

In this section we shall give two applications of Theorem 1, cf. also [2,4,5 and 7]. Let $x(t) \equiv 0$ and $A(t) \equiv A$; from now on we shall write L_t^A instead of $L_t^{A,0}$.

The stochastic process $(Z_t) := (W_t, L_t^A)$ is the unique solution to the stochastic differential equation

$$dZ_t = \sigma_A(Z_t)dW_t, \quad Z_0 = 0,$$

where $\sigma_A(z)$, $z = (\theta,\omega) \in \mathbb{R}^d \times \mathbb{R}$, denotes the $(d+1)\times d$ - dimensional matrix

$$\sigma_A(z) = \begin{pmatrix} Id \\ (A\omega)^* \end{pmatrix}$$

Put $\alpha(z) = \sigma_A(z)\sigma_A^*(z)$; the generator of (Z_t) is thus given by

$$\frac{1}{2}\Delta_A = \frac{1}{2}\sum_{i,j=1}^{d+1} \alpha_{ij}(z)\frac{\partial^2}{\partial z_i \partial z_j} = \frac{1}{2}\sum_{i=1}^{d}\left(\frac{\partial}{\partial \omega_i} + \sum_{j=1}^{d} A_{ij}\omega_j\frac{\partial}{\partial\theta}\right)^2.$$

Let $p_0(\tau;z), \tau>0$, denote the fundamental solution to the equation

$$\left(\frac{\partial}{\partial\tau}p - \Delta_A p\right) = 0 \quad \text{with pole at zero.}$$

Taking the Fourier transform of Eq. 6 with respect to γ first we obtain the following result.

Theorem 2. *Put* $n = [d/2]$, *then*

$$p_0(\tau;(\omega,\theta)) = (2\pi\tau)^{-(1+d/2)} \int_{\mathbb{R}} \prod_{k=1}^{n} \frac{\rho a_k}{sinh(\rho a_k)} exp\left[\frac{i\langle\theta,\rho\rangle}{\tau} - \frac{\rho a_k}{2\tau}[(\mathcal{O}\gamma)_{2k-1}^2 + (\mathcal{O}\gamma)_{2k}^2]coth(\rho a_k)\right] d\rho.$$

Since in dimension $d \geq 3$ Brownian motion is not recurrent we can derive from Theorem 2 a formula for the fundamental solution for the operator Δ_A. For $d = 2$ the integral can be easily calculated.

Corollary 4. The fundamental solution for Δ_A is given by

$$q_0(z) = \int_0^\infty p_0(\tau;z)d\tau.$$

If A = J, *then*

$$q_0(z) = \frac{(2\pi)^{-1}}{\sqrt{(\omega_1^2+\omega_2^2)^2 + \theta^2}}$$

Finally, let us consider a more general situation: Instead of just one matrix A we take $m\geq 1$ matrices $A^{(1)},\dots,A^{(m)}$ having the properties that each matrix $A^{(i)}$ is skew-symmetric and orthogonal and that for every pair of indices (i,j) the relation

$$A^{(i)}A^{(j)} = -A^{(j)}A^{(i)}$$

holds. We consider the stochastic process $I_t := (W_t, L_t^{(1)},\dots,L_t^{(m)})$ where

$$L_t^{(i)} := L_t^{A^{(i)}}, \quad 1\leq i\leq m.$$

Theorem 3. The characteristic function of I_t, $t>0, \gamma\in \mathbb{R}^d, \vec{\Lambda}\in \mathbb{R}^m$, is

$$E\left[\exp\{i<(\gamma,\vec{\theta})I_t>\}\right] = \frac{1}{\cosh(\Lambda t)^{[d/2]}} \exp\{-\frac{|\gamma|^2\tanh(\Lambda t)}{2\Lambda}\} ,$$

where $\Lambda := |\vec{\Lambda}|$. The fundamental solution to $(\partial/\partial t - \Delta_{A^1,\dots}) = 0$ with pole at zero is given by

$$p_0(\tau;\omega,\vec{\theta}) = \frac{i^m}{(2\pi\tau)^{1+m+d/2}} \int_{\mathbb{R}^m} \left(\frac{\Lambda}{\sinh(2\Lambda)}\right)^{[d/2]} \exp\left[-\frac{i<\vec{\Lambda},\vec{\theta}>}{\tau} - \right.$$

$$\left. - \frac{|\omega|^2\Lambda\coth(\Lambda)}{2\tau}\right] d\vec{\Lambda} .$$

Proof. Let $0 \neq \vec{\Lambda}\in \mathbb{R}^m$ be given. Put

$$A := \Lambda^{-1}\sum_{i=1}^m \Lambda_i A^{(i)} ;$$

Since $i<(\gamma,\vec{\theta}),I_t> = i<\gamma,W_t> + i<\Lambda,L_t^A>$ and A is a skew-symmetric and orthogonal matrix the first assertion follows from Corollary 2 since, in that case, all the numbers a_k are +1 or -1 only. The second formula is derived from the first one by taking Fourier transform on both sides of the equation.

REFERENCES

[1] BREIMAN, L.: *Probability*, Addison Wesley, New York, 1968.

[2] CYGAN, J.: *Heat kernels for class 2 nilpotent groups*, Studia Math. 64 (1979), pp. 227-238.

[3] FRIEDMAN, A.: *Stochastic differential equations and applications*, Academic Press, New York, 1976.

[4] GAVEAU, B.: *Principe de moindre action, propagation de la chaleur et estimées sous elliptiques sur certains groupes nilpotents*, Acta Math. 139 (1977), pp. 95-153.

[5] HULANICKI, A.: *The distribution of energy in the Brownian motion in the Gaussian field and analytic-hypoellipticity of certain subelliptic operators on the Heisenberg group*, Studia Math. 56 (1976), pp. 165-173.

[6] IKEDA, N. and WATANABE, S.: *Stochastic differential equations and diffusion processes*, North-Holland Publ., Amsterdam, 1981.

[7] KAPLAN, A.: *Fundamental solutions for a class of hypoelliptic pde generated by composition of quadratic forms*, Trans Amer. Math. Soc. 258 (1980), pp. 147-153.

[8] LÉVY, P.: *Le mouvement Brownien plan*, Amer. Jour. Math. 62 (1940), pp. 487-550.

[9] ——— : *Processus stochastiques et mouvement Brownien*, Gauthier-Villars, Paris, 1948.

[10] ——— : *Calcul des probabilités-fonctions aléatoires Laplaciennes*, C. R. Acad, Sci. 229 (1949), pp. 1057-1058.

[11] ——— : *Calcul des probabilités - sur l'aire comprise entre un arc de la courbe du mouvement Brownien plan et sa corde*, C. R. Acad. Sci. 230 (1950), pp. 432-434; errata p. 689.

[12] ——— : *Wiener's random function, and other Laplacian random functions*, Proc. 2nd Berkeley Symp., pp. 171-187, 1951.

[13] LIPTSER. R. and SHIRYAYEV, A.: *Statistics of random processes*, vols. 1, 2, Springer-Verlag, New York, 1977.

[14] STROOCK, D. and VARADHAN S.: *Multidimensional diffusion processes*, Springer-Verlag, New York, 1979.

[15] WILLIAMS, D.: *On a stopped Brownian motion formula of H.M. Taylor*, Séminaire de Probabilités X, Lect. Notes in Maths. 511, pp. 235-239, Springer-Verlag, Berlin, 1976.

[16] YOR, M.: *Remarques sur une formule de Paul Lévy*, Séminaire de Probabilités XIV, Lect. Notes in Maths. 784, pp. 343-346, Springer-Verlag, Berlin, 1980.

ASYMPTOTIC NONLINEAR FILTERING AND LARGE DEVIATIONS

Omar Hijab

Mathematics and Statistics
Case Western Reserve University
Cleveland, Ohio 44106

0. Introduction. Consider a diffusion $t \to x^\varepsilon(t)$ evolving on R^n and governed by a generator of the form

$$A^\varepsilon = f + \frac{\varepsilon}{2}(g_1^2 + \dots + g_m^2) \tag{1}$$

corresponding to a given set of vector fields $f, g_1, \dots, g_m$ on R^n. It is of interest to study the asymptotic behavior of the probability distributions P^ε on $\Omega^n \equiv C([0,T];R^n)$ of the diffusions $t \to x^\varepsilon(t)$ as $\varepsilon \downarrow 0$. It turns out that the asymptotic properties of P^ε depend strongly on properties of the associated control system

$$\dot{x} = f(x) + g_1(x)u_1 + \dots + g_m(x)u_m. \tag{2}$$

Indeed, it turns out that in some sense

$$P^\varepsilon(dx(\cdot)) = \exp(-\frac{1}{2\varepsilon}\int_0^T u(t)^2 dt)dx(\cdot)$$

as $\varepsilon \downarrow 0$. More precisely, suppose that the diffusions $t \to x^\varepsilon(t)$ satisfy $x^\varepsilon(0) = x_o$ almost surely and suppose that to each u in $L^2([0,T];R^m)$ there is a well-defined solution x_u of (2) in Ω^n satisfying $x_u(0) = x_o$. Then the asymptotic behavior of P^ε is given by the following estimates: For any open set G in Ω^n and closed set C in Ω^n,

$$\begin{aligned} \underline{\lim}_{\varepsilon \downarrow 0}\ \varepsilon \log P^\varepsilon(G) &\geq -\inf\{\frac{1}{2}\int_0^T u^2 dt \mid x_u \text{ in } G\} \\ \overline{\lim}_{\varepsilon \downarrow 0}\ \varepsilon \log P^\varepsilon(C) &\leq -\inf\{\frac{1}{2}\int_0^T u^2 dt \mid x_u \text{ in } C\}. \end{aligned} \tag{3}$$

In 1966 S.R.S. Varadhan set down a general framework [1] for dealing with the asymptotic behavior of families of measures and certain associated expectations, and in particular derived the above estimates for processes with independent increments [1]. Subsequently, he derived these estimates for the case of drift-free nondegenerate diffusions (i.e., $f = 0$) [2]. Later Glass [3] and Ventsel and Freidlin [4] established these estimates for nondegenerate diffusions with drift.

In 1978 Azencott [5] established these estimates in a general case; Azencott's results imply that if $f, g_1, \ldots, g_m$ are C^2, if for each $\varepsilon > 0$ there is a solution to the martingale problem on Ω^n corresponding to A^ε and if for each u in $L^2([0,T];R^m)$ the solution x_u of (2) starting at x_o exists in Ω^n, then the above estimates hold.

Suppose that the diffusions $t \to x^\varepsilon(t)$ are observed in the presence of an independent Brownian motion $t \to b(t)$,

$$y^\varepsilon(t) = \int_o^t h(x^\varepsilon(s))ds + \sqrt{\varepsilon}\, b(t) \;,\; t \geq 0,$$

where $h : R^n \to R^p$ is a given map. Then the unnormalized conditional distribution $Q^\varepsilon_{x|y}$ of $t \to x^\varepsilon(t)$ given $t \to y^\varepsilon(t)$ is well-defined. In this paper we show that if h is C^3 and $h, f(h), g_1(h), \ldots, g_m(h), g_1^2(h), \ldots, g_m^2(h)$ are all bounded on R^n then for any open set G in Ω^n and closed set C in Ω^n,

$$\begin{aligned}
&\varliminf_{\varepsilon \downarrow 0} \varepsilon \log Q^\varepsilon_{x|y}(G) \geq -\inf \{\tfrac{1}{2}\int_o^T u^2 + h(x_u)^2 dt - \int_o^T h(x_u)dy \,|\, x_u \text{ in } G\} \\
&\varlimsup_{\varepsilon \downarrow 0} \varepsilon \log Q^\varepsilon_{x|y}(C) \leq -\inf \{\tfrac{1}{2}\int_o^T u^2 + h(x_u)^2 dt - \int_o^T h(x_u)dy \,|\, x_u \text{ in } C\}
\end{aligned} \tag{4}$$

for almost all y in $\Omega^p \equiv C([0,T];R^p)$.

1. Large Deviations. Throughout Ω^n will denote $C([0,T];R^n)$ with Ω^m and Ω^p defined analogously, where $[0,T]$ is a fixed time interval. The topology on Ω^n is that of uniform convergence on $[0,T]$. We suppose we are given

(i) C^2 vector fields $g_1, \ldots, g_m$ on R^n and a time-varying vector field f in $C^{o,2}([0,T] \times R^n, R^n)$.

If g is any vector field on R^n, let $g(\varphi)(x)$ denote the directional derivative of φ in the direction of g at the point x. The vector field g can then be thought of as a first order differential operator taking φ in $C^\infty(R^n)$ to $g(\varphi)$. If $g^2(\varphi)$ is short for $g(g(\varphi))$, then (1) defines a second order (possibly time-varying) differential operator A^ε. Let $C_o^\infty(R^n)$ denote the space of smooth compactly supported functions on R^n.

Let $b(t) : \Omega^m \to R^m$ be given by $b(t,\omega) = \omega(t)$ and impose Wiener measure on Ω^m. Then $t \to b(t) = (b_1(t), \ldots, b_m(t))$ is an R^m-valued Brownian motion.

One way to construct diffusions on R^n governed by A^ε is to pick a point x_o in R^n and to let $t \to x^\varepsilon(t)$ be the unique process $\Omega^m \to \Omega^n$ satisfying

$$\varphi(x^\varepsilon(t)) - \varphi(x^\varepsilon(s)) - \int_s^t A^\varepsilon(\varphi)(x^\varepsilon(r))dr = \sqrt{\varepsilon} \int_s^t g(\varphi)(x^\varepsilon(r))db(r) \tag{5}$$

for all φ in $C_o^\infty(R^n)$, $0 \leq s \leq t \leq T$, and $x^\varepsilon(0) = x_o$, almost surely on Ω^m. Here $g(\varphi)db$ is short for $g_1(\varphi)db_1 + \ldots + g_m(\varphi)db_m$ where $g_i(\varphi)$ is defined above.

Using the standard exixtence and uniqueness theorem for stochastic differential equations and Ito's differential rule, it is easy to show that there is a unique such process defined up to an explosion time $\zeta^\varepsilon \leq \infty$ characterized by the fact that $t \to x^\varepsilon(t)$ leaves every compact subset of R^n as $t \uparrow \zeta^\varepsilon$, almost surely on $\zeta^\varepsilon < \infty$.

The merit of the above definition of $t \to x^\varepsilon(t)$ is that it makes sense on any manifold X. Indeed, the Whitney embedding theorem allows one to embed any such X into some R^N and by extending $f, g_1, \ldots, g_m$ to R^N one can derive the result described above on any manifold. Of course in R^n $t \to x^\varepsilon(t)$ is the "Stratonovitch solution". In any event, as $\varepsilon \downarrow 0$ the "correction factor" disappears and so estimates (3) are expected to hold just as well for the diffusions $t \to x^\varepsilon(t)$ constructed here. In what follows we are careful to state everything in such a way as to make sense on any manifold X.

If $T \leq \zeta^\varepsilon$ then the probability distribution P^ε of $t \to x^\varepsilon(t)$ exists on Ω^n and is the unique probability measure on Ω^n satisfying $P^\varepsilon(x(0) = x_o) = 1$ and

$$E^\varepsilon(\varphi(x(t)) - \varphi(x(s)) - \int_s^t A^\varepsilon(\varphi)(x(r))dr \,|\, \mathcal{F}_s) = 0 \tag{6}$$

for all φ in $C_o^\infty(R^n)$ and $0 \leq s \leq t \leq T$. Here $x(t) : \Omega^n \to R^n$ is the canonical map and $\mathcal{F}_s$ is the σ-algebra generated by the maps $x(r)$, $0 \leq r \leq s$.

Conversely, if one assumes that

(ii) for each $\varepsilon > 0$ there is a probability measure P^ε on Ω^n satisfying $P^\varepsilon(x(0) = x_o) = 1$ and (6) for all φ in $C_o(R^n)$ and $0 \leq s \leq t \leq T$,

then one can show that the solution $t \to x^\varepsilon(t)$ of (5) explodes after time T, i.e., $\zeta^\varepsilon \geq T$, almost surely.

In what follows we shall assume (ii) and

(iii) to each u in $L^2([0,T];R^m)$ there is a path x_u in Ω^n satisfying (2) and $x_u(0) = x_o$. In other words, the solution of (2) starting at x_o has escape time greater or equal to T, for all u in L^2.

Under assumptions (i),(ii) and (iii) estimates (3) hold for the measures $\{P^\varepsilon\}$ constructed here [5]. To understand these estimates from a more general perspective consider the following definition [1].

Definition. Let Ω be a completely regular topological space and let P^ε, $\varepsilon > 0$, be a family of probability measures on Ω. We say that $\{P^\varepsilon\}$ admits large deviation if there is a function I on Ω satisfying

(i) $0 \leq I \leq +\infty$.

(ii) I is lower semicontinuous on Ω.

(iii) $\{\omega | I(\omega) \leq M\}$ is a compact subset of Ω for all finite M.

(iv) For any open set G in Ω, $\underline{\lim}_{\varepsilon\downarrow 0} \varepsilon \log P^{\varepsilon}(G) \geq -\inf\{I(\omega)|\omega \text{ in } G\}$.

(v) For any closed set C in Ω, $\overline{\lim}_{\varepsilon\downarrow 0} \varepsilon \log P^{\varepsilon}(C) \leq -\inf\{I(\omega)|\omega \text{ in } C\}$.

The function I is then referred to as the corresponding "I-functional".

Estimates (3) then state that (iv) and (v) hold for the probability distributions of the diffusions $t \to x^{\varepsilon}(t)$, where I is given by

$$I(\omega) = \inf\{\tfrac{1}{2} \int_0^T u(t)^2 dt | x_u = \omega\}$$

for all ω in Ω^n, with the understanding that the infimum of an empty set of real numbers is $+\infty$. Since (ii) is easy to derive and (iii) is the statement that $u \to x_u$ is a compact map from $L^2([0,T];R^m)$ into Ω^n, we have

Theorem 1.1. The probability distributions $\{P^{\varepsilon}\}$ corresponding to A^{ε} admit large deviation as $\varepsilon \downarrow 0$.

A consequence of the above abstract definition is the following theorem, which is a summary of results appearing in section 3 of [1].

Theorem 1.2. Let $\{P^{\varepsilon}\}$ admit large deviation with corresponding I-functional I and let Φ_{ε} be a bounded continuous function on Ω such that Φ_{ε} converges uniformly to Φ as $\varepsilon \downarrow 0$. Let Q^{ε} be given by

$$dQ^{\varepsilon} = e^{-\frac{1}{\varepsilon}\Phi_{\varepsilon}} dP^{\varepsilon}.$$

Then $\{Q^{\varepsilon}\}$ satisfies

$$\underline{\lim}_{\varepsilon\downarrow 0} \varepsilon \log Q^{\varepsilon}(G) \geq -\inf\{I(\omega) + \Phi(\omega)|\omega \text{ in } G\}.$$

$$\overline{\lim}_{\varepsilon\downarrow 0} \varepsilon \log Q^{\varepsilon}(C) < -\inf\{I(\omega) + \Phi(\omega)|\omega \text{ in } C\}$$

for G open in Ω and C closed in Ω.

We note that for the results of theorem 1.2 to hold it is not necessary that Φ_{ε} be bounded: That is required is that the tail estimate

$$\lim_{R\to\infty} \overline{\lim}_{\varepsilon\downarrow 0} \varepsilon \log E^{\varepsilon}(1_{[\Phi_{\varepsilon}\geq R]} \times \exp(-\Phi_{\varepsilon}/\varepsilon)) = -\infty$$

holds [1].

2. Nonlinear Filtering. Let $h : R^n \to R^p$ be a locally bounded measurable map and let $t \to b(t)$ denote an R^p-valued Brownian motion independent of given processes $t \to x^{\varepsilon}(t)$ on R^n. Let

$$y^{\varepsilon}(t) = \int_0^t h(x^{\varepsilon}(s))ds + \sqrt{\varepsilon}b(t), \quad 0 \leq t \leq T. \tag{7}$$

In this section, we study the conditional distribution $P^{\varepsilon}_{x|y}$ on Ω^n of $t \to x^{\varepsilon}(t)$ given $t \to y^{\varepsilon}(t)$. We shall use Bayes' rule to compute $P^{\varepsilon}_{x|y}$.

Let W denote Wiener measure on Ω^p, let P^{ε}_x denote the probability distribution of $t \to x^{\varepsilon}(t)$ on Ω^n, let P^{ε}_y denote the probability distribution of $t \to y^{\varepsilon}(t)$ on Ω^p, let $P^{\varepsilon}_{(x,y)}$ denote the probability distribution of $t \to (x^{\varepsilon}(t), y^{\varepsilon}(t))$ on $\Omega^n \times \Omega^p$, and let $P^{\varepsilon}_{x|y}$ denote the conditional distribution of $t \to x^{\varepsilon}(t)$ on Ω^n given $t \to y^{\varepsilon}(t)$. Let $y(t) : \Omega^p \to R^p$ denote the canonical map. Let W^{ε} denote the Wiener measure on Ω^p "of variance ε".

For $0 \leq t \leq T$ set

$$\Lambda(t) = \frac{1}{2} \int_0^t h(x(s))^2 ds - \int_0^t h(x(s))dy(s).$$

$\Lambda(t)$ is then a measurable function on $\Omega^n \times \Omega^p$ for each t. Using (7) and invoking the Cameron-Martin formula it is easy to see that

$$dP^{\varepsilon}_{(x,y)} = e^{-\frac{1}{\varepsilon}\Lambda} d(P^{\varepsilon}_x \times W^{\varepsilon})$$

where $\Lambda = \Lambda(T)$. Here and elsewhere, $h^2 = h_1^2 + \ldots + h_p^2$, $hdy = h_1 dy_1 + \ldots + h_p dy_p$, etc.

Using Bayes' rule, the conditional distribution is given by

$$\begin{aligned} dP_{x|y} &= \frac{dP^{\varepsilon}_{(x,y)}}{d(P^{\varepsilon}_x \times P^{\varepsilon}_y)} \cdot dP^{\varepsilon}_x \\ &= \frac{dP^{\varepsilon}_{(x,y)}}{d(P^{\varepsilon}_x \times W^{\varepsilon})} \cdot \frac{d(P^{\varepsilon}_x \times W^{\varepsilon})}{d(P^{\varepsilon}_x \times P^{\varepsilon}_y)} \cdot dP^{\varepsilon}_x \\ &= e^{-\frac{1}{\varepsilon}\Lambda} dP^{\varepsilon}_x / E^{\varepsilon}(e^{-\frac{1}{\varepsilon}\Lambda}). \end{aligned} \tag{8}$$

Equation (8) is the formula of Kallianpur-Striebel [6]. We rewrite it as

$$= dQ^{\varepsilon}_{x|y} / Q^{\varepsilon}_{x|y}(\Omega^n),$$

and refer to $Q^{\varepsilon}_{x|y}$ as the unnormalized conditional distribution.

So far equation (8) holds for <u>any</u> processes $t \to x^{\varepsilon}(t)$. Now suppose that P^{ε}_x is governed by A^{ε} in the sense of equation (6), where A^{ε} is given by (1). For any bounded measurable φ let

$$\mu^{\varepsilon}_t(\varphi) = E^{\bar{P}^{\varepsilon}}(\varphi(x(t))\exp(-\Lambda(t)/\varepsilon)) \tag{9}$$

the "unnormalized conditional expectation of $\varphi(x^{\varepsilon}(t))$ given $y^{\varepsilon}(s)$, $0 \leq s \leq t$". We derive the equation governing the time evolution of $\mu^{\varepsilon}_t(\varphi)$.

Ito's rule guarantees that $z^{\varepsilon}(t) = \exp(-\Lambda(t)/\varepsilon)$ satisfies

$$z^{\varepsilon}(t) - z^{\varepsilon}(s) = \frac{1}{\varepsilon} \int_s^t h(x(r)) z^{\varepsilon}(r) dy(r)$$

for $0 \le s \le t \le T$. This last equation together with equations (6) and (8) and the Ito product rule then yield

Theorem 2.1. For all φ in $C_0^\infty(R^n)$ and $0 \le s \le t \le T$,

$$\mu_t^\varepsilon(\varphi) - \mu_s^\varepsilon(\varphi) - \int_s^t \mu_r^\varepsilon(A^\varepsilon\varphi)dr = \frac{1}{\varepsilon}\int_s^t \mu_r^\varepsilon(h\varphi)dy(r).$$

We emphasize that this proof is valid for any locally bounded measurable h and any generator A^ε of the form (1). This equation is well-known and appears in various forms in the literature.

In the next section we study the asymptotic behaviour of $Q^\varepsilon_{x|y}$ as $\varepsilon \downarrow 0$.

3. Asymptotic Filtering. In this section, we shall assume (i),(ii),(iii) and

(iv) h is C^3 and $f(h)$, $g_i(h)$, $g_i^2(h)$ and h, $i = 1, \ldots, m$ are bounded on $[0,T] \times R^n$.

Let I be the I-functional given by theorem 1.1,

$$I(\omega) = \inf\{\tfrac{1}{2}\int_0^T u(t)^2 dt \,|\, x_u = \omega\}.$$

In this section we shall prove

Theorem 3.1. Let $Q^\varepsilon_{x|y}$ denote the unnormalized conditional distribution on Ω^n. Then estimates (4) hold for almost all y in Ω^p.

Note that for $h = 0$ this theorem reduces to estimates (3). The idea of the proof is simple enough; apply theorem 1.2 to $Q^\varepsilon_{x|y}$ using the representation given by equation (8). This however does not work directly because the exponent Λ is not a continuous function on Ω^n for each y in Ω^p. We therefore have to make a slight detour and integrate by parts the stochastic integral appearing in Λ.

For each $\varepsilon > 0$ let Φ_ε on Ω^n be given by

$$\Phi_\varepsilon(\omega) = -y(T)h(\omega(T)) + y(0)h(\omega(0)) + \int_0^T [+yA^\varepsilon(h)(\omega) + \tfrac{1}{2}h(\omega)^2 - \tfrac{1}{2}y^2 g(h)(\omega)^2]dt.$$

Then $\Phi_\varepsilon \to \Phi_0 = \Phi$ as $\varepsilon \downarrow 0$ uniformly on Ω^n, for each y in Ω^p. Referring to (8) and performing an integration by parts in the stochastic integral appearing in Λ and invoking Girsanov's theorem we see that

$$dP^\varepsilon_{x:y} \equiv e^{\frac{1}{\varepsilon}\Phi_\varepsilon} dQ^\varepsilon_{x|y} \tag{10}$$

satisfies equation (6) with A^ε replaced by

$$A^\varepsilon - yg_1(h)g_1 - \cdots - yg_m(h)g_m.$$

We wish to apply theorem 1.1 to $\{P^\varepsilon_{x:y}\}$. To do so we must check that assumptions (i), (ii), (iii) of section 1 hold for the vector fields

$$f_y = f - yg_1(h)g_1 - \cdots - yg_m(h)g_m,\ g_1, \ldots, g_m$$

for all y in Ω^p, given that they hold for $y = 0$. For (i) this is obvious. For (ii) this is also obvious and for (iii) this is so because $g_1(h), \ldots, g_m(h)$ are bounded feedback terms. Thus theorem 1.1 applies to $\{P^{\varepsilon}_{x:y}\}$ and hence theorem 1.2 applies to $Q^{\varepsilon}_{x|y}$ via equation (10).

Thus let $x_{u:y}$ denote the unique path in Ω^n satisfying

$$\dot{x} = f_y(x) + g_1(x)u_1 + \ldots g_m(x)u_m \quad \text{and} \quad x(0) = x_o.$$

Let I_y be the I-functional corresponding to $P^{\varepsilon}_{x:y}$ according to theorem 1.1:

$$I_y(\omega) = \inf\{\tfrac{1}{2}\int_0^T u(t)^2 dt \,|\, x_{u:y} = \omega\}.$$

The theorem 1.2 implies that for any G open and C closed in Ω^n

$$\begin{aligned} &\underline{\lim}_{\varepsilon \downarrow 0}\, \varepsilon \log Q^{\varepsilon}_{x|y}(G) \geq -\inf\{I_y(\omega) + \Phi(\omega) \,|\, \omega \text{ in } G\} \\ &\overline{\lim}_{\varepsilon \downarrow 0}\, \varepsilon \log Q^{\varepsilon}_{x|y}(C) \leq -\inf\{I_y(\omega) + \Phi(\omega) \,|\, \omega \text{ in } C\}. \end{aligned} \tag{11}$$

Now a little algebraic manipulation in (11) using the fact that

$$I_y(\omega) = \inf\{\tfrac{1}{2}\int_0^T (u + yg(h)(\omega))^2 dt \,|\, x_u = \omega\}$$

yields theorem 3.1. Applications of theorem 3.1 will appear elsewhere.

REFERENCES

[1] S.R.S. Varadhan, "Asymptotic Probabilities and Differential Equations," Comm. Pure & Applied Math., Vol. XIX, 261-286 (1966).

[2] ______________, "Diffusion Processes in a Small Time Interval", Comm. Pure & Applied Math., Vol. XX, 659-685 (1967).

[3] M. Glass, "Perturbation of a First Order Equation by a Small Diffusion", Ph.D. Disseration, New York University, 1969.

[4] A.D. Ventsel and M.I. Friedlin, "Small Random Perturbations of Dynamical Systems, Russian Math. Surveys, 25 (1970) 1-56 [Uspehi Mat. Nauk. 25 (1970) 3-55].

[5] R. Azencott, Lecture Notes in Math. #774, Springer 1978.

[6] G. Kallianpur and C. Striebel, "Estimation of Stochastic Processes", Annals Math. Statistics, 39 (1968) 785-801.

Representation and approximation of counting processes

Thomas G. Kurtz
Department of Mathematics
University of Wisconsin - Madison
Madison, Wisconsin 53706 USA

1. Introduction

By a counting process we mean a stochastic process N whose sample paths are constant except for jumps of $+1$. The simplest example is, of course, the Poisson process. Recall that the distribution of the Poisson process is determined by specifying the intensity parameter λ which gives

(1.1) $$P\{N(t+\Delta t) - N(t) > 0 \mid F_t\} = \lambda \Delta t + o(\Delta t)$$

where F_t is the history for the process up to time t, i.e. $F_t = \sigma(N(s): s \le t)$. More general counting processes are determined by specifying an intensity function $\lambda(t,N)$ which, as in (1.1), gives

(1.2) $$P\{N(t+\Delta t) - N(t) > 0 \mid F_t\} = \lambda(t,N)\Delta t + o(\Delta t).$$

Of course for (1.2) to make sense, $\lambda(t,N)$ can depend only on the values of N up to time t. To be precise, let $\mathbb{Z}^+$ be the nonnegative integers and $\mathbb{Z}^{+\infty} = \mathbb{Z}^+ \cup \{+\infty\}$. (Topologically think of $\mathbb{Z}^{+\infty}$ as being the one-point compactification of $\mathbb{Z}^+$.) Let $J[0,\infty)$ denote the right continuous, nondecreasing $\mathbb{Z}^{+\infty}$-valued functions x such that $x(0) = 0$ and $x(t) - x(t-)$ is 0 or 1. In particular if $x(t) = \infty$, then $x(s) = \infty$ for $s > t$. (We give $J[0,\infty)$ the Skorohod topology when a topology is needed.) Let $\tau_n(x)$ denote the time of the nth jump of x and define x_t by $x_t(s) = x(s \wedge t)$. A (Borel)-measurable function $\lambda: [0,\infty) \times J[0,\infty) \to [0,\infty)$ is an <u>intensity function</u>, if for all $x \in J[0,\infty)$

(1.3) $$\lambda(t,x) = \lambda(t,x_t), \quad t \ge 0,$$

and

(1.4) $$\int_0^{\tau_m(x)} \lambda(t,x)dt < \infty, \quad m = 1, 2, 3 \ldots .$$

Given an intensity function λ, the problem then becomes to associate with it a counting process N satisfying (1.2). There are a variety of ways of accomplishing this. Here we will specify a stochastic equation for which N is the unique solution. For other approaches see the books by Brémaud (1981) and Snyder (1975). All these approaches are essentially equivalent. This equivalence is discussed in Kurtz (1982).

Let Y be a Poisson process with parameter 1. Then the equation for the counting process N corresponding to a given λ is

(1.5) $$N(t) = Y(\int_0^t \lambda(s,N)ds).$$

Existence and uniqueness of the solution follows by using (1.3) and (1.4) to solve the equation "from one jump to the next". This is discussed in detail in Kurtz (1982). The uniqueness implies $\int_0^t \lambda(s,N)ds$ is a stopping time for Y and observing that on the set where $N(t + \Delta t) = N(t)$

(1.6) $$\int_0^{t+\Delta t} \lambda(s,N)ds = \int_0^{t+\Delta t} \lambda(s,N(\cdot\wedge t))ds$$

we have, on the event $\{N(t) < \infty\}$,

(1.7) $$\begin{aligned} &P\{N(t+\Delta t) - N(t) > 0|F_t\} \\ &= 1 - P\{N(t+\Delta t) - N(t) = 0|F_t\} \\ &= 1 - P\{Y(\int_0^{t+\Delta t} \lambda(s,N(\cdot\wedge t))ds) - Y(\int_0^t \lambda(s,N(\cdot\wedge t))ds) = 0|F_t\} \\ &= 1 - \exp\{-\int_t^{t+\Delta t} \lambda(s,N(\cdot\wedge t))ds\} \end{aligned}$$

which is a precise version of (1.2). The fact that $\int_0^t \lambda(s,N)ds$ is a stopping time also give us the relation between the stochastic equation (1.5) and the martingale approach described in Brémaud (1981). Since $Y(u) - u$ is a martingale, the optional sampling theorem implies

(1.8) $$\begin{aligned} &N(t\wedge\tau_m) - \int_0^{t\wedge\tau_m} \lambda(s,N)ds \\ &= Y(\int_0^{t\wedge\tau_m} \lambda(s,N)ds) - \int_0^{t\wedge\tau_m} \lambda(s,N)ds \end{aligned}$$

is a martingale, where $\tau_m = \tau_m(N)$ is the time of the mth jump of N.

Equations of the form (1.5) also can be specified for systems of counting processes. Here we require that $Y_1, Y_2 \dots$ are independent Poisson processes with unit intensity. Letting $N = (N_1, N_2 \dots)$, we have the system of equations

(1.9) $$N_k(t) = Y_k(\int_0^t \lambda_k(s,N)ds)$$

where for each k, $\lambda_k: [0,\infty) \times (J[0,\infty))^d \to [0,\infty)$ (d may be finite or infinite) and for each $x \in (J[0,\infty))^d$ and each k

(1.10) $$\lambda_k(t,x) = \lambda_k(t, x_t)$$

and

$$(1.11) \qquad \int_0^{\tau_m(x)} \sum_k \lambda_k(t,x)\,dt < \infty .$$

Here

$$(1.12) \qquad \tau_m(x) = \inf\{t: \sum_k x_k(t) \geq m\}.$$

Examples

(a) Counter model. Let $\rho \geq 0$ and $\lim_{u\to\infty} \rho(u) = 0$. The equation

$$N(t) = Y(\int_0^t \lambda e^{-\int_0^t \rho(t-s)dN(s)}\,ds)$$

models the number of counts registered on a counter in a Poisson stream of particles where the sensitivity of the counter is reduced by each count but recovers in time.

(b) Birth and death process in a random environment. We can easily introduce "external" randomness. Let A, B, and C be positive stochastic processes independent of the (independent) Poisson processes Y_1, Y_2. Then the equation

$$(1.13) \quad Z(t) = Z(0) + Y_1(\int_0^t A(s)Z(s)ds) - Y_2(\int_0^t (B(s)Z(s) + C(s)Z(s)^2)ds)$$

determines a birth and death process in a random environment. Note that the counting processes are just the number of births

$$(1.14) \qquad N_1(t) = Y_1(\int_0^t A(s)Z(s)ds)$$

and the number of deaths

$$(1.15) \qquad N_2(t) = Y_2(\int_0^t (B(s)Z(s) + C(s)Z(s)^2)\,ds).$$

Since $Z = Z(0) + N_1 - N_2$, except for the additional complication of random coefficients, (1.14) and (1.15) form a system of the type in (1.9).

(c) Markov chain in $\mathbb{Z}^d$. Let

$$(1.16) \qquad X(t) = X(0) + \sum_{\ell\in \mathbb{Z}^d} \ell Y_\ell(\int_0^t \beta_\ell(X(s))ds)$$

where the Y_ℓ are independent Poisson processes and $\sum_\ell \beta_\ell(k) < \infty$, $k \in \mathbb{Z}^d$. Here the counting processes

$$(1.17) \qquad N_\ell(t) = Y_\ell(\int_0^t \beta_\ell(X(s))ds)$$

count the jumps of X of type ℓ, and X is the minimal Markov chain associated with the intensities $q_{k,k+\ell} = \beta_\ell(k)$. See Karlin (1966), page 228.

(d) Controlled counting process. The intensity $\lambda(s,x,u)$ may depend on a control parameter u. Then the controlled counting process is given by

(1.18) $$N(t) = Y(\int_0^t \lambda(s,N,u(s,N))ds)$$

provided $u(s,x) = u(s,x_t)$(cf. (1.3)).

In Sections 2 and 3 we use the stochastic equations described above to prove limit theorems for counting processes. In Section 4, in order to give another example of the type of argument used in Section 3, we consider the asymptotics of a simple fiber bundle model. The model is expressed as a solution of a stochastic equation similar to (1.5) but using the empirical process rather than the Poisson process.

2. Continuous dependence on λ

In this section we use the stochastic equations to show that the distribution of a counting process depends continuously on its intensity in a very strong sense.

2.1 Theorem Let λ, $\lambda^{(n)}$ $n = 1, 2 \ldots$ be intensity functions, N the counting process corresponding to λ and $N^{(n)}$ the counting process corresponding to $\lambda^{(n)}$. If for each $T > 0$, m, and $x \in J[0,\infty)$

(2.1) $$\lim_{n\to\infty} \int_0^{\tau_m(x)\wedge T} |\lambda^{(n)}(s,x) - \lambda(s,x)|ds = 0,$$

then $N^{(n)} \Rightarrow N$.

Proof We actually prove a stronger result than that stated. Furthermore we use a different representation of $N^{(n)}$ than that given by (1.5). Specifically let Y_1, Y_2 be independent Poisson processes with intensity one and let $N^{(n)}$ satisfy

(2.2) $$N^{(n)}(t) = Y_1(\int_0^t \lambda\wedge\lambda^{(n)}(s,N^{(n)})ds) + Y_2(\int_0^t(\lambda^{(n)}-\lambda\wedge\lambda^{(n)})(s,N^{(n)})ds)$$

and

(2.3) $$\tilde{N}^{(n)}(t) = Y_1(\int_0^t \lambda\wedge\lambda^{(n)}(s,\tilde{N}^{(n)})ds) + Y_2(\int_0^t(\lambda-\lambda\wedge\lambda^{(n)})(s,\tilde{N}^{(n)})ds).$$

Note that $\lambda^{(n)} = \lambda\wedge\lambda^{(n)} + \lambda^{(n)} - \lambda\wedge\lambda^{(n)}$ and $\lambda = \lambda\wedge\lambda^{(n)} + \lambda - \lambda\wedge\lambda^{(n)}$, and it follows by the multiparameter optional sampling theorem, Kurtz (1980), that

(2.4) $$N^{(n)}(t\wedge\tau_m(N^{(n)})) - \int_0^{t\wedge\tau_m(N^{(n)})} \lambda^{(n)}(s,N^{(n)})ds$$

and

(2.5) $$\tilde{N}^{(n)}(t\wedge\tau_m(\tilde{N}^{(n)})) - \int_0^{t\wedge\tau_m(\tilde{N}^{(n)})} \lambda(s,\tilde{N}^{(n)})ds$$

are martingales. Consequently, by the martingale characterization of counting processes (See Brémaud (1981)) $N^{(n)}$ has intensity $\lambda^{(n)}$ and $\tilde{N}^{(n)}$ has intensity λ. In particular all the $\tilde{N}^{(n)}$ have the same distribution, namely that of N satisfying

$$(2.6)\qquad N(t) = Y(\int_0^t \lambda(s,N)ds).$$

Note that $N^{(n)}(t) = \tilde{N}^{(n)}(t)$ for

$$(2.7)\qquad t < \gamma_n \equiv \inf\{t: Y_2(\int_0^t (\lambda^{(n)} - \lambda\wedge\lambda^{(n)})(s,N^{(n)})ds) > 0$$

$$\text{or}\quad Y_2(\int_0^t (\lambda - \lambda\wedge\lambda^{(n)})(s,\tilde{N}^{(n)})ds) > 0\}.$$

Consequently for any Borel set $\Gamma \subset J[0,\infty)$

$$(2.8)\qquad |P\{N(\cdot\wedge\tau_m(N)\wedge T) \in \Gamma\} - P\{N^{(n)}(\cdot\wedge\tau_m(N^{(n)})\wedge T) \in \Gamma\}|$$

$$= |P\{\tilde{N}^{(n)}(\cdot\wedge\tau_m(\tilde{N}^{(n)})\wedge T) \in \Gamma\} - P\{N^{(n)}(\cdot\wedge\tau_m(N^{(n)})\wedge T) \in \Gamma\}|$$

$$\leq P\{\tilde{N}^{(n)}(t) \neq N^{(n)}(t),\quad \text{some}\quad t \leq \tau_m(\tilde{N}^{(n)})\wedge T)\}$$

$$\leq P\{\gamma_n \leq \tau_m(\tilde{N}^{(n)})\wedge T\}$$

$$\leq \{Y_2(\int_0^{T\wedge\tau_m(\tilde{N}^{(n)})} |\lambda - \lambda^{(n)}|(s,\tilde{N}^{(n)})ds) > 0\}$$

$$\leq E[1 - \exp\{-\int_0^{T\wedge\tau_m(\tilde{N}^{(n)})} |\lambda(s,\tilde{N}^{(n)}) - \lambda^{(n)}(s,\tilde{N}^{(n)})|ds\}]$$

$$= E[1 - \exp\{-\int_0^{T\wedge\tau_m(N)} |\lambda(s,N) - \lambda^{(n)}(s,N)|ds\}].$$

Note that since $\tilde{N}^{(n)}(t) = N^{(n)}(t)$ for $t < \gamma_n$

$$Y_2(\int_0^{T\wedge\tau_m(N^{(n)})} (\lambda^{(n)} - \lambda\wedge\lambda^{(n)})(s,N^{(n)})ds) > 0$$

if and only if

$$Y_2(\int_0^{T\wedge\tau_m(\tilde{N}^{(n)})} (\lambda^{(n)} - \lambda\wedge\lambda^{(n)})(s,\tilde{N}^{(n)})ds) > 0.$$

Finally as $n\to\infty$ the right side of (2.8) goes to zero by (2.1).

3. Density dependent Markov families

We now consider a sequence of processes X_n, with values in $E_n = \{n^{-1}k : k \in \mathbb{Z}^d\}$, satisfying

$$(3.1)\qquad X_n(t) = X_n(0) + \sum_{\ell\in\mathbb{Z}^d} \ell n^{-1} Y_\ell(n\int_0^t \beta_\ell(X_n(s))ds)$$

where the Y_d are independent Poisson processes. As noted in example (c) in Section 1, X_n is a Markov chain. If we set $\tilde{Y}_\ell(u) = Y_\ell(u) - u$

and

(3.2) $$F(x) = \sum_{\ell} \ell\beta_{\ell}(x)$$

then

(3.3) $$X_n(t) = X_n(0) + \sum_{\ell\in\mathbb{Z}^d} \ell n^{-1} \tilde{Y}_{\ell}\left(n\int_0^t \beta_{\ell}(X_n(s))ds\right) + \int_0^t F(X_n(s))ds.$$

Since the law of large numbers gives

(3.4) $$\lim_{n\to\infty} \sup_{u\le u_0} n^{-1}|\tilde{Y}(nu)| = 0, \quad u_0 > 0,$$

if $X_n(0) \to X(0)$ we expect that $X_n(t) \to X(t)$ where

(3.5) $$X(t) = X(0) + \int_0^t F(X(s))ds$$

i.e. $\dot{X} = F(X)$. Similarly, if we define $W_{\ell}^{(n)}(u) = n^{-1/2}\,\tilde{Y}_{\ell}(nu)$, then $W_{\ell}^{(n)} \Rightarrow W_{\ell}$ where W_{ℓ} is a standard Brownian motion. Defining $V_n(t) = n^{1/2}(X_n(t) - X(t))$, we have

(3.6) $$\begin{aligned} V_n(t) &= V_n(0) + \sum_{\ell} \ell\, W_{\ell}^{(n)}\left(\int_0^t \beta_{\ell}(X_n(s))ds\right) \\ &\quad + \int_0^t n^{1/2}(F(X_n(s)) - F(X(s)))ds \\ &= V_n(0) + \sum_{\ell} \ell W_{\ell}^{(n)}\left(\int_0^t \beta_{\ell}(X_n(s))ds\right) \\ &\quad + \int_0^t n^{1/2}(F(X(s) + n^{-1/2}V_n(s)) - F(X(s))ds \end{aligned}$$

and if $V_n(0) \to V(0)$, we expect that $V_n \Rightarrow V$ where

(3.7) $$V(t) = V(0) + \sum_{\ell} \ell\, W_{\ell}\left(\int_0^t \beta_{\ell}(X(s))ds\right) + \int_0^t \partial F(X(s))V(s)ds.$$

Both of the suggested limits hold in considerable generality. See Kurtz (1978) or Kurtz (1981) for details. In particular the limits hold if F is continuously differentiable, the solution of (3.5) exists for all $t > 0$, the β_{ℓ} are continuous and $\sum |\ell|^2 \sup_{x\in K} \beta_{\ell}(x) < \infty$ for each compact K. We will assume these conditions in what follows.

In this section we would like to show how the above results can be used to obtain asymptotic results for hitting distributions of the X_n.

<u>3.1</u> <u>Theorem</u> Let φ be continuously differentiable on $\mathbb{R}^d$ with

$\varphi(X(0)) > 0$. Let

(3.8) $$\tau_n = \inf\{t : \varphi(X_n(t)) \leq 0\}$$

and

(3.9) $$\tau_\infty = \inf\{t : \varphi(X(t)) \leq 0\}.$$

Suppose $\tau_\infty < \infty$ and $\partial\varphi(X(\tau_\infty)) \cdot F(X(\tau_\infty)) < 0$.

Then

(3.10) $$\sqrt{n}(\tau_n - \tau_\infty) \Rightarrow -\frac{\partial\varphi(X(\tau_\infty))\cdot V(\tau_\infty)}{\partial\varphi(X(\tau_\infty)\cdot F(X(\tau_\infty))}$$

and

(3.11) $$\sqrt{n}(X_n(\tau_n) - X(\tau_\infty))$$
$$\Rightarrow V(\tau_\infty) - \frac{\partial\varphi(X(\tau_\infty))\cdot V(\tau_\infty)}{\partial\varphi(X(\tau_\infty))\cdot F(X(\tau_\infty))} F(X(\tau_\infty)).$$

Proof Note that

(3.12) $$\frac{\partial}{\partial t}\varphi(X(t)) = \partial\varphi(X(t))\cdot F(X(t))$$

so the condition that $\partial\varphi(X(\tau_\infty))\cdot F(X(\tau_\infty)) < 0$ implies that for $\epsilon > 0$, $\varphi(X(\tau_\infty - \epsilon)) > 0$ and $\varphi(X(\tau_\infty + \epsilon)) < 0$. Since the convergence of X_n to X is uniform on bounded time intervals $\varphi(X_n(t)) \to \varphi(X(t))$ uniformly on bounded time intervals and it follows that $\tau_n \to \tau_\infty$ in probability.

Next note that $\varphi(X(\tau_\infty)) = 0$ and, since $\varphi(X_n(\tau_n)) \leq 0$ and $\varphi(X_n(\tau_{n-})) \geq 0$

(3.13) $$|\sqrt{n}\,\varphi(X_n(\tau_n))| \leq |\sqrt{n}\,(\varphi(X_n(\tau_n)) - \varphi(X_n(\tau_{n-})))|$$
$$= |\partial\varphi(\theta_n)\cdot(V_n(\tau_n) - V_n(\tau_{n-}))|$$

for some θ_n on the line between $X_n(\tau_{n-})$ and $X_n(\tau_n)$. The right side of (3.13) goes to zero since $\tau_n \to \tau_\infty$, $V_n \Rightarrow V$ and V is continuous. (Note we are appealing to the continuous mapping theorem here, Billingsley (1968), page 30, but see the Remark below.) Consequently

(3.14) $$\sqrt{n}\,(\varphi(X(\tau_\infty)) - \varphi(X(\tau_n)))$$
$$= \sqrt{n}\,(\varphi(X_n(\tau_n)) - \varphi(X(\tau_n))) - \sqrt{n}\,\varphi(X_n(\tau_n))$$
$$= \sqrt{n}\,(\varphi(X(\tau_n) + n^{-1/2}\,V_n(\tau_n)) - \varphi(X(\tau_n)) - \sqrt{n}\,\varphi(X_n(\tau_n))$$
$$\Rightarrow \partial\varphi(X(\tau_\infty))\cdot V(\tau_\infty).$$

Since the left side of (3.14) is asymptotic to

(3.15) $$-\partial\varphi(X(\tau_\infty))\cdot F(X(\tau_\infty))\sqrt{n}\,(\tau_n-\tau_\infty),$$ (3.10) follows.

Finally, noting that $\sqrt{n}(X(\tau_n)-X(\tau_\infty)) \sim F(X(\tau_\infty))\sqrt{n}\,(\tau_n-\tau_\infty)$

(3.16)
$$\begin{aligned}\sqrt{n}\,(X_n(\tau_n) - X(\tau_\infty)) &= \sqrt{n}\,(X_n(\tau_n) - X(\tau_n)) + \sqrt{n}\,(X(\tau_n) - X(\tau_\infty)) \\ &\Rightarrow V(\tau_\infty) - \frac{\partial\varphi(X(\tau_\infty))\cdot V(\tau_\infty)}{\partial\varphi(X(\tau_\infty))\cdot F(X(\tau_\infty))} F(X(\tau_\infty)).\end{aligned}$$

Remark Although the continuous mapping theorem is sufficient to justify the above conclusions, it may be helpful to recall the Skorohod representation theorem which states that if $V_n \Rightarrow V$ then there is a probability space and versions $\tilde{V}_n$ and $\tilde{V}$ of V_n and V such that $\tilde{V}_n \to \tilde{V}$ a.s., which in our case means

(3.17) $$\lim_{n\to\infty} \sup_{s\leq t} |\tilde{V}_n(s) - \tilde{V}(s)| = 0 \qquad \text{a.s.}$$

See Billingsley (1971), page 7. In Kurtz (1978) a construction based on work of Komlos, Mayor, and Tusnady (1975) is given which provides a rate of convergence for (3.17).

Epidemic Model We consider the simplest stochastic epidemic model. In this example, the parameter n can be thought of as the total population size. Two subclasses of the population are considered, the susceptibles and the infectives. Assuming that the probability that a particular susceptible is infected during a short interval of time is proportional to the fraction of the population that is infective, we have transition intensities

(3.18) $$\lambda s \frac{i}{n} = n\lambda \frac{s}{n}\frac{i}{n}$$

for the infection of a susceptible and

(3.19) $$\mu i = n\,\mu\,\frac{i}{n}$$

for the recovery of an infective, where i is the number of infectives and s the number of susceptibles. Assuming recovered infectives are immune, we have the Markov chain given by

(3.20) $$I_n(t) = I_n(0) + n^{-1} Y_1(n\int_0^t \lambda S_n(s)I_n(s)\,ds) - n^{-1} Y_2(n\int_0^t \mu I_n(s)\,ds)$$

and

(3.21) $$S_n(t) = S_n(0) - n^{-1} Y_1(n\int_0^t \lambda S_n(s)I_n(s)\,ds)$$

where I_n and S_n are the fractions of the population that are infective and susceptible. Letting $n \to \infty$ the limiting deterministic model is

(3.22) $$I(t) = I(0) + \int_0^t (\lambda S(s)I(s) - \mu I(s))ds$$

(3.23) $$S(t) = S(0) - \int_0^t \lambda S(s)I(s)ds.$$

Note that the Markov chain eventually absorbs on the boundary with $I_n = 0$. We want to obtain results of Nagaev and Startsev (1970) on the asymptotics of the number of susceptibles remaining after absorption. Unfortunately, Theorem 3.1 is not directly applicable since $\tau_\infty = \infty$. However, since we are only interested in where the process absorbs we can speed up the process without affecting where it hits the boundary. In particular let γ_n satisfy

(3.24) $$\int_0^{\gamma_n(t)} I_n(s)ds = t, \quad 0 < t \leq \int_0^\infty I_n(s)ds$$

and define $\tilde{I}_n(t) = I_n(\gamma_n(t))$ and $\tilde{S}_n(t) = S_n(\gamma_n(t))$. Then, substituting in (3.20) and (3.21)

(3.25) $$\tilde{I}_n(t) = I_n(0) + n^{-1} Y_1(n \int_0^t \lambda\tilde{S}_n(s)ds) - n^{-1} Y_2(n\mu t)$$

and

(3.26) $$\tilde{S}_n(t) = S_n(0) - n^{-1} Y_1(n \int_0^t \lambda\tilde{S}_n(s)ds)$$

for $t \leq \int_0^\infty I_n(s)ds$. Of course (3.25) and (3.26) have a unique solution for all $t \geq 0$, so we will assume $\tilde{I}_n$ and $\tilde{S}_n$ are defined for all $t \geq 0$. Let

(3.27) $$\tau_n = \inf\{t:\tilde{I}_n(t) = 0\}$$

and note that $\tilde{S}_n(\tau_n)$ is the number of susceptibles remaining in the original epidemic model after absorption.

Theorem 3.1 is applicable to (I_n, S_n). The limiting deterministic model is now given by

(3.28) $$\tilde{I}(t) = I(0) + \int_0^t \lambda\tilde{S}(s)ds - \mu t$$

and

(3.29) $$\tilde{S}(t) = S(0) - \int_0^t \lambda\tilde{S}(s)ds,$$

so

(3.30) $$\tilde{S}(t) = S(0)e^{-\lambda t}$$

and

(3.31) $$\tilde{I}(t) = I(0) + S(0)(1 - e^{-\lambda t}) - \mu t.$$

Consequently τ_∞ is the solution of

(3.32) $$I(0) + S(0)(1 - e^{-\lambda\tau_\infty}) - \mu\tau_\infty = 0.$$

We also have $(\sqrt{n}(\tilde{I}_n - \tilde{I}), \sqrt{n}(\tilde{S}_n - \tilde{S})) \Rightarrow (V_I, V_S)$

where

(3.33) $$V_I(t) = V_I(0) + W_1(S(0)(1 - e^{-\lambda t})) - W_2(\mu t) + \int_0^t \lambda V_S(s)ds$$

and

(3.34) $$V_S(t) = V_S(0) - W_1(S(0)(1 - e^{-\lambda t})) - \int_0^t \lambda V_S(s)ds.$$

Then V_S and V_I can be written

(3.35) $$V_S(t) = e^{-\lambda t} V_S(0) + \int_0^t e^{-\lambda(t-s)} \sqrt{S(0)(1-e^{-\lambda s})}\, d\tilde{w}(s).$$

and

(3.36) $$V_I(t) = V_I(0) + V_S(0)(1 - e^{-\lambda t}) - \int_0^t e^{-\lambda(t-s)} \sqrt{S(0)(1-e^{-\lambda s})}\, d\tilde{w}(s) - W_2(\mu t),$$

where $\tilde{w}$ is a standard Brownian motion.

Finally we have

(3.37) $$\sqrt{n}(\tilde{S}_n(\tau_n) - \tilde{S}(\tau_\infty)) \Rightarrow V_S(\tau_\infty) + \lambda\tilde{S}(\tau_\infty)V_I(\tau_\infty)/(\lambda\tilde{S}(\tau_\infty) - \mu).$$

The mean and variance for the limiting Gaussian distribution can be calculated easily in terms of τ_∞.

The limit (3.37) is a special case of results of Nagaev and Startsev (1970). Because of the time substitution, the asymptotic behavior of τ_n tells essentially nothing about the duration of the epidemic. Some asymptotic results for the duration are given in Nagaev and Mukhomor (1975). For more information on epidemic models see Bailey (1975).

Repairman Model We now consider a repair facility model of Iglehart and Lemoine (1974). (See also Lemoine (1978).) The model involves an operating facility and two repair facilities. When operating at full capacity, the operating facility requires n operating units. There are $n + m_n$ operating units available if none are being repaired (i.e. there are m_n spare units.) The time until failure of a unit is exponentially distributed with parameter λ and a failure is of type 1 and is sent to the first repair facility with probability p and is of type two and sent to the second repair facility with probability $1-p$. The repair facilities have s_n^1 and s_n^2 repairmen respectively, and the service times in each facility are exponentially distributed with parameters μ_1 and μ_2 respectively. Let X_n^1 and X_n^2 be the numbers of units in the repair facilities normalized by n. Then setting $\lambda_1 = \lambda p$ and $\lambda_2 = \lambda(1-p)$

$$(3.38)\qquad X_n^i(t) = X_n^i(0) + n^{-1}Y_{1i}\left(\int_0^t \lambda_i n\wedge(n+m_n-nX_n^1(u)-nX_n^2(u))du\right)$$
$$- n^{-1}Y_{2i}\left(\int_0^t \mu_i\, S_n^i\wedge(nX_n^i(u))du\right)$$
$$= X_n^i(0) + n^{-1}Y_{1i}\left(n\int_0^t\lambda_i\, 1\wedge(1+n^{-1}m_n-X_n^1(u)-X_n^2(u))du\right)$$
$$- n^{-1}Y_{2i}\left(n\int_0^t \mu_i(n^{-1}S_n^i)\wedge X_n^i(u)du\right)$$

where Y_{11}, Y_{21}, Y_{12}, Y_{22} are independent Poisson processes. Assume that $n^{-1}m_n = m + O(1/n)$ and $n^{-1}S_n^i = S^i + O(1/n)$. Then the limiting deterministic model is

$$(3.39)\qquad X^i(t) = X^i(0) + \int_0^t (\lambda_i 1\wedge(1+m-X^1(u)-X^2(u))-\mu_i S^i\wedge X^i(u))du.$$

Since $F_i(x) = \lambda_i\, 1\wedge(1+m-x_1-x_2) - \mu_i S^i\wedge x_i$ is Lipschitz the "law of large numbers" (for example in Kurtz (1981)) applies and assuming $X_n^i(0) \to X^i(0)$, $i = 1,2$ we have

$$(3.40)\qquad \sup_{u\le t} |X_n^i(u) - X^i(u)| \to 0 \quad \text{in probability.}$$

The "central limit theorem" does not quite apply since F_i is not continuously differentiable. But the same proof gives the following. Let

$$(3.41)\qquad V_n^i(t) \equiv \sqrt{n}(X_n^i(t) - X^i(t))$$
$$= V_n^i(0) + W_{1i}^{(n)}\left(\int_0^t \lambda_i 1\wedge(1+n^{-1}m_n-X_n^1(u)-X_n^2(u))du\right)$$
$$- W_{2i}^{(n)}\left(\int_0^t \mu_i(n^{-1}S_n^i)\wedge X_n^i(u)du\right)$$
$$+ \int_0^t \lambda_i\sqrt{n}(1\wedge(1+n^{-1}m_n-X_n^1(u)-X_n^2(u))-1\wedge(1+m-X^1(u)-X^2(u)))du$$
$$- \int_0^t \mu_i\sqrt{n}((n^{-1}S_n^i)\wedge X_n^i(u)-S^i\wedge X^i(u))du \quad .$$

Assuming $V_n^i(0) \to V^i(0)$, $i = 1,2$ the weak convergence of $W_{1i}^{(n)}$ and $W_{2i}^{(n)}$ and (3.40) imply $V_n^i \Rightarrow V^i$ where

$$(3.42) \qquad V^i(t) = V^i(0) + W_{1i}\left(\int_0^t \lambda_i 1\wedge(1+m-X^1(u)-X^2(u))du\right)$$
$$- W_{2i}\left(\int_0^t \mu_i S^i \wedge X^i(u)du\right)$$
$$- \int_0^t \lambda_i(\chi_{\{m\}}(X^1(u)+X^2(u))(V^1(u)+V^2(u)) \vee 0$$
$$+ \chi_{(m,\infty)}(X^1(u)+X^2(u))(V^1(u)+V^2(u)))du$$
$$- \int_0^t \mu_i(\chi_{[0,S^i)}(X^i(u))V^i(u) + \chi_{\{S^i\}}(X^i(u))V^i(u)\wedge 0)du$$

($\chi_A(z) = 1$ if $z \in A$ and $= 0$ otherwise). Since the integral term is not linear in V, V may not be Gaussian. But, except for very special cases of the parameters, the Lebesgue measure of $\{u: X^1(u) + X^2(u) = m$ or $X^i(u) = S^i\}$ is zero and the nonlinear terms can be dropped. We do not want to go into an exhaustive case by case analysis here (see Lemoine (1978)), but if $S^1 \neq \lambda_1/\mu_1$, $S^2 \neq \lambda_2/\mu_2$ and $S^2\mu_2/\lambda_2 \neq S^1\mu_1/\lambda_1$ then the deterministic model has a unique asymptotically stable fixed point (x_1, x_2) and, assuming $x_i \neq S_i$ $i = 1,2$, a theorem of Norman (1974) (see Kurtz (1981)) implies that the convergence in distribution $(V_n^1(t), V_n^2(t)) \Rightarrow (V^1(t), V^2(t))$ is uniform for all $t \geq 0$. In particular the stationary distribution for (V_n^1, V_n^2) converges to the (Gaussian) stationary distribution for (V^1, V^2).

4. Fiber Bundles

In order to give another example of an argument similar to those used in Section 3, we consider the simplest example of the fiber bundle models studied by Phoenix and Taylor (1973) and Phoenix (1979). In fact the particular result we give goes back to Daniels (1945).

We consider a bundle of n fibers under a load nL. We assume that all fibers share the load equally (i.e. initially each fiber is subjected to a load L). Under this load a number of fibers $N_n(L)$ will break leaving $n - N_n(L)$ fibers to support the load, and hence the load on each remaining fiber is $nL/(n-N_n(L)) = L/(1-X_n(L))$ where $X_n(L) = n^{-1}N_n(L)$ is the fraction of fibers that have broken. Finally assume that a fiber subjected to a load ℓ breaks with probability $F(\ell)$ and that the fibers break independently of each other. We can construct this model by associating with each fiber an independent random variable ξ_k, uniformly distributed on $[0,1]$. If the kth fiber is subjected to a load ℓ, then it breaks if $\xi_k \leq F(\ell)$. Define the empirical process

$$(4.1) \qquad Y^{(n)}(u) = \#\{k:\xi_k \le u\}.$$

Then $N_n(L)$ must satisfy

$$(4.2) \qquad N_n(L) = Y^{(n)}(F(\frac{nL}{n-N_n(L)}))$$

or, equivalently, $X_n(L)$ satisfies

$$(4.3) \qquad X_n(L) = n^{-1}Y^{(n)}(F(\frac{L}{1-X_n(L)})).$$

Unfortunately (4.3) may not have a unique solution so we must specify that $X_n(L)$ is the smallest solution, if it exists, of (4.3). The analogy with the limit theorems of Section 3 is clear. We know that

$$(4.4) \qquad \lim_{n\to\infty} \sup_{0\le u\le 1} |n^{-1} Y^{(n)}(u) - u| = 0$$

and, defining $W_n(u) = n^{-1/2}(Y^{(n)}(u) - nu)$, that $W_n \Rightarrow W_B$ where W_B is Brownian bridge (see for example Billingsley (1968) page 64). The limiting deterministic model is

$$(4.5) \qquad X(L) = F(\frac{L}{1-X(L)})$$

(again take $X(L)$ to be the smallest solution, if one exists, of (4.5)). Assume that F is continuously differentiable. Then it is not difficult to see that $X_n(L) \to X(L)$ for each L such that $X(L)$ exists and

$$(4.6) \qquad (1 - X(L))^2 - LF'(\frac{L}{1-X(L)}) > 0.$$

Assuming (4.6), then setting $\tilde{Y}^{(n)}(u) = Y^{(n)}(u)-nu$

$$(4.7) \qquad V_n(L) \equiv \sqrt{n}\,(X_n(L) - X(L))$$
$$= n^{-1/2}\tilde{Y}^{(n)}(F(\frac{L}{1-X_n(L)}))$$
$$+ \sqrt{n}(F(\frac{L}{1-X_n(L)}) - F(\frac{L}{1-X(L)}))$$

and it follows that $V_n(L) \Rightarrow V(L)$ where $V(L)$ satisfies

$$(4.8) \qquad V(L) = W_B(F(\frac{L}{1-X(L)}))$$
$$+ F'(\frac{L}{1-X(L)}) \frac{L}{(1-X(L))^2} V(L),$$

that is

$$(4.9) \qquad V(L) = \frac{(1-X(L))^2 W_B(F(\frac{L}{1-X(L)}))}{(1-X(L))^2 - LF'(\frac{L}{1-X(L)})}.$$

Finally consider the maximum load the bundle will support, that is the maximum L for which (4.3) has a solution. Rewriting (4.3) we see that

(4.10) $$L = \frac{L}{1-X_n(L)} \left(1 - n^{-1}\, Y^{(n)} \left(F\left(\frac{L}{1-X_n(L)}\right)\right)\right)$$

and if L_n^* is the maximum load, then

(4.11) $$L_n^* = \sup_u u(1 - n^{-1}\, Y^{(n)}(F(u))).$$

Similarly define

(4.12) $$L^* = \sup_u u(1 - F(u)).$$

Then

(4.13) $$\sqrt{n}\,(L_n^* - L^*)$$

$$= \sup_u [-un^{-1/2}\, \tilde{Y}^{(n)}(F(u)) - \sqrt{n}\,(L^* - u(1-F(u)))].$$

Noting that $L^* \geq u(1-F(u))$ we conclude that

(4.14) $$\sqrt{n}\,(L_n^* - L^*) \Rightarrow \sup_{u\in\Gamma}(-uW_B(F(u)))$$

where $\Gamma = \{u : u(1-F(u)) = L^*\}$.

References

1. Bailey, Norman T. J. (1975). The Mathematical Theory of Infectious Diseases. Griffin, London.

2. Billingsley, Patrick (1968). Convergence of Probability Measures: John Wiley, New York.

3. Billingsley, Patrick (1971). Weak Convergence of Measures: Applications in Probability. SIAM, Philadephia.

4. Brémaud, Pierre (1981). Point Processes and Queues: Martingale Dynamics. Springer-Verlag, New York.

5. Daniels, H. E. (1945). The statistical theory of strength of bundles of threads. Proc. R. Soc. London A 183, 405-435.

6. Iglehart, Donald L. and Austin J. Lemoine (1974). Approximations for the repairman problem with two repair facilities, II: Spaces. Adv. App. Prob. 6, 147-158.

7. Karlin, Samuel (1966). A First Course in Stochastic Processes. Academic Press, New York.

8. Komlos, J., P. Major, and G. Tusnady (1975). An approximation of partial sums of independent random variables and the sample distribution function, I. Z. Wahr und Verw. Gebiete 32, 111-131.

9. Kurtz, Thomas G. (1978). Strong approximation theorems for density dependent Markov chains. Stochastic Processes Appl. 6, 223-240.

10. Kurtz, Thomas G. (1980). The optional sampling theorem for martingales indexed by directed sets. Ann. Probability, 8, 675-681.

11. Kurtz, Thomas G. (1981). Approximation of Population Processes. SIAM, Philadelphia.

12. Kurtz, Thomas G. (1982). Counting processes and multiple time changes. (in preparation)

13. Lemoine, Austin J. (1978). Networks of queues - A survey of weak convergence results. Management Science, 24, 1175-1193.

14. Nagaev, A. V. and T. P. Mukhomor (1975). A limit distribution of the duration of an epidemic. Theory Prob. Applications 20, 805-818.

15. Nagaev, A. V. and A. N. Startsev (1970). The asymptotic analysis of a stochastic model of an epidemic. Theory Prob. Applications; 15, 98-107.

16. Norman, M. Frank (1974). A central limit theorem for Markov processes that move by small steps. Ann. Probability, 2, 1065-1074.

17. Phoenix, S. Leigh (1979). The asymptotic distribution for the time to failure of a fiber bundle. Adv. Appl. Prob 11, 153-187.

18. Phoenix, S. Leigh and Howard M. Taylor (1973). The asymptotic strength distribution of a general fiber bundle. Adv. Appl. Prob. 5, 200-216.

19. Snyder, Donald L. (1975). Random Point Processes. John Wiley, New York.

APPROXIMATE INVARIANT MEASURES FOR THE ASYMPTOTIC DISTRIBUTIONS OF DIFFERENTIAL EQUATIONS WITH WIDE BAND NOISE INPUTS*

Harold J. Kushner
Division of Applied Mathematics
Brown University
Providence, Rhode Island 02912

ABSTRACT

Diffusion models are useful and of widespread use in many areas of control and communication theory. The models are frequently used as approximations to continuous or discrete parameter systems which are not quite diffusions but are, hopefully, close to a diffusion in some sense. For example, the input noise might be 'wide-band'--but not 'white-Gaussian'. Many approximation techniques have been developed and the typical results are of a weak convergence nature. The physical process $x^{\varepsilon}(\cdot)$ is parameterized by ε, and one tries to show that $\{x^{\varepsilon}(\cdot)\}$ converges weakly to some diffusion $x(\cdot)$ as $\varepsilon \to 0$. The limit process $x(\cdot)$ is then used to study various properties of $x^{\varepsilon}(\cdot)$ for small ε. Frequently, in applications, we are concerned with asymptotic properties, as $t \to \infty$ (for small ε), as well as with weak convergence. Such information is not normally provided by the weak convergence theory. We discuss the problem of approximating functionals on the 'tail' of $x^{\varepsilon}(\cdot)$ for small ε, by such functionals on the 'tail' of $x(\cdot)$; e.g., approximating the measures of $\{x^{\varepsilon}(t),$ large $t\}$, for small ε, by an invariant measure of $x(\cdot)$. This is particularly useful in problems in communication theory, where the (say) detection system is often supposed to be in operation for a very long time.

*This research was supported in part by the Air Force Office of Scientific Research under AFOSR-76-3063D, in part by the National Science Foundation under NSF-Eng.77-12946-A02, and in part by the Office of Naval Research under N00014-76-C-2079-P0004.

1. Introduction

Let $\{x^{\varepsilon}(\cdot)\}$ be solutions to ordinary differential equations with random right hand sides, e.g., where $\dot{x}^{\varepsilon} = F_{\varepsilon}(x^{\varepsilon},\xi^{\varepsilon})$ for some function $F_{\varepsilon}(\cdot,\cdot)$ and a "wide band" noise process $\xi^{\varepsilon}(\cdot)$. Many results are available concerning the weak convergence of $\{x^{\varepsilon}(\cdot)\}$ to a diffusion $x(\cdot)$. The weak convergence basically gives us information on the approximation of $x^{\varepsilon}(\cdot)$ by $x(\cdot)$ on arbitrarily large but still finite time intervals. In applications to control and communication theory, information concerning the closeness of the distributions of the $x^{\varepsilon}(t)$ for large t, small ε, to $\mu(\cdot)$, the invariant measure of $x(\cdot)$, is of considerable interest. Some results along this line were obtained in [2, Section 6] for the system

$$(1.1)\qquad \dot{x}^{\varepsilon} = F(x,\xi^{\varepsilon})/\varepsilon + G(x,\xi^{\varepsilon}) + \overline{G}(x),\quad x^{\varepsilon}(t) \in R^{r},$$

where $EF(x,\xi)=EG(x,\xi)=0$, $\xi^{\varepsilon}(t)=(\xi\, t/\varepsilon^{2})$, and $\xi(\cdot)$ is a Markov jump process. The weak limit of $\{x^{\varepsilon}(t)\}$ is a diffusion with differential generator

$$(1.2)\qquad \mathscr{L}f(x) = f_x'(x)\overline{G}(x) + \int_0^{\infty} E(f_x'(x)F(x,\xi(t)))_x' F(x,\xi(0))dt.$$

Suppose that $x(\cdot)$ has a unique invariant measure $\mu(\cdot)$, and let there be a smooth Liapunov function satisfying $0 \leq V(x) \to \infty$ as $|x| \to \infty$, and a $\gamma > 0$ such that

$$(1.3)\qquad \mathscr{L}V(x) \leq -\gamma V(x),\ \text{for large } |x|.$$

Then, for small ε, $(x^{\varepsilon}(\cdot), \xi^{\varepsilon}(\cdot))$ has an invariant measure $\nu^{\varepsilon}(\cdot)$ whose x-marginals $\mu^{\varepsilon}(\cdot)$ converge weakly to $\mu(\cdot)$ as $\varepsilon \to 0$ [2, Section 6].

In this paper, $\xi^{\varepsilon}(\cdot)$ need not be Markov, F or G might not be smooth and (1.3) is replaced by a weaker condition. Equation (1.3) would not normally hold if $\overline{G}$ were bounded, for example. The basic techniques used here are similar to those in [2]; both heavily depending on the use of "averaged Liapunov functions."

Section 2 contains the basic approximation theorem, but using the condition (A5) which is not usually directly verifiable. A verifiable condition for (A5) is then given.

We present only some of the theorems and their assumptions. A fuller development is in [6] which contains all proofs, examples, and extensions to the cases of unbounded noise, discontinuous dynamical terms and state dependent noise.

2. The Basic Convergence Theorem

Assumptions

A1. For each $x(0)$, $x(\cdot)$ is a Feller Markov diffusion process with continuous coefficients.

A2. $x(\cdot)$ has a unique invariant measure $\mu(\cdot)$ and $\mu(x,t,\cdot)$, the measure of $x(t)$ when $x(0) = x$, converges weakly to $\mu(\cdot)$ for each x, as $t \to \infty$.

A3. The convergence in (A2) is uniform in compact x-sets; i.e., for each $f(\cdot) \in \mathscr{C}_r$, the space of continuous bounded functions on R^r, $E_x f(x(t)) \to E_\mu f(x(0))$ uniformly in x in compact sets, where E_μ denotes expectation with respect to the stationary measure of $x(\cdot)$.

A4. If $t_\varepsilon \to t_0 \leq \infty$ and $x^\varepsilon(t_\varepsilon) \to x(0)$ weakly, then $x^\varepsilon(t_\varepsilon+\cdot) \to x(\cdot)$ weakly. (The usual conditions which imply this for $t_0=0$, also imply it as stated.)

A5. There is an $\varepsilon_0 > 0$ such that $\{x^\varepsilon(t),\ 0 < \varepsilon \leq \varepsilon_0,\ t \geq 0\}$ is tight. (See Theorem 3 for a verifiable criterion for (A5).)

Lemma 1. For any integer m, let $f(\cdot) \in \mathscr{C}_{mr+r}$. Assume (A1)-(A3) and let $0 = \Delta_0 < \Delta_1 \ldots < \Delta_m$. Let S denote a tight set of R^r valued random variables. Then

$$E_{x(0)} f(x(t + \Delta_i),\ i \leq m) \to E_\mu f(x(\Delta_i),\ i \leq m)$$

uniformly for $x(0) \in S$, as $t \to \infty$.

Theorem 1 is the basic convergence theorem.

Theorem 1. Assume (A1)-(A5). Then for each $f(\cdot) \in \mathscr{C}_{rm+r}$, $T < \infty$ and $\delta > 0$, there are $t_0(f,\delta) < \infty$ and $\varepsilon_0(f,\delta) > 0$ such that for all $t \geq t_0(f,\delta)$, and $\Delta_i \leq T$, and $\varepsilon \leq \varepsilon_0(f,\delta)$, and any sequence $\{x^\varepsilon(\cdot)\}$ which converges weakly to $x(\cdot)$,

$$|Ef(x^\varepsilon(t+\Delta_i),\ i \leq m) - E_\mu f(x(\Delta_i),\ i \leq m)| < \delta. \tag{2.1}$$

Let $(x^\varepsilon(\cdot), \xi^\varepsilon(\cdot))$ be Markov and have an invariant measure $\nu^\varepsilon(\cdot)$.

Replace (A5) by (A5'): There is a sequence $T_\varepsilon \to \infty$ (T_ε can depend on the initial condition $x^\varepsilon(0), \xi^\varepsilon(0)$) such that (A5) holds for $t \geq T_\varepsilon$. Then $\{\mu^\varepsilon(\cdot)\}$, the x-marginals of $\{v^\varepsilon(\cdot)\}$ are tight. In addition, let (A4) hold when $(x^\varepsilon(0), \xi^\varepsilon(0))$ has the distribution $v^\varepsilon(\cdot)$. Then $\{\mu^\varepsilon(\cdot)\}$ converges weakly to $\mu(\cdot)$ as $\varepsilon \to 0$.

Remark. The theorem implies that the convergence as $t \to \infty$ in (2.1) is uniform in ε for small ε, a fact which is important in applications. In applications, it is often possible to prove results such as $\overline{\lim_{t\to\infty}}\ E_{x,\xi}|x^\varepsilon(t)| \leq K$, where for small ε, K does not depend on $\varepsilon, \xi = \xi^\varepsilon(0)$, or x. Then the replacement (A5') for (A5) in the last paragraph holds. Under (A5'), $(x^\varepsilon(\cdot),\xi^\varepsilon(\cdot))$ has an invariant measure $v^\varepsilon(\cdot)$ for small ε, if for some $\xi^\varepsilon(0)$, $x^\varepsilon(0)$, $\{\xi^\varepsilon(t), t \geq 0\}$ is tight for each small ε.

Proof. Suppose that (2.1) is false. Then there is a subsequence $\varepsilon \to 0$ and a sequence $\{t_\varepsilon\} \to \infty$ such that

$$(2.2)\qquad |Ef(x^\varepsilon(t_\varepsilon+\Delta_i), i \leq m) - E_\mu f(x(\Delta_i), i \leq m)| \geq \delta > 0.$$

We will find a further subsequence, also indexed by ε, which violates (2.2). Fix $T > 0$. By (A5), we can choose a further subsequence such that $\{x^\varepsilon(t_\varepsilon - T)\}$ converges weakly to a random variable $x(0)$. By (A4), $\{x^\varepsilon(t_\varepsilon - T + \cdot)\}$ converges weakly to $x(\cdot)$ with initial condition $x(0)$ and

$$(2.3)\qquad Ef(x^\varepsilon(t_\varepsilon - T + T + \Delta_i), i \leq m) \to EE_{x(0)}f(x(T + \Delta_i), i \leq m).$$

By (A5), the set S of all possible $x(0)$ (over all $T > 0$ and weakly convergent subsequences) is tight. By Lemma 1, we can take T large enough such that

$$(2.4)\qquad |EE_{x(0)}f(x(T + \Delta_i), i \leq m) - E_\mu f(x(\Delta_i), i \leq m)| < \delta/2.$$

Equations (2.3) and (2.4) contradict (2.2).

The proof of the last assertion is similar to the last part of the proof of Theorem 4 of [6] and is omitted.

Q.E.D.

3. A Liapunov Function Criterion for Tightness of $\{x(t), t \geq 0\}$

Here, we state conditions which guarantee (3.1).

(3.1) $\{x(t),\ t \geq 0,\ x(0) \in B\}$ is tight for each compact B. To prove (3.1), condition (A6) is required.

A6. There is a continuous Liapunov function $0 \leq V(x) \to \infty$ as $|x| \to \infty$ and a λ_0 and $\alpha_0 > 0$ such that $\mathscr{L}V(x) \leq -\alpha_0$ for $x \notin Q_0 = \{x : V(x) \leq \lambda_0\}$. The partial derivatives of $V(\cdot)$ up to order 2 are continuous.

The proof of Theorem 2 is a prototype of the technique used to verify (A5). See [6].

Theorem 2. Under (A1) and (A6), condition (3.1) holds.

Comment on the proof. (A6) implies that Q_0 is a recurrence set for $x(\cdot)$. Let τ_0 and τ_1 be stopping times such that $V(x(\tau_0)) \in \partial Q_1 = \{x : V(x) = \lambda_1 > \lambda_0\}$, and let $\tau_1 > \tau_0$ be the next return time of $x(\cdot)$ to ∂Q_0. For $x(t) \notin Q_0$, $\{V(x(t))\}$ is a 'supermartingale' and probability estimates can be obtained of the maximum excursion of $V(x(\cdot))$ on any such interval $[\tau_0, \tau_1]$. Given $\delta > 0$, we find numbers k, λ (not depending on t) such that for each t there are k 'return intervals' satisfying

$$P\{\ t \notin \text{these } k \text{ return intervals}\ \} \leq \delta/2$$
$$P\{\ V(x(t)) \geq \lambda \text{ on these } k \text{ return intervals}\} \leq \delta/2.$$

This yields the desired tightness.

4. An Averaged Liapunov Function Criterion for (A5).

In this section, we use the model (1.1), smooth F, G, $\bar{G}$ and a strong mixing condition on $\xi(\cdot)$. The result and techniques should be viewed as an illustration of the general possibilities. The mixing condition is too strong for many applications, and other conditions on $\xi^{\varepsilon}(\cdot)$ and on F, G, $\bar{G}$ are considered in [6]. In order to get the necessary inequalities for any Liapunov function based approach, an assumption such as (B4) seems to be required. The conditions hold in numerous cases of interest.

B1. $\xi(\cdot)$ is a bounded, right continuous, stationary ϕ-mixing process [7] with $\int_0^\infty \phi^{1/2}(t)dt < \infty$.

B2. $F(\cdot,\cdot)$, $G(\cdot,\cdot)$ and $\bar{G}(\cdot)$ are continuous, R^r valued functions whose growth (as $|x| \to \infty$) is $0(|x|)$. The partial derivatives of $F(\cdot,\xi)$ up to order 2 (and of $G(\cdot,\xi)$ up to order 1) are bounded uniformly in x,ξ, and $EF(x,\xi) \equiv 0 \equiv EG(x,\xi)$.

B3. There is a diffusion process $x(\cdot)$ with differential generator defined by (1.2), and which satisfies (A1)-(A3). Also, (A6) holds, but the partial derviatives of $V(\cdot)$ up to order 3 are continuous.

B4. There are constants K such that, uniformly in x,ξ,

(4.1a) $$|V_x'(x)G(x,\xi)| + |V_x'(x)F(x,\xi)| \le K(1 + V(x))$$

(4.1b) $$|(V_x'(x)F(x,\xi))_x'F(x,\xi)| \le K(1 + V(x))$$

(4.2) $$|(V_x'(x)G(x,\xi))_x'U(x,\xi)| \le K(1 + |\mathscr{L}V(x)|) \quad \text{for} \quad U = F,G,\bar{G}.$$

(4.3) $$|(V_x'(x)F(x,\xi))_x'U(x,\xi)| \le K(1 + |\mathscr{L}V(x)|), \quad U = G,\bar{G}$$

(4.4) $$|((V_x'(x)F(x,\xi))_x'F(x,\xi))_x'U(x,\xi)| \le K(1 + |\mathscr{L}V(x)|), \quad U = F,G,\bar{G}.$$

Theorem 3. Under (B1)-(B4) and the tightness of $\{x^\varepsilon(0)\}$, condition (A5) holds.

Remarks on Theorem 3. (B4) fits many examples. In a sense it is a prototype condition for two typical cases, where the orders would often be as required by (B4): (a) $F,G,\bar{G}$ increase roughly (at most) linearly in x and V increases roughly quadratically in x as $|x| \to \infty$; (b) $F,G,\bar{G}$ are bounded and V increases roughly linearly in $|x|$ for large x.

In the proof of Theorem 3, an 'averaged' Liapunov function $V^\varepsilon(\cdot)$ is obtained from $V(\cdot)$ for small $\varepsilon > 0$, $V^\varepsilon(x^\varepsilon(t),t)$ is a 'supermartingale' on the time intervals during which $x^\varepsilon(t) \notin Q_0$. This is used to prove recurrence of $x^\varepsilon(\cdot)$, and to get probability estimates on the path excursions of $x^\varepsilon(\cdot)$ on trips from $\partial Q_1 = \{x: V(x) = \lambda_1 > \lambda_0\}$ to ∂Q_0. Then we apply a technique similar to that used to complete the tightness proof in Theorem 2.

REFERENCES

[1] R. Z. Khazminskii, "A limit theorem for solutions of differential equations with a random right hand side." Theory of Probability and Applic., 11, 1966, p. 390-406.

[2] G. Blankenship, G. C. Papanicolaou, "Stability and control of stochastic systems with wide-band noise disturbances." SIAM J. on Appl. Math., 34, 1978, pp. 437-476.

[3] G. C. Papanicolaou, W. Kohler, "Asymptotic theory of mixing ordinary stochastic differential equations." Comm. Pure and Applied Math., 27, 1974, pp. 641-668.

[4] H. J. Kushner, "Jump-diffusion approximations for ordinary differential equations with wide-band random right hand sides." SIAM J. on Control and Optimization, 17, 1979, p. 729-744.

[5] H. J. Kushner, "A martingale method for the convergence of a sequence of processes to a jump-diffusion process." Z. Wahrscheinlichkeitsteorie, 53, 1980, p. 209-219.

[6] H. J. Kushner, "Asymptotic distributions of solutions of ordinary differential equations with wide band noise inputs; approximate invariant measures." To appear in Stochastics, early 1982.

[7] P. Billingsley, Convergence of Probability Measures, John Wiley, New York, 1968.

OPTIMAL STOCHASTIC CONTROL OF DIFFUSION TYPE PROCESSES

AND HAMILTON-JACOBI-BELLMAN EQUATIONS

P.L. Lions
Ceremade , Paris IX University
Place de Lattre de Tassigny
75775 Paris Cedex 16
France

I. Introduction :

In this paper we present a general approach and several results (obtained in particular by the author) concerning general optimal stochastic control problems and more precisely the associated Hamilton-Jacobi-Bellman equations (also called the dynamic programming equations).

Let us first describe briefly the type of problems we consider : the state of the system we want to control is given by the solution of the following stochastic differential equation :

$$(1)\qquad \begin{cases} dy_x(t) = \sigma(y_x(t),v(t,\omega))dW_t + b(y_x(t),v(t,\omega))dt \\ y_x(0) = x \in \overline{\mathcal{O}} \end{cases}$$

where $\mathcal{O}$ is a smooth domain in R^N ; W_t is a Brownian motion in R^p ; $\sigma(x,v)$ is a matrix-valued function from $R^N \times V$; $b(x,v)$ is a vector-valued function from $R^N \times V$, V is a separable metric space. We assume that (1) takes place in a probability space (Ω,F,F_t,P) having the usual properties. And $v(t,\omega)$ (called the control process) is any progressively measurable process with values in a compact subset of V (which may of course depend on v). We will call an admissible system the collection : $\left(\Omega,F,F_t,P,W_t,v(\cdot),(y_x(\cdot))_{x\in\overline{\mathcal{O}}}\right) = A$.

For each admissible system A , we define a cost function :

$$(2)\qquad J(x,A) = E\int_0^{\tau_x} f(y_x(t),v(t)) \exp\left(-\int_0^t c(y_x(s),v(s))ds\right)$$

where $f(x,v)$, $c(x,v)$ are real-valued functions from $\bar{\mathcal{O}} \times V$ and τ_x is the first exit time of the process $y_x(t,\omega)$ from $\bar{\mathcal{O}}$. To simplify the presentation, we will assume throughout this paper : for all $\varphi = \sigma, b, c, f$

$$(3) \qquad \begin{cases} \varphi(\cdot,v) \in W^{2,\infty}(R^N) \quad , \quad \sup\limits_{v \in V} \|\varphi(\cdot,v)\|_{W^{2,\infty}} < \infty \\ \forall\, x \in R^N \quad , \quad \varphi(x,\cdot) \in C(V) \quad ; \end{cases}$$

$$(4) \qquad \lambda = \inf\,\{c(x,v) \,/\, x \in R^N \,,\, v \in V\} > 0 \,.$$

In particular (4) insures that $J(x,A)$ has a meaning.

We want to minimize $J(x,A)$ over all possible admissible systems A that is we consider the minimum cost function - also called the value function or the criter rion :

$$(5) \qquad u(x) = \inf_{A} \, J(x,A) \,.$$

It is a well-known consequence of the dynamic programming principle (due to R. Bellman [2]) that u should be "related to the solution" of the following non-linear second-order elliptic equation :

$$(6) \qquad \sup_{v \in V} \,\{A^v u(x) - f(x,v)\} = 0 \qquad \text{in } \mathcal{O}$$

and u should vanish on $\partial\mathcal{O}$ or on some portion of $\Gamma = \partial\mathcal{O}$:

$$(7) \qquad u = 0 \qquad \text{on } \Gamma \;.$$

Here and below A^v denotes the 2nd order elliptic operator (eventually degenerate) defined by :

$$(8) \qquad A^v = -\sum_{i,j} a_{ij}(x,v)\,\partial_{ij} - \sum_i b_i(x,v)\,\partial_i + c(x,v)$$

and the matrix $a(x,v)$ is given by : $a = \frac{1}{2}\sigma\,\sigma^T$.

The equation (6) is called the Hamilton-Jacobi-Bellman equation associated with the above optimal stochastic control problem : in some sense it is an extension of the classical first-order Hamilton-Jacobi equations occuring in the Calculus of Variations (see P.L. Lions [26]). Let us also point out that in the litterature (6) is sometimes called the Bellman equation, or the dynamic programming equation.

A more precise relation between (5) and (6) is the following (see W.H. Fleming and R. Rishel [16], A. Bensoussan and J.L. Lions [3], N.V. Krylov [21]) : i) If $u \in C^2(\mathcal{O})$ then u solves (6), ii) If $\tilde{u} \in C^2(\mathcal{O}) \cap C(\bar{\mathcal{O}})$ satisfies (6) and (7) then $\tilde{u}(x) = u(x)$ in $\bar{\mathcal{O}}$. Unfortunately this classical theory (consisting of verification theorems) is not convenient since i) u is not in general C^2, it may even happen on simple examples that u is not continuous ! ; ii) no classical tools can take care of (6) and this for several reasons : first it is a fully nonlinear equation that is the nonlinearity acts on second derivatives of the unknown and second it is a degenerate equation since a may not be positive definite.

To solve these difficulties, we propose here a notion of weak solution of (6) that we call viscosity solution (since it is an extension of the notion introduced in first-order Hamilton-Jacobi equations by M.G. Crandall and P.L. Lions [7], [8] - see also M.G. Crandall, L.C. Evans and P.L. Lions [9]). This notion is briefly discussed in section II. Since this notion requires continuity, we give in section III a few results concerning, in particular, the continuity of u. In section IV we give a general uniqueness result for viscosity solutions of (6). Next (section V) we present various regularity results which, combined with the notion of viscosity solution, immediately yield that (6) holds in elementary ways (such that a.e. for example). Finally in section VI we mention several related topics that may be treated by the same methods.

Finally we want to mention a problem that we do not consider here : we will not give any result concerning optimal controls. Let us just mention that using results due to N.V. Krylov [21], [22] and a method due to S.R.S. Varadhan [48], it is possible to give under very general assumptions the existence of ε-optimal (or even optimal) markovian controls (i.e. controls in the so-called feedback form).

II. Viscosity solutions of Hamilton-Jacobi-Bellman equations.

Remarking that (6) may be rewritten :

$$H(D^2u, Du, u, x) = 0 \quad \text{in } \mathcal{O} , \tag{9}$$

where H is specified to be :

$$H(u^{ij}, u^i, u, x) = \sup_{v \in V} \left[-\sum_{i,j} a_{ij}(x,v)u^{ij} - \sum_i b_i(x,v)u^i + c(x,v)u - f(x,v) \right]$$

we are going to define weak solutions of (9) where H is a function on L^N_S (= {A : N x N symmetric matrix}) x R^N x R x $\mathcal{O}$ satisfying :

$$(10) \quad \begin{cases} H \in C(L^N_S \times R^N \times R \times \mathcal{O}) \quad ; \quad \text{for all } A,B \in L^N_S ,\ (p,t,x) \in R^N \times R \times \mathcal{O} : \\ H(A,p,t,x) \leqslant H(B,p,t,x) \quad \text{if } A \geqslant B \quad (\text{in } L^N_S) \ . \end{cases}$$

(the second part of (10) express the fact that (9) is elliptic).

For any continuous function $\varphi \in C(\mathcal{O})$, we define generalized derivatives as follows : we denote by $T_\pm$ the sets

$$T_+ = \Big\{ x \in \mathcal{O} \ / \ \exists A \in L^N_S ,\ \exists \xi \in R^N$$

$$\limsup_{y \to x} \{\varphi(y)-\varphi(x)-(\xi,y-x) + \tfrac{1}{2}(y-x,A(y-x))\} \, |y-x|^{-2} \leqslant 0 \Big\}$$

$$T_- = \Big\{ x \in \mathcal{O} \ / \ \exists A \in L^N_S ,\ \exists \xi \in R^N$$

$$\liminf_{y \to x} \{\varphi(y)-\varphi(x)-(\xi,y-x) + \tfrac{1}{2}(y-x,A(y-x))\} \, |y-x|^{-2} \geqslant 0 \Big\}.$$

And for $x \in T_+$ (resp. T_-) we set :

$$D_+\varphi(x) = \Big\{ (A,\xi) \in L^N_S \times R^N \ /$$

$$\limsup_{y \to x} \{\varphi(y)-\varphi(x)-(\xi,y-x) - \tfrac{1}{2}(y-x,A(y-x))\} \, |y-x|^{-2} \leqslant 0 \Big\}$$

(resp. $$D_-\varphi(x) = \Big\{ (A,\xi) \in L^N_S \times R^N \ /$$

$$\liminf_{y \to x} \{\varphi(y)-\varphi(x)-(\xi,y-x) - \tfrac{1}{2}(y-x,A(y-x))\} \, |y-x|^{-2} \geqslant 0 \Big\}).$$

Remark II.1 :

T_+ is the set of points where φ has, roughly speaking, an upper expansion up to order 2 ; and $D_+\varphi(x)$ consists in all possible couples (A,ξ) such that the expansion holds. Let us point out that for a general continuous function φ , T_+ and T_- are dense in $\mathcal{O}$ and that if $(A,\xi) \in D_+\varphi(x)$ and $B \geq A$ then $(B,\xi) \in D_+\varphi(x)$.

Definition II.1 :

$u \in C(\mathcal{O})$ is said to be a viscosity subsolution (resp. supersolution, resp. solution) of (9) if u satisfies :

(11) $\forall\, x \in T_+ \;,\; \forall\, (A,\xi) \in D_+u(x) \qquad H(A,\xi,u(x),x) \leq 0$

(resp. (12) $\forall\, x \in T_- \;,\; \forall\, (A,\xi) \in D_-u(x) \qquad H(A,\xi,u(x),x) \geq 0$;

resp. (11) and (12)).

Remark II.2.

This definition is the extension to second-order equations of the notion of viscosity solutions of first-order Hamilton-Jacobi equations introduced in M.G. Crandall and P.L. Lions [7], [8]. It is also related to some notions of accretivity considered by L.C. Evans [10], [11] (see also M.G. Crandall, P.L. Lions and L.C. Evans [9]) ; and to a notion introduced for linear elliptic equations by E. Calabi [5].

Let us give without proof a few elementary results :

Proposition II.1 :

The following are equivalent for any $u \in C(\mathcal{O})$:

i) u *is a viscosity solution of* (9) ;

ii) u *satisfies for all* $\varphi \in C^2(\mathcal{O})$:

(13) $H(D^2\varphi(x),D\varphi(x),u(x),x) \leq 0$ *at any local maximum* x *of* $u-\varphi$,

(14) $H(D^2\varphi(x),D\varphi(x),u(x),x) \geq 0$ *at any local minimum* x *of* $u-\varphi$.

Remark II.3 :

A similar result holds for viscosity subsolutions or supersolutions : (11)

is equivalent to (13) and (12) is equivalent to (14). In addition we may take $\varphi \in C^\infty(\bar{\mathcal{O}})$ in (13)-(14), we may replace any by some and we may **add** "strict extremum".

Proposition II.2 (Consistency) :

i) *If* $u \in C^2(\mathcal{O})$ *is a solution of* (9) *then* u *is a viscosity solution of* (9).

ii) *If* $u \in C(\mathcal{O})$ *is a viscosity solution of* (9), u *is differentiable near* $x_o \in \mathcal{O}$ *and* u *is twice differentiable at* x_o *then we have :*

$$H(D^2u(x_o),Du(x_o),u(x_o),x_o) = 0 .$$

A fundamental application of this notion is the following easy :

Proposition II.3 (Stability) :

Let H_n *be a sequence of functions satisfying* (10) *and converging uniformly on compact sets of* $L_S^N \times R^N \times R \times \mathcal{O}$ *to some function* H *(thus satisfying* (10)*). Let* u_n *be a sequence in* $C(\mathcal{O})$ *of viscosity solutions of :* $H_n(D^2u_n,Du_n,u_n,x) = 0$ *in* $\mathcal{O}$. *We assume that* u_n *converges on compact sets to* $u \in C(\mathcal{O})$. *Then* u *is a viscosity solution of* (9).

Finally, in the case when we specialize H to correspond to the Hamilton-Jacobi Bellman equation (6), the connection between the optimal stochastic control problem and the above considerations is illustrated by :

Theorem II.1 (Dynamic Programming) :

Let u *be the minimum cost function given by* (5). *If* $u \in C(\mathcal{O})$ *then* u *is a viscosity solution of* (6).

We briefly sketch the proof since we believe it to be enlightening : let $x_o \in \underline{T}$ (for example), let $(A,\xi) \in D\_u(x_o)$. It is an easy exercise to build $\varphi \in C_b^2(\bar{\mathcal{O}})$ such that : $u(x_o) = \varphi(x_o)$, $\xi = D\varphi(x_o)$, $A = D^2\varphi(x_o)$ and $u \geqslant \varphi$ in $\bar{\mathcal{O}}$. Writing now the mathematical formulation of the dynamic programming principle (see K. Itô [19], or next section) : $\forall\, T > 0$,

$$u(x_o) = \inf_{\mathcal{A}} \quad E \int_0^{\tau_{x_o} \wedge T} f\left(y_{x_o}(t),v(t)\right) \exp\left\{- \int_0^t c\left(y_{x_o}(s),v(s)\right) ds\right\} dt\, +$$

$$+\, u\left(y_{x_o}(\tau_{x_o} \wedge T)\right) \exp\left\{- \int_0^{\tau_{x_o} \wedge T} c\left(y_{x_o}(t),v(t)\right) dt\right\} .$$

and this yields :

$$\sup_{A} \Big[\frac{1}{T}\Big\{\varphi(x_o) - E\Big[\varphi\big(y_{x_o}(\tau_{x_o} \wedge T)\big) \exp\Big\{-\int_o^{\tau_{x_o} \wedge T} c\big(y_{x_o}(t),v(t)\big)dt\Big\}\Big]\Big\} +$$

$$- \frac{1}{T} E \int_o^{\tau_{x_o} \wedge T} f\big(y_{x_o}(t),v(t)\big) \exp\Big\{-\int_o^t c\big(y_{x_o}(s),v(s)\big)ds\Big\}\, dt\Big] \geq 0 .$$

Using Itô's formula, we deduce easily :

$$\sup_{A} \frac{1}{T} E \int_o^{\tau_{x_o} \wedge T} \{A^{v(t)} \varphi(x_o) - f(x_o,v(t))\}\, dt \geq - \varepsilon(T)$$

where $\varepsilon(T) \to 0$ if $T \to 0_+$. And we conclude remarking that :

$$\sup_{A} P\,[\tau_{x_o} < T\,] \leq \sup_{A} P\,[\sup_{0 \leq t \leq T} |y_{x_o}(t) - x_o)| \geq \delta]$$

$$\leq \frac{1}{\delta^2} \sup_{A} E\,[\sup_{0 \leq t \leq T} |y_{x_o}(t) - x_o|^2] \leq \frac{C}{\delta^2} T \qquad \text{for } T \leq 1 ,$$

$\delta = \mathrm{dist}(x_o,\Gamma)$ (we used at that point that the coefficients are bounded independently of v).

Combining Theorem II.1 and Proposition II.2 we deduce the :

Corollary II.1 :

Let u *be given by* (5) :

i) *We have* : $\forall\, v \in V$, $A^v u \leq f(\cdot,v)$ *in* $\mathcal{D}'(\mathcal{O})$;

ii) *If* $u \in W^{2,p}_{loc}(\mathcal{O})$ *for some* $p > N$ *then we have* :

(6') $$\sup_{v \in V} \{A^v u(x) - f(x,v)\} = 0 \qquad a.e.\ in\ \mathcal{O} .$$

iii) *If* u *belongs to the vector space generated by the cone* X *defined by* :

$$X = \Big\{u \in C(\mathcal{O}) ,\ \exists\, \varphi \in W^{2,p}_{loc}(\mathcal{O})\ (p > N) ,\ D^2 u \leq D^2\varphi \text{ in } \mathcal{D}'(\mathcal{O})\Big\}$$

then $u \in W^{1,\infty}_{loc}(\mathcal{O})$, $D^2 u \in \mathcal{M}(\mathcal{O})$ *and* $\sup_{v \in V} \{A^v u - f(\cdot,v)\}$ *is a negative measure on* $\mathcal{O}$ *orthogonal to the Lebesgue measure.*

iv) *If* $u \in X$ *then* : $\forall\, v \in V$, $h \leq A^v u \leq C$ *for some* $h \in L^p_{loc}(\mathcal{O})$ $(p > N)$; *and* (6') *holds.*

Remark II.4 :

An example due to Genis and N.V. Krylov [18] shows that $\{A^{v}u - f(\cdot,v)\}$ may be a non zero measure on $\mathcal{O}$.

Remark II.5 :

ii) is a consequence of well-known differentiability properties of functions in $W^{2,p}_{loc}(\mathcal{O})$ spaces (see E. Stein [46]) ; while iii),iv) are deduced from a differentiability theorem due to Alexandrov [1], H. Busemann [5] - see for more details P.L. Lions [27].

All the results mentioned in this section are detailed and proved in P.L. Lions [27].

III. Continuity of the minimum cost function :

As we just saw, the notion of viscosity solutions requires that some continuity of the value function is known. On the other hand since we are dealing with eventually degenerate diffusion processes the question whether the value function is continuous can be extremely difficult to settle (even in the case without control - see D.W. Stroock and S.R.S. Varadhan [47]). It turns out that there exists a natural assumption which gives quite general results : we assume that "the problem has a subsolution" :

(15)
$$\left\{\begin{array}{l} \exists\, w \text{ bounded meausrable on } \mathcal{O}\ ,\ \exists\, \Gamma_0 \subset \Gamma \text{ such that} \\ \text{i) } \forall\, A\ ,\ w(y_x(t \wedge \tau_x)) \exp\left[-\int_0^{t \wedge \tau_x} c\, ds\right] + \int_0^{t \wedge \tau_x} f \exp\left(-\int_0^s c\, d\lambda\right) ds \\ \quad \text{is a } F_{t \wedge \tau_x} \text{ strong submartingale for all } x \in \overline{\mathcal{O}}\ ; \\ \text{ii) } w = 0 \text{ on } \Gamma_0\ ,\ 1_{(\tau_x < \infty)} y_x(\tau_x) \in \Gamma_0 \text{ a.s. } \forall\, A\ ,\ \forall\, x \in \overline{\mathcal{O}} \text{ and} \\ \quad \liminf_{\operatorname{dist}(y,\Gamma_0) \to 0} w(y) \geq 0\ . \end{array}\right.$$

Remark III.1 :

To explain and motivate this complicated condition, let us make a few remarks :

i) If $\mathcal{O} = R^N$ of if, more generally, $\tau_x = +\infty$ a.s. $\forall x$, A then we choose $\Gamma_0 = \emptyset$; $w = \inf_{x,v} \frac{f(x,v)}{c(x,v)}$ and (15) holds. This is the case when $\sigma \equiv 0$ on Γ and $-b(x,v) \cdot n(x) \geq \alpha > 0$ for all $(x,v) \in \Gamma \times V$ (where $n(x)$ is the unit outward normal to $\partial\mathcal{O}$).

ii) If there exists $w \in W^{1,\infty}(\mathcal{O})$ satisfying :

$$A^v w \leq f(\cdot,v) \quad \text{in } \mathcal{D}'(\mathcal{O}) \quad , \quad \forall v \in V \quad ; \quad w = 0 \quad \text{on } \Gamma \quad ; \tag{16}$$

then (15) holds.

This is the case for example when $f(x,v) \geq 0$ ($\forall x,v$) : take $\Gamma_0 = \Gamma$, $w = 0$. It is also the case when all the processes "cross the boundary" that is when we assume :

$$\exists \alpha > 0 \text{ , } \forall (x,v) \in \Gamma \times V \quad \text{either} \quad (n(x),a(x,v)n(x)) \geq \alpha \tag{17}$$
$$\text{or} \quad b(x,v) \cdot n(x) + \sum_{i,j} a_{ij}(x,v)\, \partial_{ij}\, \delta(x) \geq \alpha$$

(where $\delta(x) = \text{dist}\,(x,\Gamma)$).

It is also possible to combine the two cases i) and ii) above.

Under assumption (15), we have the following :

<u>Theorem III.1</u> :

Under assumption (15), *we have :*

i) $J(\cdot,A)$, $u(x)$ *are u.s.c. on* $\bar{\mathcal{O}}$ *for all* A ; $u \geq w$, $u \geq \underline{u}$ *in* $\bar{\mathcal{O}}$ *where* $\underline{u}(x) = \inf_A E \int_0^{\tau'_x} f(y_x(t),v(t)) \exp\left(-\int_0^t c(y_x(s),v(s))ds\right) dt$ *and* τ'_x *is the first exit time of* $y_x(t)$ *from* $\mathcal{O}$. *In particular:* $u \geq 0$ *on* Γ;

ii) $1_{(\tau_x < \infty)}\, u(y_x(\tau_x)) = 0$ *a.s. for all* A, $x \in \bar{\mathcal{O}}$;

iii) *For all* A *and* $x \in \bar{\mathcal{O}}$, $u(y_x(t \wedge \tau_x)) \exp\left(-\int_0^{t \wedge \tau_x} c\right) + \int_0^{t \wedge \tau_x} f \exp\left(-\int_0^s c\right)$ *is a* F_t *strong submartingale. In particular, we have :*

$$u(x) = \inf_A E \int_0^{\theta \wedge \tau_x} f(y_x(t),v(t)) \exp\left(-\int_0^t c(y_x(s),v(s))ds\right) dt \; + \tag{18}$$
$$+ \; u(y_x(\theta \wedge \tau_x)) \exp\left(-\int_0^{\theta \wedge \tau_x} c(y_x(t),v(t))dt\right) ,$$

where θ *is a stopping time (which may depend on* A*).*

iv) *In* (5) , *the infimum may be restricted to admissible systems where the probability space* (Ω,F,F_t,P) *and the Brownian motion* W_t *is prescribed.*

v) *If* $w \in W^{1,\infty}_{loc}(\mathcal{O})$ *satisfies* : $\limsup_{y \to x} (w-u)(y) \leq 0$, $\forall x \in \Gamma$; $A^v w \leq f(\cdot,v)$ *in* $\mathcal{D}'(\mathcal{O})$, $\forall v \in V$; *then* $w(x) \leq u(x)$ *in* $\bar{\mathcal{O}}$.

Remark III.2 :

Let us mention that assumption (15) is discussed in P.L. Lions [26] for deterministic control problems and that if (15) (or some analogue of (15)) does not hold, everything may happen.

Remark III.3 :

This result improves various previous results obtained by J.L. Menaldi and P.L. Lions [37].

Corollary III.1 :

Under assumption (15) , *the following assertions are equivalent* :

i) $u \in C(\bar{\mathcal{O}})$; ii) $u_{|\Gamma} \in C(\Gamma)$; iii) $\Gamma_1 = \{x \in \Gamma \ , \ u(x) = 0\}$ *is closed.*

In particular if $u = 0$ *on* Γ *then* $u \in C(\bar{\mathcal{O}})$ *and* $u = \underline{u}$ *in* $\bar{\mathcal{O}}$.

Corollary III.2 :

If (15) *holds and if we assume* : $\exists \alpha \in (0,1]$

(19) $$|u(x)| \leq C \text{ dist } (x,\Gamma_1)^\alpha \quad ;$$

then denoting by λ_0 *the following constant* :

$$\lambda_0 = \sup \left\{ \frac{1}{2} \text{Tr} \left(\frac{\partial\sigma}{\partial\xi} \cdot \frac{\partial\sigma}{\partial\xi}^T \right) + \frac{\partial b}{\partial\xi} \cdot \xi \ / \ (x,v,\xi) \in R^N \times V \times R^N \ , \ |\xi| = 1 \right\}$$

we have : $u \in C^{0,\beta}(\bar{\mathcal{O}})$ *with* $\beta = \min\left(\alpha, \frac{\lambda}{\lambda_0}\right)$ *if* $0 < \lambda < \lambda_0$; $\beta = \alpha$ *if* $\alpha < 1$, $\lambda = \lambda_0$; $\beta < 1$ *if* $\alpha = 1$, $\lambda = \lambda_0$; $\beta = \alpha$ *if* $\lambda > \lambda_0$.

All these results are detailed and proved in P.L. Lions [27] : they extend various results obtained by J.L. Menaldi and P.L. Lions [37]. Let us finally mention that some extensions of (15) are given in [27].

IV. Uniqueness of viscosity solutions.

Again, the boundary and the eventual degeneracy of the processes create technical difficulties. A typical result that we may obtain is the following :

Theorem IV.1 :

Let u *be given by* (5). *We assume :*

(20) $$u \in C_b(\bar{O}) \quad , \quad u = 0 \quad \text{on } \Gamma \; ;$$

(21) $$\forall\, \varepsilon > 0 \;,\; \exists w_\varepsilon \in C^1(O) \cap C(\bar{O}) \;,\; \forall\, v \in V \quad \check{A} w_\varepsilon \leq f(\cdot,v) + \varepsilon \quad \text{in } O \;,$$
$$|w_\varepsilon| \leq \varepsilon \quad \text{on } \Gamma \; .$$

Then if $\tilde{u}$ *is a viscosity solution of* (6) *satisfying* (20), *we have :*

$$\tilde{u}(x) = u(x) \quad , \quad \forall\, x \in \bar{O} \; .$$

Remark IV.1 :

In P.L. Lions [27], the proof of this result is given together with several variants or extensions. Let us mention that if $O = R^N$ then conditions (20)-(21) are vacuous ; let us also indicate that it is possible to replace (21) by : there exist Γ_1,Γ_2 relatively open subsets of Γ such that $\Gamma = \Gamma_1 \cup \Gamma_2$ and on Γ_2 $\sigma(x,v) = 0$ $(\forall\, v)$ and either $b(x,v) = 0$ $(\forall\, v)$ or $b(x,v)\cdot n(x) \leq -\alpha < 0$ $(\forall\, v)$; and for each $\varepsilon > 0$ there exists $w_\varepsilon \in C^1(\bar{O})$ satisfying $\check{A} w_\varepsilon \leq f(\cdot,v) + \varepsilon$ in O, $|w_\varepsilon| \leq \varepsilon$ on Γ_1. Then we may replace (20) by : $u \in C_b(\bar{O})$, $u = 0$ on Γ_1.

V. Regularity results.

We will use below the following formulation of the fact that the controlled processes really cross the boundary (or some part of it) :

(22) $$\begin{cases} \exists\, \Gamma_1,\Gamma_2 \text{ relatively open in } \Gamma \text{ , disjoint, eventually empty such that :} \\ \Gamma = \Gamma_1 \cup \Gamma_2 \text{ and} \\ \forall\; (x,v) \in \Gamma_2 \times V \;,\; \sigma(x,v) = b(x,v) = 0 \\ \forall\; (x,v) \in \Gamma_1 \times V \text{ , then either } (a(x,v)n(x),n(x)) \geq \alpha > 0 \text{ or} \\ D_x^j\, \sigma(x,v) = 0 \text{ for } |j| \leq 1 \text{ and } b(x,v)\cdot n(x) \geq \alpha > 0 \; . \end{cases}$$

Theorem V.1.

Under assumption (22), *there exists* λ_1 *depending only on bounds on* $\|D^{\alpha}\varphi\|_{L^{\infty}(R^N x V)}$ *for* $|\alpha| = 1,2$, $\varphi = b,\sigma$ ($\lambda_1 = 0$ *if* b *and* σ *do not depend on* x) *such that we have for* λ $(= \inf_{x,v} c(x,v)) > \lambda_1$:

(23) $$u \in W^{1,\infty}(\mathcal{O}) \quad , \quad u = 0 \quad on\ \Gamma_1 \quad ;$$

(24) $$A^{v}u \in L^{\infty}(\mathcal{O}) \text{ for all } v \in V \text{ and } \sup_{v \in V} \|A^{v}u\|_{L^{\infty}(\mathcal{O})} < \infty \ ;$$

(25) $$\sup_{v \in V} [A^{v}u - f(x,v)] = 0 \quad a.e.\ in\ \mathcal{O} \ ;$$

(26) $$\frac{\partial^2 u}{\partial \chi^2} \leqslant C \quad in\ \mathcal{D}'(\mathcal{O}) \quad , \quad \forall\, \chi \in R^N : |\chi| = 1 \ .$$

Theorem V.2 :

Let ω *be an open subset of* $\mathcal{O}$. *Let* $p \in \{1,\ldots,N\}$.

i) *Under the assumptions of* Theorem V.1 *and if we have* :

(27) $$\begin{cases} \exists\, \alpha > 0 \ , \ \forall\, x \in \omega \ , \ \exists\, m \geqslant 1 \ , \ \exists\, v_1,\ldots,v_m \in V \ , \ \exists\, \theta_1,\ldots,\theta_m \in \,]0,1[\\ \text{such that : } \sum_{i=1}^{m} \theta_i = 1 \ , \ \sum_{i=1}^{m} \theta_i\, a_{k\ell}(x,v_i)\xi_k\, \xi_\ell \geqslant \alpha \sum_{k=1}^{p} |\xi_k|^2 \\ \text{for all } \xi \in R^N \end{cases}$$

then $\partial_{ij}u \in L^{\infty}(\omega)$ *for* $1 \leqslant i,j \leqslant p$.

ii) *Under assumption* (15) *and if we have* :

(28) $$\inf \{(a(x,v)\xi,\xi) \ / \ v \in V \ , \ \xi \in R^N \ , \ |\xi| = 1\} > 0 \quad , \quad \forall\, x \in \omega$$

then $u \in C^{2,\beta}(\omega')$ *for any open set* $\omega' \subset\subset \omega$, *where* $\beta \in (0,1)$ *depends only on* ω' *and* $\{a(x,v)\}_{x,v}$.

Theorem V.3 : (uniqueness)

i) *Under assumption* (15) , *if* w *satisfies* : $w \in W^{1,\infty}_{loc}(\mathcal{O})$,

$$A^{v}w \leqslant f(\cdot,v) \ in\ \mathcal{D}'(\mathcal{O}) \ , \ \forall\, v \in V \ ; \ \limsup_{y \to x} w(y) \leqslant 0 \ , \ \forall\, x \in \Gamma$$

then $w \leqslant u$ *in* $\bar{\mathcal{O}}$.

ii) *If* $u \in C(\bar{O})$ *and* $u = 0$ *on* Γ *; then for any* w *satisfying :*

(23') $$w \in W^{1,\infty}_{loc}(O) \cap C(\bar{O}) \quad , \quad w = 0 \quad on\ \Gamma \ ;$$

(24') $$\begin{cases} A^v u \in L^\infty_{loc}(O) \ , \ \forall\, v \in V \ ; \ \sup\limits_{v \in V} \|A^v u\|_{L^\infty(\omega)} < \infty \ \text{for any bounded} \\ \text{open set } \omega \subset\subset O \ ; \end{cases}$$

(25) $$\sup_{v \in V} \ [A^v u(x) - f(x,v)] = 0 \quad a.e.\ in\ O \ ;$$

(26') $$\exists\, g \in L^N_{loc}(O) \quad , \quad \Delta w \leqslant g \ in\ \mathcal{D}'(O) \ ;$$

we have : $w \equiv u$ *in* $\bar{O}$.

Remark V.1 :

It is easily seen on simple examples that those results are essentially optimal : in particular the assumption $\lambda > \lambda_1$ is in general necessary (see Genis and N.V. Krylov [18]). Nevertheless in the uniformly non-degenerate case, i.e. when the matrices $a(x,v)$ are uniformly in (x,v) positive definite $\lambda > 0$ is enough (this is proved in P.L. Lions [28], L.C. Evans and P.L. Lions [15]) and if O is bounded then the precise range of admissible λ in R such that u solves (25) is given in P.L. Lions [29].

Remark V.2 :

Many variants of Theorems V.1-3 are given in [27]. Let us just mention that if $O = R^N$, (22) becomes vacuous.

Remark V.3 :

Previous regularity results (included in Theorems V.1-3) are given in N.V. Krylov [23], [24], [25], [21]; H. Brézis and L.C. Evans [4]; L.C. Evans and A. Friedman [14]; M.V. Safonov [44], [45]. The non degenerate case is treated in [28], [15] by purely p.d.e. methods. The case $O = R^N$ was proved independently by N.V. Krylov [22] and P.L. Lions [30], [31] by slightly different probabilistic methods. Those results are taken from P.L. Lions [27] (see also [32], [33], [34]) and part ii) of Theorem V.2 is an easy consequence of deep regularity results due to L.C. Evans [12], [13].

VI. Related problems :

We indicate briefly a list of problems where the above results and methods are used.

1. Similar problems : Let us mention that similar results hold for problems where i) there is a pay-off when the process reach Γ ; or ii) the coefficients are unbounded in x ; or iii) we add optimal stopping time problems ; or iv) we consider time-dependent diffusion processes and time-dependent HJB equations. Analogous results hold and are proved by similar methods - see P.L. Lions [27] for more details.
Using similar methods, the case of impulse control problems has been treated by Perthame [42].

2. Numerical approximation : We refer the interested reader to P.L. Lions and B. Mercier [38], J.P. Quadrat [43].

3. Bifurcation and optimal stochastic control : in P.L. Lions [29], some bifurcation phenomena are interpreted in terms of optimal stochastic control. In addition the analogues of eigenvalues are introduced for HJB equations.

4. Differential Games : The main open questions in stochastic differential games concern the uniqueness result and regularity results. We hope to come back on these questions in a future study. Let us also mention that deterministic differential games are covered by the results in M.G. Crandall and P.L. Lions [7], [8]; P.L. Lions [26].

5. Nonlinear semi-groups : It is possible to adapt with the methods of sections II-IV the method of M. Nisio [41] to build nonlinear semi-groups associated with HJB equations. Then the results of section IV yield easily a very general uniqueness result concerning those nonlinear semi-groups (see P.L. Lions and M. Nisio [39]).

6. Asymptotic problems : The preceding results and methods can be used in the proofs of various asymptotic results such as the simplification of large-scale systems (see R. Jensen and P.L. Lions [20]), the homogeneization of optimal stochastic control problems (see P.L. Lions, G. Papanicolaou and S.R.S. Varadhan [40]).

7. Monge-Ampère equations : The results and methods given above have enabled us recently to solve the classical Monge-Ampère equations (arising in Differential Geometry) - see P.L. Lions [35], [36]. The remark that Monge-Ampère equations are indeed HJB equations is due to B. Gaveau [17].

Bibliography :

[1] A.D. ALEXANDROV : Almost everywhere existence of the second differential of a convex function and some properties of convex surfaces connected with it. Ucen. Zap. Leningrad Gos. Univ. No. 37 (Ser. Mat., vyp 6) (1939), p. 3-35 (in Russian).

[2] R. BELLMAN : Dynamic programming . Princeton Univ. Press, Princeton, N.J. (1957).

[3] A. BENSOUSSAN and J.L. LIONS : Applications des inéquations variationnelles en contrôle stochastique. Dunod, Paris (1978).

[4] H. BREZIS and L.C. EVANS : A variational approach to the Bellman-Dirichlet equation for two elliptic operators. Arch. Rat. Mech. Anal., 71 (1979), p. 1-14.

[5] H. BUSEMANN : Convex surfaces, Interscience, New-York (1958).

[6] E. CALABI : Improper affine hyperspheres of convex type and a generalization of a theorem by K. Jörgens. Michigan Math. J., 5 (1958), p.105-126.

[7] M.G. CRANDALL and P.L. LIONS : Condition d'unicité pour les solutions généralisées des équations de Hamilton-Jacobi du 1er ordre. Comptes-Rendus Paris, 252 (1981), p. 183-186.

[8] M.G. CRANDALL and P.L. LIONS : Viscosity solutions of Hamilton-Jacobi equations. To appear in Trans. Amer. Math. Soc.

[9] M.G. CRANDALL , L.C. EVANS and P.L. LIONS : Some properties of viscosity solutions of Hamilton-Jacobi equations. Preprint.

[10] L.C. EVANS : A convergence theorem for solutions of nonlinear second-order elliptic equations. Ind. Univ. Math. J., 27 (1978), p. 875-887.

[11] L.C. EVANS : On solving certain nonlinear partial differential equations by accretive operator methods. Israel J. Math., 36 (1981), p. 225-247.

[12] L.C. EVANS : Classical solutions of fully nonlinear, convex, second-order elliptic equations. To appear in Comm. Pure Appl. Math.

[13] L.C. EVANS : Classical solutions of the Hamilton-Jacobi-Bellman equation for uniformly elliptic operators. Preprint.

[14] L.C. EVANS and A. FRIEDMAN : Optimal stochastic switching and the Dirichlet problem for the Bellman equation. Trans. Amer. Math. Soc., 253 (1979), p. 365-389.

[15] L.C. EVANS and P.L. LIONS : Résolution des équations de Hamilton-Jacobi-Bellman pour des opérateurs uniformément elliptiques. Comptes-Rendus Paris, 290 (1980), p. 1049-1052.

[16] W.H. FLEMING and R. RISHEL : Deterministic and stochastic optimal control. Springer, Berlin (1975).

[17] B. GAVEAU : Méthodes de contrôle optimal en analyse complexe. I. Résolution d'équations de Monge-Ampère. J. Funct. Anal., 25 (1977), p. 391-411.

[18] I.L. GENIS and N.V. KRYLOV : An example of a one dimensional controlled process. Th. Proba. Appl., 21 (1976), p. 148-152.

[19] K. ITO : personnal communication.

[20] R. JENSEN and P.L. LIONS : Some asymptotic problems in optimal stochastic control and fully nonlinear elliptic equations. Preprint.

[21] N.V. KRYLOV : Control of diffusion type processes, Moscou (1979) - in Russian.

[22] N.V. KRYLOV : Control of the diffusion type processes. Proceedings of the International Congress of Mathematicians, Helsinki, 1978.

[23] N.V. KRYLOV : On control of the solution of a stochastic integral equation. Th. Proba. Appl., 17 (1972), p. 114-131.

[24] N.V. KRYLOV : On control of the solution of a stochastic integral equation with degeneration. Math. USSR Izv., 6 (1972), p. 249-262.

[25] N.V. KRYLOV : On equations of minimax type in the theory of elliptic and parabolic equations in the plane. Math. USSR Sbornik, 10 (1970), p. 1-19.

[26] P.L. LIONS : Generalized solutions of Hamilton-Jacobi equations. Pitman, London (1982).

[27] P.L. LIONS : Optimal control of diffusion processes and Hamilton-Jacobi-Bellman equations. To appear.

[28] P.L. LIONS : Résolution analytique des problèmes de Bellman-Dirichlet, 146 (1981), p. 151-166.

[29] P.L. LIONS : Bifurcation and optimal stochastic control. Preprint ; see also MRC report, University of Wisconsin-Madison (1982).

[30] P.L. LIONS : Contrôle de diffusions dans R^N. Comptes-Rendus Paris. 288 (1979), p. 339-342.

[31] P.L. LIONS : Control of diffusion processes in R^N. Comm. Pure Appl. Math., 34 (1981), p. 121-147.

[32] P.L. LIONS : Equations de Hamilton-Jacobi-Bellman dégénérées. Comptes-Rendus, 289 (1979), p. 329-332.

[33] P.L. LIONS : Equations de Hamilton-Jacobi-Bellman. In Séminaire Goulaouic-Schwartz 1979-1980, Ecole Polytechnique, Palaiseau.

[34] P.L. LIONS : Optimal stochastic control and Hamilton-Jacobi-Bellman equations. In Mathematical Control Theory, Banach Center Publications, Warsaw.

[35] P.L. LIONS : Une méthode nouvelle pour l'existence de solutions régulières de l'équation de Monge-Ampère. Comptes-Rendus Paris, 293 (1981), p. 589-592.

[36] P.L. LIONS : Sur les équations de Monge-Ampère. I, II and III. Preprint.

[37] P.L. LIONS and J.L. MENALDI : Optimal control of stochastic integrals and Hamilton-Jacobi-Bellman equations. I, II. SIAM J. Control Optim., 20 (1982), p. 58-95.

[38] P.L. LIONS and B. MERCIER : Approximation numérique des équations de Hamilton-Jacobi-Bellman. R.A.I.R.O., 14 (1980), p. 369-393.

[39] P.L. LIONS and M. NISIO : A uniqueness result for the semi-group associated with the Hamilton-Jacobi-Bellman operator. Preprint.

[40] P.L. LIONS , G. PAPANICOLAOU and S.R.S. VARADHAN : work in preparation.

[41] M. NISIO : On stochastic optimal controls and envelope of Markovian semi-groups. Proc. Intern. Symp. SDE, Kyoto (1976).

[42] PERTHAME : Thèse de 3e cycle, Université P. et M. Curie, 1982-83.

[43] J.P. QUADRAT : Analyse numérique de l'équation de Bellman stochastique. Rapport Laboria n° 140 (1975), INRIA, Rocquencourt.

[44] M.V. SAFONOV : On the Dirichlet problem for Bellman's equation in a plane domain. Math. USSR Sbornik, 31 (1977), p. 231-248.

[45] M.V. SAFONOV : On the Dirichlet problem for Bellman's equation in a plane domain. II. Math. USSR Sbornik, 34 (1978), p. 521-526.

[46] E. STEIN : Singular integrals and differentiability properties of functions. Princeton Univ. Press, Princeton (1970).

[47] D.W. STROOCK and S.R.S. VARADHAN : On degenerate elliptic- parabolic operators of second-order and their associated diffusions. Comm. Pure Appl. Math., 25 (1972), p. 651-713.

[48] S.R.S. VARADHAN : personnal communication.

ON REDUCING THE DIMENSION OF CONTROL PROBLEMS
BY DIFFUSION APPROXIMATION

Petr Mandl
Department of Probability and Mathematical Statistics, Charles University
Sokolovská 83, 186 00 Prague 8, Czechoslovakia

1. The control problem

The topic was dealt by several authors in Prague ([2] - [5]). They apply a martingale method explained here on a specific model.

Let $\{X_t, t \geq 0\}$ be a controlled Markov process with finite state space I whose dynamics is defined by the transition rates

$$rq(i,j;z),\ i,j \in I,\quad q(i,i;z) = -\sum_{j \neq i} q(i,j;z),\quad i \in I.$$

z denotes the control parameter taking values in a set J. r is a time scale parameter. We shall study the limit behaviour of the model as $r \longrightarrow \infty$. Let the reward arise from the process in the following way. The reward rate in state i is a random variable with distribution function ${}^rF_i(y)$. The rate is constant as long as the system stays in i. After the transition to the next state, say j, a new reward rate realizes independently according to distribution function ${}^rF_j(y)$, etc. We assume

$$\int y \, d\, {}^rF_i(y) = a_i, \qquad \int y^2 \, d\, {}^rF_i(y) \sim rb_i,$$

$$\int y^4 d\, {}^rF_i(y) = O(r^2), \quad i \in I, \quad \text{as } r \longrightarrow \infty .$$

Let Y_t, Z_t denote the reward rate and the control parameter at time t, respectively. Further, let C_t be the capital accumulated up to time t,

$$C_t = C_o + \int_0^t Y_s \, ds , \quad t \geq 0.$$

The initial capital C_o is nonrandom.

If the aim is to maximize EC_T, the expected capital at time T, $0 < T < \infty$, the controller can limit himself to the controls

(1) $$Z_t = u(t, X_t^-), \quad t \in [0,T] ,$$

where u is mapping from $[0,T]\times I$ to J . $\{X_t^-, t\geq 0\}$ denotes the left continuous version of $\{X_t, t\geq 0\}$. However, if <u>risk sensitivity</u> is desirable, the optimality problem is usually formulated as

(2) $$E\, g(C_T) = \max,$$

where $g(y)$ is a given function. For $g(y) = -\exp\{-hy\}$, with constant $h>0$, the controls (1) are a sufficiently broad class. For general $g(y)$ the controls of the form

(3) $$Z_t = \bar{u}(t, C_t, X_t^-), \quad t\in[0,T] \quad ,$$

are to be used.

Similarly, when maximizing the expected discounted reward

(4) $$E \int_0^\infty e^{-lt} d\, C_t, \qquad l > 0,$$

the controls

$$Z_t = u(X_t^-), \quad t\geq 0,$$

suffice. If the risk of the ruin is to be taken into account, (4) is changed into

(5) $$E\left(\int_0^\tau e^{-lt}\, d\, C_t - h\, e^{-l\tau} \right),$$

where

$$\tau = \inf\{t : C_t < 0\}$$

is the time of the ruin, and h a positive constant. To maximize (5), more general controls

$$Z_t = \bar{u}(C_t, X_t^-), \quad t\geq 0,$$

are needed.

A simplification is achieved in diffusion approximation as $r\to\infty$. The original problem reduces to the optimization of a one-dimensional process $\{\tilde{C}_t, t\geq 0\}$ satisfying

$$d\, \tilde{C}_t = \theta(U_t)dt + \sigma(U_t)d\, W_t, \quad t\geq 0,$$

where $\{W_t, t\geq 0\}$ is a Wiener process, and the control U_t takes values in the set $\mathcal{U}$ of stationary controls u represented by mappings $u(i)$ from I to J.

The model considered here is a simple generalization of the service system investigated in [5] (see Example 1). In [2] , [3] discrete time controlled finite Markov chains with criterion (5) are treated. [3] introduces the aggregation of Markov chains (see Example 2). In [4] various kinds of the coupling of two Markov chains with different time scales are studied.

Example 1. V.Lánská ([5]) considers $\{X_t, t\geq 0\}$ to be the model of an M/M/1/1 service system to which customers of n types arrive for service. The type i customers arrive with rate rq_i, and have the service completion rate rd_i. $X_t=0$ means that the system is idle, $X_t=i$ means that a customer of type i is in service. The random reward rates introduced above are the payments of the customers for service. The control consists in deciding which types of customers will be accepted. The blocking of the line by unprofitable customers is to be avoided. Thus, $z=(z_1,\dots,z_n)$ where $z_i=1$ or 0 says whether the type i customers are accepted or not. Hence, the rates are

$$q(0,i;z) = z_i q_i, \quad q(0,0;z) = -\sum_{j\neq 0} z_j q_j,$$

$$q(i,0;z) = d_i = -q(i,i;z), \qquad i\neq 0,$$

$$q(i,j;z) = 0, \qquad 0\neq i\neq j\neq 0.$$

For the diffusion approximation one has

$$\theta(z) = (a_0 + \sum_{j\neq 0} \frac{z_j q_j a_j}{d_j}) \;/\; (1 + \sum_{j\neq 0} \frac{z_j q_j}{d_j}) ,$$

$$\sigma(z)^2 = 2 \sum_{j\neq 0} \frac{z_j q_j b_j}{d_j^2} \;/\; (1 + \sum_{j\neq 0} \frac{z_j q_j}{d_j}) .$$

Since the decisions are effective only in state 0, stationary controls u can be identified with control parameter values z . The goal is to maximize (5).

Example 2. Consider an aggregate of n independent processes of the kind defined in the introduction. Let its trajectory be

$$X_t = ({}^1X_t, \dots, {}^nX_t), \quad t\geq 0.$$

The total capital at time t equals

$$\bar{C}_t = \sum_k {}^k c_t \quad .$$

To maximize $Eg(C_T)$, one has to employ in general the controls

$${}^k z_t = {}^k\bar{u}(t, c_t, \bar{X}_t^-), \quad t \in [0,T] \;, \quad k=1,\dots,n.$$

Again, it is advantageous to simplify the problem by means of diffusion approximation. This can be done provided that

$$na_i \longrightarrow \bar{a}_i, \qquad nb_i \longrightarrow \bar{b}_i, \qquad i \in I, \qquad \text{as} \qquad r \longrightarrow \infty \quad .$$

2. Diffusion approximation

Theorem. Let $J \subset R^m$ be closed and bounded, and let $q(i,j;z)$, $i,j \in I$, be continuously differentiable. For each $u \in \mathcal{U}$ let the matrix

$$\| q(i,j;u(i)) \|_{i,j \in I}$$

be indecomposable. Assume (3) with $\bar{u}(t,y,i)$ having bounded derivatives. Then, as $r \longrightarrow \infty$, the probability distribution of $\{ C_t, t \in [0,T] \}$ converges weakly to the probability distribution of a diffusion process $\{ \tilde{C}_t, t \in [0,T] \}$ satisfying

$$d\, \tilde{C}_t = \theta(U_t)dt + \sigma(U_t)d\, W_t, \qquad t \in [0,T] \quad ,$$

where

$$U_t = \bar{u}(t, \tilde{C}_t, \cdot), \qquad t \in [0,T] \quad .$$

The coefficients $\theta(u)$, $\sigma(u)$ are obtained by solving

$$a_i + \sum_j q(i,j;u(i))\bar{w}(j,u) - \theta(u) = 0, \qquad i \in I, \tag{6}$$

for unknowns $\theta(u)$, $\bar{w}(j,u)$, $j \in I$, and

$$\sum_{j \neq i} q(i,j;u(i))\Big(\frac{b_i}{q(i,i;u(i))^2} + \frac{b_j}{q(j,j;u(j))^2} + \bar{w}_2(j,u)\Big) + \tag{7}$$

$$+ \; q(i,i;u(i))\bar{w}_2(i,u) - \sigma(u)^2 = 0, \qquad i \in I,$$

for unknowns $\sigma(u)^2$, $\bar{w}_2(j,u)$, $j \in I$.

The proof of the Theorem will be briefly sketched. Let u be a stationary control. Solve (6), which is a known system of equations defining the average reward $\theta(u)$ and auxiliary constants $\bar{w}(j,u)$ for stationary control u in a controlled Markov process with transition rates $q(i,j;z)$ and reward rates a_i (see e.g. [1]). Further, let

$$w(i,y,u) = \frac{\bar{w}(i,u)}{r} - \frac{y - a_i}{rq(i,i;u(i))} \quad . \tag{8}$$

Then

$$\begin{aligned} & y + r \sum_{j\neq i} q(i,j;u(i)) \int w(j,s,u)\, d\, {}^rF_j(s) + \\ & + rq(i,i;u(i))w(i,y,u) - \theta(u) = 0, \qquad i \in I. \end{aligned} \tag{9}$$

Let (3) hold. Set

$$w(t) = w(X_t,Y_t,\bar{u}(t,C_t,.)), \qquad \theta_t = \theta(\bar{u}(t,C_t,.)).$$

Introduce the counting process of the state changes in X, $\{N_t, t\geq 0\}$. Using (9) it can be shown that the process

$$\begin{aligned} M_t &= C_t - \int_0^t \theta_s\, ds + \int_0^t (w(s) - w(s-))\,d\, N_s = \\ &= C_t - \int_0^t \theta_s\, ds + w(t)-w(0) - \int_0^t \frac{d}{ds}\, w(s)ds, \qquad t\geq 0, \end{aligned} \tag{10}$$

is a martingale. Namely, from (9) it follows that

$$\int_0^t \theta_s\, ds - C_t \, , \qquad t\geq 0,$$

is the compensator of

$$\int_0^t (w(s)-w(s-))\,d\, N_s \, , \qquad t\geq 0.$$

(8) indicates that

$$w(t) = O(\tfrac{1}{r}), \qquad r \longrightarrow \infty \quad .$$

Hence it is derived that

$$\{M_t, \quad t\geq 0\} \qquad \text{and} \qquad \{C_t - \int_0^t \theta_s\, ds, \quad t\geq 0\}$$

have identical limiting distributions.

The second part of the proof consists in verifying that the limiting quadratic variation of M is

$$\langle M\rangle_t = \int_0^t \sigma(\bar{u}(s,C_s,.))^2 ds, \qquad t\geq 0.$$

To do so, one in fact repeats the above reasoning. It holds

$$\langle M\rangle_t = \int_0^t (w(s)-w(s-))^2 d\,N_s, \qquad t\geq 0,$$

because only jumps, and not the absolutely continuous terms in (10), contribute to the quadratic variation.

Further, one solves for stationary controls u

$$\text{(11)}\qquad \begin{aligned} & r\sum_{j\neq i} q(i,j;u(i))\int[(w(j,s,u)-w(j,y,u))^2 + w_2(j,s,u)]\, d\,{}^rF_j(s) + \\ & + rq(i,i;u(i))w_2(i,y,u) - \zeta(u) = 0, \qquad i\in I. \end{aligned}$$

$w_2(j,y,u)$, $j\in I$, are a new set of auxiliary functions. Introducing

$$w_2(t) = w_2(X_t,Y_t,\bar{u}(t,C_t,.)), \qquad \zeta_t = \zeta(\bar{u}(t,C_t,.)),$$

we define a martingale

$$\begin{aligned} L_t = & \int_0^t (w(s) - w(s-))^2 d\,N_s - \int_0^t \zeta_s\, ds + \\ & + w_2(t) - w_2(0) - \int_0^t \frac{d}{ds} w_2(s) ds, \qquad t\geq 0. \end{aligned}$$

The last three terms are negligible, since again

$$w_2(j,y,u) = O(\tfrac{1}{r}), \qquad r\longrightarrow\infty .$$

More over,

$$E\, L_t^2 = E \int_0^t (w(s) - w(s-))^4 d\,N_s \longrightarrow 0, \quad t\geq 0, \quad r\longrightarrow\infty .$$

We conclude that

$$\langle M\rangle_t \longrightarrow \int_0^t \sigma(\bar{u}(s,C_s,.))^2 ds, \qquad t\geq 0,$$

where

$$\sigma(u)^2 = \lim_{r\to\infty} \zeta(u), \quad u\in\mathcal{U} .$$

We have established that in the limit

$$C_t - \int_0^t \theta(\bar{u}(s,C_s,.))ds, \qquad t\geq 0,$$

is a martingale with quadratic variation

$$\int_0^t \sigma(\bar{u}(s,C_s,.))^2 ds, \qquad t \geq 0.$$

Consequently,

$$d\, C_t = \theta(\bar{u}(t,C_t,.))dt + \sigma(\bar{u}(t,C_t,.))d\, W_t, \qquad t \geq 0,$$

for a Wiener process $\{W_t,\ t \geq 0\}$. Integrating (11) with respect to $d^r F_i(y)$ and letting $r \to \infty$ equations (7) for $\sigma(u)^2$ are obtained. The analogy between (6) and (7) is apparent.

Example 3. Let us examine, what solution is obtained from the diffusion approximation for the most common risk averting criterion

$$E \exp\{-h\, \tilde{C}_T\} = \min,$$

where $h > 0$. Assume that u constant is used on $[T-\Delta, T]$. Then the increment $\tilde{C}_T - \tilde{C}_{T-\Delta}$ is Gaussian, and

$$E \exp\{-h(\tilde{C}_T - \tilde{C}_{T-\Delta})\} = \exp\{-h\, \theta(u)\Delta + \frac{h^2}{2} \sigma(u)^2 \Delta\}.$$

From here it is seen that the optimal control is constant,

$$U_t = \hat{u}, \qquad t \quad 0,T \quad ,$$

where $\hat{u}$ is the maximizer of

$$h\, \theta(u) - \frac{h^2}{2} \sigma(u)^2 = \xi(u).$$

Combining (6) and (7), we get for $\xi(u)$ the system of equations

$$d(i,u(i)) + \sum_j q(i,j;u(i))w(j,u) - \xi(u) = 0, \qquad i \in I,$$

with

$$d(i,z) = h\, a_i + \frac{h^2}{2} \sum_{j \neq i} q(i,j;z)\left(\frac{b_i}{q(i,i;z)^2} + \frac{b_j}{q(j,j;z)^2}\right) .$$

Thus, $\hat{u}$ can be found by standard methods of dynamic programming, e.g. by Howard's iteration method.

3. Optimization

Using the diffusion approximation the original problem was transformed into the optimization of a controlled Markov process $\{\tilde{C}_t, t \geq 0\}$ with differential generator

$$L^u = \frac{1}{2} \sigma(u)^2 \frac{d^2}{dy^2} + \theta(u)\frac{d}{dy} .$$

With regard to (2),(5) the aim is to achieve

(12) $$\sup E_y\, g(C_T) = V_1(T,y),$$

or

(13) $$\sup E_y \left(\int_0^{\tau} e^{-lt}\, d\,C_t - h\, e^{-l\tau} \right) = \sup E_y \left(\int_0^{\tau} e^{-lt}\, \theta(U_t)dt - h\, e^{-l\tau} \right) = V_2(y).$$

The subscript y refers to the starting position of the process. The dynamic programming equations for (12) and (13) are

(14) $$\frac{\partial}{\partial T} V_1 = \sup_{u \in \mathcal{U}} L^u V_1, \quad V_1(y,0) = g(y),$$

and

(15) $$0 = \sup_{u \in \mathcal{U}} \{ L^u V_2 + \theta(u) - l\, V_2 \}, \quad V_2(0) = -h.$$

Let $\hat{u}(T,y)$ and $\hat{u}(y)$ be the maximizers of the expression on the right hand side of (14) and of (15), respectively. The corresponding optimal feedback controls are

(16) $$U_t = \hat{u}(T-t, C_t), \quad t \in [0,T] \quad \text{and} \quad U_t = \hat{u}(C_t), \quad t \geq 0.$$

It should however be noted that the controls (16) often do not fulfil the smoothness assumptions of the Theorem. Of course, they can be approximated arbitrarily closely by smooth controls. A method for solving (15) is presented in [3], [5].

For illustration let us return to Example 3. We have

$$V_1(T,y) = \inf E_y \exp\{-h\, \tilde{C}_t\} = \exp\{-hy - (h\theta(\hat{u}) - \frac{h^2}{2}\sigma(\hat{u})^2)T\}.$$

Hence,

$$\frac{\partial}{\partial T} V_1 = -(h\theta(\hat{u}) - \frac{h^2}{2}\sigma(\hat{u})^2)V_1 = \inf_{u \in \mathcal{U}} [-(h\theta(u) - \frac{h^2}{2}\sigma(u)^2)V_1] =$$

$$= \inf_{u \in \mathcal{U}} L^u V_1, \quad V_1(y,0) = \exp\{-hy\}.$$

4. References

[1] R.A. Howard : Dynamic Programming and Markov Processes, Cambridge (Mass.)-New York, 1960.

[2] Pham Van Kieu : A diffusion approximation in the ruin problem for a controlled Markov chain. Kybernetika (Prague) 10 (1974), 125-132.

[3] P. Mandl : On aggregating controlled Markov chains. Contributions to Statistics (ed.J.Jurečková) pp.137-156, Prague 1979.

[4] D.Kuklíková-Jarušková: Approximation of bivariate Markov chains by one-dimensional diffusion processes. Aplikace matematiky 23 (1978), 267-279.

[5] V. Lánská : Diffusion approximation for a controlled service system. Kybernetika (Prague) 18 (1982) to appear.

LIE ALGEBRAIC AND APPROXIMATION METHODS FOR SOME NONLINEAR FILTERING PROBLEMS

S.I. Marcus*, C.-H. Liu**, and G.L. Blankenship***

ABSTRACT

State estimation problems for systems involving small parameters are treated by both Lie algebraic and analytical approximation techniques. An asymptotic expansion for the unnormalized conditional density corresponding to the case of observations of a Gauss-Markov process through a (weak) polynomial nonlinearity is computed and a convergence result is derived. The convergence result is based on arguments used recently to prove existence and uniqueness and to estimate the tail behavior of solutions to nonlinear filtering problems with unbounded coefficients. The expansion is related to certain approximations of the associated estimation Lie algebra. Lie algebraic methods are used to compute finite dimensional filters for the terms in the asymptotic expansion.

---

*Department of Electrical Engineering, University of Texas at Austin; Austin, Texas. Supported in part by NSF Grant ECS-8022033.

**Department of Electrical Engineering, National Taiwan Institute of Technology, Taipei, Taiwan. Supported in part by the Joint Services Electronics Program at the University of Texas under Contract F49620-77-C-0101.

***Department of Electrical Engineering, University of Maryland, College Park, Maryland. Supported in part by ONR Contract N00014-79-C-0808.

1. INTRODUCTION

This paper is concerned with the problem of estimating a signal x_t, $t \geq 0$, from noisy observations y_s, $s \leq t$, based on the model

$$dx_t = f(x_t)dt + g(x_t)dw_t$$
$$dy_t = h(x_t)dt + dv_t, \quad 0 \leq t \leq T < \infty \quad . \tag{1.1}$$

Here, f, g, h are smooth functions, w_t, v_t are independent standard Wiener processes, and x_0 is a random variable independent of (w_t, v_t) for all t. We assume that $x_t \in \mathbb{R}$, $y_t \in \mathbb{R}$ for convenience. It is well known that the problem of recursively estimating x_t given $\mathcal{Y}_t = \sigma\{y(s), S \leq t\}$, the σ-algebra generated by the observations, involves in an essential way the (Ito) stochastic partial differential equation (Duncan-Mortensen-Zakai) (DMZ) equation) [1], [2], for the unnormalized conditional density u(t,x)

$$du(t,x) = (\mathcal{L}^*u)(t,x)dt + h(x)u(t,x)dy_t$$
$$\mathcal{L}^*u = \frac{1}{2}\partial_{xx}(g^2(x)u) - \partial_x(f(x)u) \tag{1.2}$$
$$u(0,x) = p_0(x), \text{ the density of } x_0.$$

Specifically, if (1.2) has a "nice" solution, then the conditional density of x_t given $\mathcal{Y}_t$ is $p(t,x) = u(t,x)/\int u(t,z)dz$.

Over the past five years or so a considerable effort has been devoted to the search for (recursive) finite dimensional "representations" in terms of various "sufficient statistics" of either the solution of (1.2) or other conditional statistics (such as the conditional mean). This effort, which has produced few such estimators [3]-[6], has nevertheless led to a major improvement in the understanding of the estimation problem and to the tools available for its treatment. The latter include algebraic methods [7]-[10] and the use of certain transformations which simplify the analysis of the existence and uniqueness of solutions of (1.2) (see, e.g., [11]-[14] and the collection [1]). The algebraic methods are especially useful for classifying equivalences of finite dimensional filters, for indicating when no finite dimensional filters exist, and for facilitating the computation of conditional statistics. In those cases where no finite dimensional representations exist the available methods must be redirected to the construction of consistent and useful approximate filters. This is the central theme of this paper.

We consider in this paper the specific subclass of the problems (1.1) in which f is linear, g is constant, and h is a polynomial of degree $k \geq 1$; for convenience of notation, we will usually consider the system:

$$\begin{aligned} dx_t &= ax_t dt + dw_t \\ dy_t^\varepsilon &= [x_t + \varepsilon(x_t)^k]dt + dv_t, \; k \geq 1 \\ y_0^\varepsilon &= 0; \; p_0(x) \text{ Gaussian} \end{aligned} \tag{1.3}$$

where ε is a small positive parameter; as $\varepsilon \downarrow 0$, we recover the Kalman filtering problem. As is well known, the Kalman filter computes, via a finite dimensional recursion, a sufficient statistic for the associated estimation problem.

The filtering problem (1.3) does not in general admit a finite dimensional solution. However, it is reasonable to expect, as has been argued before [15]-[17], that its solution may be well approximated by the solution (conditional density, moments, etc.) of the Kalman problem when ε is small. The structure of such an approximation is discussed first by means of the natural Lie algebraic structure associated with the operators $\mathscr{L}^* - \frac{1}{2} h^2$ and h in (1.2) (h acts as multiplication by $h(x)$), specialized to (1.3). The Lie algebra associated with (1.3) is infinite dimensional for $k \geq 2$. Approximation structures are defined by identifying certain finite dimensional projections of these Lie algebras by a "mod ε^n" device. This idea was used by Hazewinkel in [17], and we consider it further in Section 2.

This approximation structure corresponds to an analytical approach in which one simply assumes a formal power series in ε for the unnormalized conditional density associated with (1.3), and obtains a formal asymptotic expansion by substituting this expansion into (1.2) and equating coefficients of like powers of ε in the usual way. This simple procedure, used previously in [15], [16], reveals clearly that terms in the formal asymptotic expansion can be evaluated by computations not much more complex than those in the Kalman problem.

In Section 3, we use Lie algebraic methods (in particular, the Wei-Norman representation [18]-[20]) to derive recursive finite dimensional filters for the terms in the asymptotic expansion. This procedure shows clearly that the problem of whether the expansion is a true (not merely formal) asymptotic expansion is equivalent to that of existence, uniqueness, and regularity of the conditional density in the original problem. These points were also made in [15]; here, however, we are able to get a little further since existence and uniqueness questions for nonlinear filtering problems are now better understood than was the case four years ago (see, e.g., [11], [13], [14]). The existence of the asymptotic expansion in an appropriate norm is stated and discussed in Section 4.

A preliminary version of this work appeared in [27]. Related results have been obtained independently with totally different methods by Sussmann [28].

2. LIE ALGEBRAIC METHODS

A Lie algebra $L(\Sigma)$ can be associated with each filtering problem (1.1), (1.2), and it is now widely recognized that the realizability of $L(\Sigma)$ or its quotients with vector fields on a finite dimensional manifold is related to the existence of finite dimensional recursive filters (this idea is originally due to Brockett [7]). The problems (1.3) do not in general admit such realizations, suggesting that for these problems no statistic of the conditional density can be computed with a finite dimensional recursive filter.

Specifically, consider the problem (1.3). The unnormalized conditional density u^ε satisfies (formally -- see [14] for results on existence and uniqueness)

$$\begin{aligned} du^\varepsilon(t,x) &= \mathscr{L}^* u^\varepsilon(t,x)dt + (x+\varepsilon x^k)u^\varepsilon(t,x)dy_t^\varepsilon \\ u^\varepsilon(0,x) &= p_0(x) \\ \mathscr{L}^* u &= \frac{1}{2}\partial_{xx}u - a\partial_x(xu) \end{aligned} \tag{2.1}$$

or in Fisk-Stratonovich form,

$$\begin{aligned} du^\varepsilon(t,x) &= [\mathscr{L}^* - \tfrac{1}{2}(x+\varepsilon x^k)^2]u^\varepsilon(t,x)dt + (x+\varepsilon x^k)u^\varepsilon(t,x)\circ dy_t^\varepsilon \\ u^\varepsilon(0,x) &= p_0(x) \end{aligned} \tag{2.2}$$

where $\circ$ denotes the Fisk-Stratonovich differential. The associated Lie algebra

$$L_{k\varepsilon} = \{\mathscr{L}^* - \frac{1}{2}(x+\varepsilon x^k)^2,\ x+\varepsilon x^k\}_{LA} \tag{2.3}$$

is the smallest vector space of differential operators containing the two operators in (2.3) and closed under the Lie bracket $[D_1, D_2] = D_1D_2 - D_2D_1$.

As a particular case, consider the "weak cubic sensor" when $k = 3$. In [17] it is shown that $L_{3\varepsilon}$ is isomorphic to the Weyl algebra $W_1 = \mathbb{R} < x, \frac{d}{dx} >$ of all differential operators (of any order) with polynomial coefficients (the same as the Lie algebra for the cubic sensor with $h(x) = x^3$ [9]). A basis for W_1 consists of the operators $e_{ij} = x^i \frac{d^j}{dx^j}$, $i, j=0,1,2,\ldots$. W_1 is an infinite dimensional Lie algebra under the commutation $[D_1, D_2] = D_1D_2 - D_2D_1$; its properties are examined in detail in [9] where the following is proved.

<u>Proposition 1</u>

Let M be a finite dimensional manifold, and let V(M) be the Lie algebra of smooth vector fields on M. There is no non-zero homomorphism of Lie algebras from W_1 into V(M).

Combined with the results of [21], [22], this result implies that for k=3 and $\varepsilon \neq 0$, fixed, no non-zero statistic of the conditional distribution for (2.1) can be computed (exactly) with a recursive finite dimensional filter. Of course, for $\varepsilon=0$, the Lie algebra $L_{30} \triangleq L_0$ has basis $\{\mathscr{L}^* - \frac{1}{2}x^2, x, d/dx, 1\}$, the Lie algebra for the Kalman problem; and in this case there is a two-dimensional sufficient statistic that can be evaluated recursively. Thus, as ε passes from zero to $\varepsilon \neq 0$, the filtering problem moves from the simplest to the most difficult class. Hazewinkel [17] has shown similar behavior for k=2; it is suspected that the behavior of the other problems (1.3) with $k \geq 4$ is similar.

To treat the case $\varepsilon \neq 0$ small, it is natural to consider expansions of the conditional density and statistics in powers of ε. This was done for several problems in [15], [16]. Approximations from a Lie algebraic point of view were considered in [17]; this can be summarized as follows. Let $W_1(\varepsilon) = \mathbb{R}\langle x, \varepsilon, d/dx\rangle$ be the Lie algebra of differential operators with coefficients that are polynomials in x and ε. Thus, $W_1(\varepsilon)$ has a basis $\{e_{ijk} = \varepsilon^i x^j d^\ell/dx^\ell;\ i,j,\ell=0,1,\ldots\}$ (Here we regard ε as a "variable".) The associated estimation algebra $L_{k\varepsilon}$ may be regarded as a sub-algebra of $W_1(\varepsilon)$. Using the notation from [17], we define $L_{k\varepsilon} \bmod \varepsilon^n$ as the Lie algebra obtained from $L_{k\varepsilon}$ by setting $\varepsilon^i=0$ for $i \geq n$. (A more precise definition is given in [17].) In [17] it is shown that $L_{k\varepsilon} \bmod \varepsilon^n$ is finite dimensional for each k, n.

This development is related to asymptotic expansions in the following way. Assume that u^ε in (2.1) has a formal expansion

$$u^\varepsilon(t,x) = u_0(t,x) + \varepsilon u_1(t,x) + \varepsilon^2 u_2(t,x) + \ldots . \tag{2.4}$$

Substituting this in (2.1) and equating coefficients of powers of ε, gives

$$\begin{aligned} du_0(t,x) &= \mathscr{L}^* u_0(t,x)dt + x u_0(t,x)dy_t^\varepsilon \\ u_0(0,x) &= p_0(x) \end{aligned} \tag{2.5}$$

$$\begin{aligned} du_\ell(t,x) &= \mathscr{L}^* u_\ell(t,x)dt + x u_\ell(t,x)dy_t^\varepsilon \\ &\quad + x^k u_{\ell-1}(t,x)dy_t^\varepsilon,\ \ell = 1,2,\ldots \end{aligned} \tag{2.6}$$

Writing (2.5)-(2.6) in Fisk-Stratonovich form and truncating after $\ell = n$ gives

$$d\begin{bmatrix} u_0 \\ u_1 \\ u_2 \\ \vdots \\ u_n \end{bmatrix} = \begin{bmatrix} \mathscr{L}^*-\frac{1}{2}x^2 & 0 & & & 0 \\ -x^{k+1} & \mathscr{L}^*-\frac{1}{2}x^2 & & & \\ -\frac{1}{2}x^{2k} & -x^{k+1} & \mathscr{L}^*-\frac{1}{2}x^2 & & \\ & \ddots & \ddots & \ddots & \\ 0 & & -\frac{1}{2}x^{2k} & -x^{k+1} & \mathscr{L}^*-\frac{1}{2}x^2 \end{bmatrix} \begin{bmatrix} u_0 \\ u_1 \\ u_2 \\ \vdots \\ u_n \end{bmatrix} dt$$

$$+ \begin{bmatrix} x & & & 0 \\ x^k & x & & \\ & \ddots & \ddots & \\ 0 & & x^k & x \end{bmatrix} \begin{bmatrix} u_0 \\ u_1 \\ \vdots \\ u_n \end{bmatrix} \circ dy_t^{\varepsilon} \tag{2.7}$$

Define, for $i=1,\ldots,n+1$, the $(n+1) \times (n+1)$ matrix E_i which has zero entries everywhere except for 1's in the sub-diagonal positions $(i,1)$, $(i+1,2),\ldots,(n+1, n+2-i)$ (i.e., on a particular sub-diagonal); note that E_1 is the identity matrix. Also, denoting $U(t,x) = [u_0(t,x), \ldots, u_n(t,x)]'$, we can rewrite (2.7) as

$$dU(t,x) = [(\mathscr{L}^*-\tfrac{1}{2}\, x^2)E_1 - x^{k+1}\, E_2 - \tfrac{1}{2}x^{2k}E_3]U(t,x) + [xE_1 + x^kE_2]U(t,x)\circ dy_t^{\varepsilon} \tag{2.8}$$

The Lie algebra of (2.7) or (2.8) is isomorphic to $L_{k\varepsilon} \bmod \varepsilon^{n+1}$. This can be seen either by a direct calculation (see the next section) or by the observation that there is an obvious isomorphism (with E_2 corresponding to ε) between the Lie algebra of (2.2) modulo ε^{n+1} and that of (2.8), since $(E_2)^{n+1} = 0$. Thus the Lie algebraic construction of $L_{k\varepsilon} \bmod \varepsilon^{n+1}$ corresponds to the equations for the approximation (2.4) to n+1 terms.

Note that the fundamental solutions of (2.5) and (2.6) are the same for every $\ell = 1,2, \ldots$. The fundamental solution is just the unnormalized conditional transition density of the Kalman problem with y_t^{ε}, $0 \le t \le T$, appearing as a parameter. The density $u_0(t,x)$ is, of course, Gaussian. Consider (2.6) in the case $\ell = 1$; clearly, $u_1(t,x)$ can be found by convolving a Gaussian density with a polynomial (x^k) times a Gaussian. This observation has been made elsewhere for related problems [15], [16]. This integral can be computed in closed form. Moreover, the several "moments" which appear in the expression can be evaluated recursively by finite dimensional equations. Such finite dimensional filters for the solution of (2.5)-(2.8) are derived by Lie algebraic methods in the next section. In the construction of formal asymptotic expansions,

Lie algebraic methods serve to organize, simplify, and explicate the structure of the expansions, as well as to lead to finite dimensional filters. However, it is also important to prove that (2.4)-(2.6) is a true asymptotic expansion in an appropriate norm; we indicate how to do this in Section 4.

3. FINITE DIMENSIONAL FILTERS FOR THE TERMS IN THE ASYMPTOTIC EXPANSION

Denote by $L(k, n+1, \varepsilon) \triangleq \{A_0, xE_1 + x^k E_2\}_{LA}$ the Lie algebra of (2.8), where $A_0 \triangleq (\mathscr{L}^* - \frac{1}{2}x^2)E_1 - x^{k+1}E_2 - \frac{1}{2}x^{2k}E_3$. The following theorem describes the structure of $L(k, n+1, \varepsilon)$; then we use Lie algebraic methods to solve (2.8) in terms of a finite number of recursively computable statistics.

Theorem 1: The Lie algebra $L(k, n+1, \varepsilon)$, for $\varepsilon \neq 0$, consists of A_0 and elements which are linear combinations of the matrix differential operators

$$x^r \frac{d^s}{dx^s} E_i; \; i=1,2,\ldots,n+1; \; 0 \le r+s \le 1 + (i-1)(k-1).$$

The dimension of $L(k, n+1, \varepsilon)$ is less than or equal to $1 + \frac{1}{2}\sum_{i=0}^{n-1} [i(k-1)+2] \times [i(k-1)+3]$.

In addition, $L(k, n+1, \varepsilon)$ is a solvable Lie algebra [5], [20] (a Lie algebra L is solvable if the derived series of Lie algebras $L^{(0)}=L$, $L^{(m+1)} \triangleq [L^{(m)}, L^{(m)}] \triangleq \{[A,B] | A,B \in L^{(m)}\}$, $m \ge 0$, is the trivial Lie algebra $\{0\}$ for some m).

Since $L(k, n+1, \varepsilon)$ is solvable and finite dimensional, (2.8) can be solved in terms of a finite number of recursively computable statistics by the method of Wei and Norman [4], [18]-[20]. The calculation, which can be rigorously justified in this case as in [4], [20], proceeds as follows. Assume that the solution of (2.8) can be written in the form

$$U(t,x) = \left[e^{g_0(t)A_0} \cdots e^{g_d(t)A_d} U_0 \right](x) \tag{3.1}$$

where $\{A_i;\, i=0,\ldots,d\}$ is a basis for $L(k, n+1, \varepsilon)$, $U_0(x) = (p_0(x), 0, \ldots, 0)'$, and $\{g_i;\, i=0,\ldots,d\}$ are real-valued functions of t and y^ε to be determined. Substituting (3.1) into (2.8), using the identity

$$e^{tA_i} A_j = \sum_{i=0}^{\infty} \frac{t^k}{k!} \, ad_{A_i}^k \, A_j \; e^{t\,A_i}, \; 0 \le i, j \le d \tag{3.2}$$

where $ad_A^0 B = B$, $ad_A^{k+1} B = [A, ad_A^k B]$,) and equating the coefficients of each basis element A_i, one obtains a set of stochastic differential equations driven by y^ε for $\{g_i,\, i=0,\ldots,d\}$. Since $L(k, n+1, \varepsilon)$ is solvable, there is an ordering of a basis for which the representation (3.1) is globally valid. This method was used to solve the

DMZ equation for the Kalman filtering problem by Ocone [4], [20]; this is precisely equation (2.5) for u_0 (which corresponds to L(k, 1, ε)). Indeed, the representation (3.1) for the first component of U(t,x) is

$$u_0(t,x) = \left[e^{g_0(t)(\mathscr{L}^* - \frac{1}{2}x^2)} e^{g_1(t)x} e^{g_2(t)\frac{d}{dx}} e^{g_d(t)} p_0 \right](x), \tag{3.3}$$

which is the form obtained by Ocone. The other components of U(t,x) are given by

$$u_i(t,x) = b_i(t,x)u_0(t,x) \tag{3.4}$$

where b_i is a polynomial differential operator containing terms of the form $\beta_{rs}(t)\, x^r \frac{d^s}{dx^s}$, $r + s \leq i(k+1)$. The form of the equations is most easily seen by considering an example.

Example 1 (the weak quadratic sensor): Consider the problem (1.3) with k=2 and a=0, and the first order approximation (n=1 in (2.7)-(2.8)); thus (2.7)-(2.8) becomes

$$d\begin{bmatrix} u_0 \\ u_1 \end{bmatrix} = \begin{bmatrix} \mathscr{L}^* - \frac{1}{2}x^2 & 0 \\ -x^3 & \mathscr{L}^* - \frac{1}{2}x^2 \end{bmatrix} \begin{bmatrix} u_0 \\ u_1 \end{bmatrix} dt + \begin{bmatrix} x & 0 \\ x^2 & x \end{bmatrix} \begin{bmatrix} u_0 \\ u_1 \end{bmatrix} \circ dy_t^\varepsilon \tag{3.5}$$

The Lie algebra L(2,2,ε) is ten-dimensional, with basis elements

$A_0 = \mathscr{L}^* E_1 - x^3 E_2$, $A_1 = xE_1$, $A_2 = \frac{d}{dx} E_1$,

$A_3 = x^2 E_2$, $A_4 = xE_2$, $A_5 = x \frac{d}{dx} E_2$, $A_6 = \frac{d}{dx} E_2$,

$A_7 = \frac{d^2}{dx^2} E_2$, $A_8 = E_2$, $A_9 = E_1$.

Substituting the representation (3.1) into (3.5), and using (3.2) repeatedly, the following differential equations for $\{g_i;\ i=0,\ldots,9\}$ are obtained (the Lie algebraic manipulations, similar to those in [19], were performed with the aid of the MACSYMA symbolic manipulation program at M.I.T.; for details, see [30]);

$$\begin{aligned} g_0(t) &= t \\ dg_1(t) &= \cosh t \circ dy_t^\varepsilon \\ dg_2(t) &= -\sinh t \circ dy_t^\varepsilon \\ dg_3(t) &= c_1(t) \circ dy_t^\varepsilon \\ dg_4(t) &= [2c_2(t)g_1(t) - 2c_1(t)g_2(t)] \circ dy_t^\varepsilon \end{aligned} \tag{3.6}$$

$$dg_5(t) = 2c_2(t) \circ dy_t^{\varepsilon}$$
$$dg_6(t) = [2c_3(t)g_1(t) - 2c_2(t)g_2(t)] \circ dy_t^{\varepsilon}$$
$$dg_7(t) = c_3(t) \circ dy_t^{\varepsilon}$$
$$dg_8(t) = [c_1(t)[g_2(t)]^2 - 2c_2(t)g_1(t)g_2(t) + c_3(t)[g_1(t)]^2 + c_2(t)] \circ dy_t^{\varepsilon}$$
$$dg_9(t) = -(\sinh t)g_1(t) \circ dy_t^{\varepsilon}$$
$$g_0(0) = \cdots = g_9(0) = 0$$

where the deterministic coefficients $\{c_i(t);\ i=1,2,3\}$ are given by

$$\begin{aligned} c_1(t) &= \cosh 2t + \cosh t - 1 \\ c_2(t) &= \sinh t - \sinh 2t \\ c_3(t) &= \cosh 2t - 2\cosh t + 1 \end{aligned} \qquad (3.7)$$

The actual form of u_0 and u_1, as given by (3.1), is obtained by calculating $\exp(g_i(t)A_i)$, $i=0,\ldots,9$, and using (3.2); the solution (assuming the initial density is $p_0(x) = \delta(x - x_0)$) is

$$u_0(t,x) = \left[e^{t(\mathscr{L}^*-\frac{1}{2}x^2)} e^{g_1(t)x} e^{g_2(t)\frac{d}{dx}} e^{g_9(t)} p_0\right](x)$$

$$= \frac{1}{\sqrt{2\pi \sinh t}} \exp\left\{-\frac{1}{2}(\tanh t)(x_0-g_2(t))^2 + g_1(t)(x_0-g_2(t))\right.$$

$$\left. + g_9(t) - \frac{1}{2}(\tanh t)^{-1}[x-(\cosh t)^{-1}(x_0-g_2(t))]^2\right\} \qquad (3.8)$$

and $u_1(t,x)$ is $q(t,x)u_0(t,x)$, where q is a polynomial in x with coefficients expressed in terms of $g_0,\ldots,g_9$ (see [30] for the exact expression).

Thus the "zero"th order approximation of u^{ε} is just the unnormalized conditional density of the Kalman filter, and the first order approximation is $u_0(t,x) + \varepsilon u_1(t,x)$, where u_0 and u_1 are computed from the nine y^{ε}-dependent statistics $g_1,\ldots,g_9$. A straightforward but involved off-line calculation is involved to obtain these equations (this is greatly facilitated by MACSYMA), but the only on-line calculations are the simple equation (3.6) for the g_i's and the calculation of u_0 and u_1 as memoryless functions of the g_i's; this constitutes a finite dimensional recursive filter for $u_0 + \varepsilon u_1$. It is also clear that finite dimensional filters for the zero-th and first order approximations of the normalization factor $\int u^{\varepsilon}(t,z)dz$, the normalized conditional density $p(t,x)$, and the conditional mean $\hat{x}(t|t)$ can be obtained as functionals of $g_1,\ldots,g_9$. The on-line computation is the same; only the off-line calculation of the form of the functionals is different. These calculations and some numerical experiments to test the performance of these filters will be reported elsewhere.

4. ASYMPTOTIC EXPANSION RESULTS

To justify the use of algebraic methods on equation (2.2) and its formal asymptotic expansion, it must be proved that (2.2) has a unique solution within an appropriate class of functions, and that the formal expansion (2.4)-(2.6) is in fact an asymptotic expansion in an appropriate norm. Existence and uniqueness results are proved in [13], [14] by first applying certain exponential transformations to (2.2), and then using the "classical" theory of PDE's [24], [25]. Here we concentrate on outlining the proof that (2.4)-(2.6) is an asymptotic expansion.

Consider the system (2.2), (2.4)-(2.6) and let $\theta^\varepsilon(t,x) = u^\varepsilon(t,x) - u_0(t,x)$. Then θ^ε satisfies the Stratonovich stochastic PDE

$$\begin{aligned} d\theta^\varepsilon(t,x) &= [\mathscr{L}^* - \tfrac{1}{2}(h^\varepsilon(x))^2]\theta^\varepsilon(t,x)dt + (h^\varepsilon(x))\theta^\varepsilon(t,x)\circ dy_t^\varepsilon \\ &\quad + \varepsilon x^k u_0(t,x)\circ dy_t^\varepsilon - \varepsilon(x + \tfrac{1}{2}\varepsilon x^k)x^k u_0(t,x)dt \\ \theta^\varepsilon(0,x) &= p_0^\varepsilon(x) - p_0{}^0(x),\ 0 \le t \le T. \end{aligned} \tag{4.1}$$

Here we have generalized (2.2) by letting $u^\varepsilon(0,x)$ depend on ε. Also,

$$\begin{aligned} du_0(t,x) &= [\mathscr{L}^* - \tfrac{1}{2}x^2]u_0(t,x)dt + xu_0(t,x)\circ dy_t^\varepsilon \\ u_0(0,x) &= p_0{}^0(x),\ 0 \le t \le T. \end{aligned} \tag{4.2}$$

If we form the vector $W^\varepsilon(t,x) = [\theta^\varepsilon(t,x), u_0(t,x)]^T$, then

$$\begin{aligned} dW^\varepsilon(t,x) &= (\tfrac{1}{2}\begin{bmatrix} \mathscr{L}^* & 0 \\ 0 & \mathscr{L}^* \end{bmatrix} W^\varepsilon(t,x) - \tfrac{1}{2}(H^\varepsilon(x))^2 W^\varepsilon(t,x))dt \\ &\quad + H^\varepsilon(x)W^\varepsilon(t,x)\circ dy_t^\varepsilon \end{aligned} \tag{4.3}$$

where

$$H^\varepsilon(x) = \begin{bmatrix} h^\varepsilon(x) & \varepsilon x^k \\ 0 & x \end{bmatrix} \tag{4.4}$$

Introducing

$$V^\varepsilon(t,x) = \exp[-H^\varepsilon(x)y_t^\varepsilon]W^\varepsilon(t,x) \tag{4.5}$$

the first component V_1^ε of the 2-vector V^ε satisfies a forced parabolic PDE (for each Hölder continuous path of y_t^ε)

$$\frac{\partial V_1^\varepsilon}{\partial t} = \frac{1}{2}\partial_{xx}V_1{}^\varepsilon + m^\varepsilon(t,x)\partial_x V_1{}^\varepsilon + n^\varepsilon(t,x)V_1^\varepsilon + \varepsilon f^\varepsilon(t,x)u_0(t,x) \tag{4.6}$$

$$V_1^\varepsilon\ (0,x) = \theta^\varepsilon(0,x) = p_0^\varepsilon(x) - p_0{}^0(x) = \pi^\varepsilon(x)$$

where

$$m^\varepsilon(t,x) = h_x^\varepsilon y_t^\varepsilon - ax$$

$$n^\varepsilon(t,x) = \frac{1}{2}\,[h_{xx}^\varepsilon y_t^\varepsilon + (h_x^\varepsilon y_t^\varepsilon)^2 - 2axh_x^\varepsilon y_t^\varepsilon - 2a - (h^\varepsilon)^2] \tag{4.7}$$

$$\varepsilon f^\varepsilon(t,x)u_0(t,x) = e^{-h^\varepsilon y_t^\varepsilon}[\mathscr{L}^* - \frac{1}{2}\,(h^\varepsilon)^2][e^{\varepsilon x^k y_t}-1)u_0]$$

$$+ (e^{-h^\varepsilon y_t}-e^{-xy_t^\varepsilon})[\mathscr{L}^* - \frac{1}{2}\,x^2]u_0$$

$$-\varepsilon x^k(x + \frac{1}{2}\,\varepsilon x^k)e^{-h^\varepsilon y_t^\varepsilon}u_0$$

The analysis of (4.6) proceeds as follows (see [14]). Note first that the term $-\frac{1}{2}\,(h^\varepsilon(x))^2$ dominates the potential $n^\varepsilon(t,x)$ as $|x| \to \infty$. Thus, the fundamental solution $\hat{\Gamma}^\varepsilon(t,x;s,z)$ of (4.6) "falls off" like $\exp[-\hat{\beta}^\varepsilon|x|^{k+1}]$ for some $\hat{\beta}^\varepsilon \in (0,1)$ as $|x| \to \infty$, and similarly as $|z| \to \infty$. The forcing function in (4.6) has $f^\varepsilon(t,x) \sim p^\varepsilon(t,x)e^{-xy_t^\varepsilon}$ where $p^\varepsilon(t,x)$ is a polynomial in x with bounded coefficients (on $\varepsilon > 0$, $t \in [0,T]$ for each fixed path y_t^ε). Thus, fi the initial data $\pi^\varepsilon(x)$ is bounded, then (4.6) has a well-defined solution

$$V_1^\varepsilon(t,x) = \int_{-\infty}^{\infty}\hat{\Gamma}^\varepsilon(t,x;0,z)[p_0^\varepsilon(z) - p_0{}^0(z)]dz$$
$$+ \varepsilon\int_0^t \int_{-\infty}^{\infty}\hat{\Gamma}^\varepsilon(t,x;s,z)f^\varepsilon(s,z)u_0(s,z)dzds \tag{4.8}$$

which approaches zero as $|x| \to \infty$, If $\pi^\varepsilon = p_0^\varepsilon - p_0{}^0 = O(\varepsilon)$, then $V_1^\varepsilon = O(\varepsilon)$. Precisely,

<u>Theorem 2</u>: Suppose (A1) $\pi^\varepsilon(x) = p_0^\varepsilon(x) - p_0{}^0(x)$ is bounded, continuous and approaches zero as $|x| \to \infty$. Suppose (A2) further that $\pi^\varepsilon = O(\varepsilon)$; i.e.,

$$\lim_{\varepsilon\downarrow 0}\ [\sup_{x\in\mathbb{R}}|\pi^\varepsilon(x)|/\varepsilon] = c < \infty \tag{4.9}$$

and that $p_0{}^0(x) \geq 0$ is bounded, continuous and approaches zero as $|x| \to \infty$ with $\int_{\mathbb{R}} p_0{}^0(x)dx < \infty$. Then, (for almost every path y_t^ε),

$$\lim_{\varepsilon\downarrow 0}\ [\sup_{\substack{0\leq t\leq T\\ x\in\mathbb{R}}}\ |V_1(t,x)|/\varepsilon] = c < \infty \tag{4.10}$$

If, in addition, (A3) $\int_{\mathbb{R}} |\pi^\varepsilon(x)|\,dx < \infty$, then

$$\lim_{\varepsilon \downarrow 0} \left[\sup_{0 \leq t \leq T} \int_{\mathbb{R}} |V_1^\varepsilon(t,x)|\,dx/\varepsilon\right] = c < \infty \tag{4.11}$$

Proof: The assumptions (A1), (A2) guarantee that (4.6) has a unique solution $V_1^\varepsilon(t,x)$ (using the results of Besala [25] and Kryzanskii [26]), and that $|V_1^\varepsilon(t,x)| \to 0$ as $|x| \to \infty$. (The assumption on $p_0{}^0(x)$ guarantees that $u_0(s,z)$ behaves like $\exp[-\alpha z^2]$ as $|z| \to \infty$.) The estimates (4.10), (4.11), follow from Cauchy's inequality applied to (4.8). QED.

From (4.5) we see that

$$\begin{aligned}\theta^\varepsilon(t,x) &= (e^{h^\varepsilon(x)y_t^\varepsilon} - e^{xy_t^\varepsilon})V_2^\varepsilon(t,x) + e^{h^\varepsilon(x)y_t^\varepsilon}V_1^\varepsilon(t,x) \\ &= \sum_{i=1}^{\infty} \frac{1}{i!}(\varepsilon x^k y_t^\varepsilon)^i u_0(t,x) + e^{h^\varepsilon(x)y_t^\varepsilon}V_1^\varepsilon(t,x)\end{aligned} \tag{4.12}$$

Thus θ^ε is $O(\varepsilon)$ in the L^1-norm of (4.11), since the first term in (4.12) consists of powers of ε multiplied by moments of a Gaussian distribution and V_1^ε falls off like $\exp[-\hat{\beta}^\varepsilon |x|^{k+1}]$. Using a similar proof, it can also be shown that $u^\varepsilon(t,x) - \sum_{i=0}^{n} \varepsilon^i u_i(t,x) = O(\varepsilon^{n+1})$ in the same norm. In addition, it is straightforward to justify the asymptotic expansions of $p(t,x)$ and $\hat{x}(t|t)$, which are computed as indicated at the end of Section 3.

CONCLUSIONS

The asymptotic analysis of the simple filtering problem involving observations of a Gauss-Markov process through a "weak" polynomial nonlinearity carried out here using a combination of Lie algebraic and analytic methods can be readily extended to a large variety of nonlinear filtering problems containing small parameters. The conditions for use of the procedure are (i) that the limit problem (corresponding to $\varepsilon = 0$) have a simple form (e.g., a Kalman filtering problem), (ii) that the Lie algebraic structures associated with the perturbed and limit problems be of a known classified type, and (iii) that existence and uniqueness conditions and analytical bounds on the solutions of the perturbed and limited PDE's be readily available. Since each of these three points is now reasonably well understood in the theory of nonlinear filtering, the consolidation achieved here may be readily extended to many of the examples discussed in [15], [16], for example.

It remains to perform some numerical experiments to test the usefulness of the asymptotic expansions in specific contexts. Work on this is now under way and will be reported elsewhere. In this connection it is worth noting that the truncated asymptotic expansions (2.4) do not always produce densities. Starting from a different point, based on the fact that $u^\varepsilon(t,x)$ is a density, one can construct asymptotic expansions whose truncated versions are always positive. However, the resulting expansions may have very poor convergence properties. This work will be reported in [31].

ACKNOWLEDGEMENTS

The third author would like to thank W.E. Hopkins and J.S. Baras for discussions on the results in Section 4. The calculations in Section 3 were performed with the aid of the MACSYMA program of the Mathlab Group at M.I.T., which is supported, in part, by the Department of Energy and the National Aeronautics and Space Administration.

REFERENCES

1. Hazewinkel, M., Willems, J.C., eds., Stochastic Systems: The Mathematics of Filtering and Identification and Applications, NATO Advanced Study Institute Series, Reidel, Dordrecht, 1981.

2. Davis, M.H.A., Marcus, S.I., "An introduction to nonlinear filtering," in [1].

3. Benes, V.W., "Exact finite-dimensional filters for certain diffusions with non-linear drift," Stochastics, 5, (1981), pp. 65-92.

4. Ocone, D., Ph.D. Thesis, Dept. of Mathematics, M.I.T., Cambridge, Mass., 1980.

5. Marcus, S.I. and Willsky, A.S., "Algebraic structure and finite dimensional nonlinear estimation," SIAM J. Math. Anal., 9 (1978), pp. 312-327.

6. Ocone, D.L., Baras, J.S., and Marcus, S.I., "Explicit filters for diffusions with certain nonlinear drifts," to appear in Stochastics.

7. Brockett, R., "Remarks on finite dimensional nonlinear estimation," in Analyse des Systems (Bordeaux, 1978), C. Lobry, Ed., and Astérisque, 75, 76, Soc. Math. de France, 1980.

8. Mitter, S.K., "The analogy between mathematical problems of nonlinear filtering and quantum physics," Richerche di Automatica, X (1979), pp. 163-216.

9. Hazewinkel, M., Marcus, S.I., "On Lie algebras and finite dimensional filtering," to appear in Stochastics.

10. Liu, C.-H. and Marcus, S.I., "The Lie algebraic structure of a class of finite dimensional nonlinear filters," in Algebraic and Geometric Methods in Linear Systems Theory, Lectures in Applied Math., vol. 18, C.I. Byrnes and C.F. Martin (eds.), Amer. Math. Soc., Providence, 1980, pp. 277-297.

11. Pardoux, E., "Stochastic partial differential equations and filtering of diffusion processes," Stochastics, 2 (1979), pp. 127-168.

12. Davis, M.H.A., "Pathwise nonlinear filtering," in [1].

13. Baras, J.S., Blankenship, G.L., Mitter, S.K., "Nonlinear filtering of diffusion processes," Proc. IFAC Triennial Congress, Kyoto, 1981.

14. Baras, J.S., Blankenship, G.L., Hopkins, W.E., "Existence, uniqueness, and asymptotic behavior of solutions to a class of Zakai equations with unbounded coefficients," to appear in IEEE Trans. Automatic Control.

15. Blankenship, G.L., Haddad, A.H., "Asymptotic analysis of a class of nonlinear filtering problems," Proc. IRIA-IFAC Workshop Singular Perturbations in Control, Paris, 1978, pp. 1-15.

16. Blankenship, G.L., "Some approximation methods in nonlinear filtering," Proc. IEEE Conf. Decision and Control, Albuquerque, 1980, pp. 51-56.

17. Hazewinkel, M., "On deformations, approximations, and nonlinear filtering," to appear in Systems and Control Letters.

18. Wei, J. and Norman, E., "On global representations of the solutions of linear differential equations as a product of exponentials," Proc. Amer. Math. Soc., 15 (1964), pp. 327-334.

19. Steinberg, S., "Applications of the Lie algebraic formulas of Baker, Campbell, Hausdorff, and Zassenhaus to the calculation of explicit solutions of partial differential equations, J. Diff. Eq., 26 (1977), pp. 404-434.

20. Ocone, D., "Nonlinear filtering problems with finite dimensional Lie algebras," Proc. 1980 Joint Automatic Control Conf., San Francisco, 1980.

21. Hazewinkel, M., Marcus, S.I., Sussmann, H.J., "Nonexistence of exact finite dimensional filters for the cubic sensor problem," in preparation.

22. Sussmann, H.J., "Rigorous results on the cubic sensor problem," in [1].

23. Baras, J.S., Blankenship, G.L., "Nonlinear filtering of diffusion processes: a generic example," to appear.

24. Aronson, D.G., Besala, P., "Parabolic equations with unbounded coefficients," J. Diff. Eqns., 3 (1967), pp. 1-14.

25. Besala, P., "On the existence of a fundamental solution for a parabolic equation with unbounded coefficients," Ann. Polonici Math., 29 (1975), pp. 403-409.

26. Kryzanskii, M., Partial Differential Equations of Second Order, Polish Scientific Publishers, Warsaw, 1971 (Theorem 1, p. 201).

27. Marcus, S.I., Liu, C.-H., and Blankenship, G.L., "Lie algebras and asymptotic expansions for some nonlinear filtering problems," in Proc. 1981 Joint Automatic Control Conf., Charlottesville, June, 1981.

28. Sussmann, H.J., "Approximate finite dimensional filters for some nonlinear problems," preprint, June 1981.

29. Sagle, A.A. and Walde, R.E., Introduction to Lie Groups and Lie Algebras. New York: Academic Press, 1973.

30. Liu, C.-H., Ph.D. Thesis, Dept. of Electrical Engineering, University of Texas at Austin, Austin, Texas, 1981.

31. Hopkins, W.E., Ph.D. Thesis, Dept. of Electrical Engineering, University of Maryland, College Park, Maryland, to appear.

32. Fleming, W.H., Mitter, S.K., "Optimal control and nonlinear filtering for non-degenerate diffusion processes," to appear.

OPTIMAL STOPPING FOR TWO-PARAMETER PROCESSES

G. MAZZIOTTO and J. SZPIRGLAS

Centre National d'Etudes des Télécommunications
92 131 - ISSY LES MOULINEAUX - FRANCE

Summary: A formalism for the optimal stopping of two-parameter processes is developed by analogy with the classical theory. An optimality criterion is established in terms of the conditional pay-off process. The problem of optimal stopping of a process indexed by $\mathbb{N}^2$ is completely solved, in probabilistic terms, by using the notion of tactics. The method consis of searching for an optimal stopping point among the maximal stopping points up to which the Snell envelope is a martingale. On $\mathbb{R}_+^2$ the difficulties arise from the lack of information about the behaviour of two-parameter supermartingales, and particularly the Snell envelope. The optimal stopping of a Brownian sheet is solved and we present the case of the bi-Brownian process. Associated systems of variational inequalities are proposed.

I-1 Preliminaries

The processes considered here are indexed by the set $\mathbb{T}$, where $\mathbb{T} = \mathbb{N}^2$ or $\mathbb{R}_+^2$. They are extended to its one-point compactification $\overline{\mathbb{T}} = \mathbb{T} \cup \{\infty\}$ by taking the null value at the infinity. The partial ordering of $\mathbb{T}$ is defined by:

$$s = (s_1,s_2) \leq t = (t_1,t_2) \quad \text{iff} \quad s_1 \leq t_1 \quad \text{and} \quad s_2 \leq t_2.$$

For two points s and t, $s \wedge t$ (resp. $s \vee t$) denotes the point

$$(\inf(s_1,t_1),\ \inf(s_2,t_2)) \ (\text{resp. } (\sup(s_1,t_1),\ \sup(s_2,t_2))) .$$

Let $(\Omega, \underline{\underline{A}}, \mathbb{P})$ be a complete probability space. A filtration $\underline{\underline{F}}$ is a family $(\underline{\underline{F}}_t;\ t \in \mathbb{T})$ of sub-σ-algebras of $\underline{\underline{A}}$ such that $\underline{\underline{F}}$ is increasing for the partial ordering, $\underline{\underline{F}}_\circ$ contains all $\mathbb{P}$-negligible sets of $(\Omega, \underline{\underline{A}})$ and (in case $\mathbb{T} = \mathbb{R}_+^2$), $\underline{\underline{F}}$ is right-continuous.

A random variable (r.v.) T, taking its values in $\overline{\mathbb{T}}$, is a stopping point (s.p.) iff:

$$\forall t \in \mathbb{T}\ ,\quad \{ T \leq t \} \in \underline{\underline{F}}_t$$

The set of s.p. is denoted by $\underline{\underline{T}}$. To a s.p. T, we associate the σ-field $\underline{\underline{F}}_T$ of events A such that:

$$A \cap \{T \leq t\} \in \underline{\underline{F}}_t\ , \forall t \in \mathbb{T}$$

Given two s.p. S and T, one can prove (12) -as in the classical theory- that:

$$\{ S = T \} \in \underline{\underline{F}}_S \cap \underline{\underline{F}}_T,\quad \{ S \leq T \} \in \underline{\underline{F}}_T$$

$$\text{but generally}\qquad ,\ \{ S \leq T \} \notin \underline{\underline{F}}_S$$

$S \vee T$ is a s.p., but $S \wedge T$ is generally not.

Two given processes X and Y are said to be pseudo-indistinguishable iff

$$\forall T \in \underline{\underline{T}}\ ,\quad X_T = Y_T \quad \text{a.s.}\ .$$

I-2 <u>The optimal stopping problem</u>

The pay-off is a non negative optional process of class (D)- i.e. the set $(Y_T;\ T \in \underline{\underline{T}})$ is uniformly integrable. The problem consists of finding a s.p. T^* , here called optimal, such that:

$$E(Y_{T^*}) = \sup_{T \in \underline{\underline{T}}} E(Y_T)$$

As in the classical theory- of (2) and (5)- the main tool is the conditional pay-off system of r.v., $(J(T);\ T \in \underline{\underline{T}})$, defined by

$$J(T) = \operatorname*{esssup}_{\substack{S \geq T \\ S \in \underline{\underline{T}}}} E(Y_S / \underline{\underline{F}}_T)\ ,\quad \forall T \in \underline{\underline{T}}\ .$$

By generalizing the Mertens theorem (2) to the bidimensional case one can prove (3), there exists one process J - up to pseudo-indistinguishability - which agregates this system. J is the Snell enveloppe of Y, i.e. the smallest strong super-martingale greater than Y:

$$\forall T \in \underline{\underline{T}} : J_T \geq Y_T \text{ a.s.},\ J_T \text{ is } \underline{\underline{F}}_T\text{-measurable}$$

$$\forall S,T \in \underline{\underline{T}},\ S \leq T \Rightarrow E(J_T/\underline{\underline{F}}_S) \leq J_S \text{ a.s.}$$

The optimality criterion - Bellman principle - has been proved (9) to be:

<u>*Proposition*</u>*: A stopping point T^* is optimal if and only if:*
(*i*) $J_{T^*} = Y_{T^*}$ *a.s.*
(*ii*) $E(J_{T^*}) = E(J_0)$

In the classical theory, the above criterion is used -(2),(5)- in the following manner. Under some appropriate regularity conditions the smallest s.p. such that (i) holds - the debut of $\{ Y = J \}$ - also verifies (ii) and then is optimal. We suggest in (9) that it seems more tractable, in the two-parameter context, to search for a s.p., which satisfies (i) among the maximal s.p. such that (ii) holds.

A s.p. T will be called maximal if it is a maximal element with respect to the partial ordering on $\underline{\underline{T}}$, of the set:

$$\{ T \in \underline{\underline{T}} : E(J_T) = E(J_0) \}$$

As a matter of fact, the debut of the random set $\{ Y = J \}$ is a weak stopping line (10) which has no a priori reason to contain any s.p.. Moreover those s.p. could be either optimal or not, as in the following example:

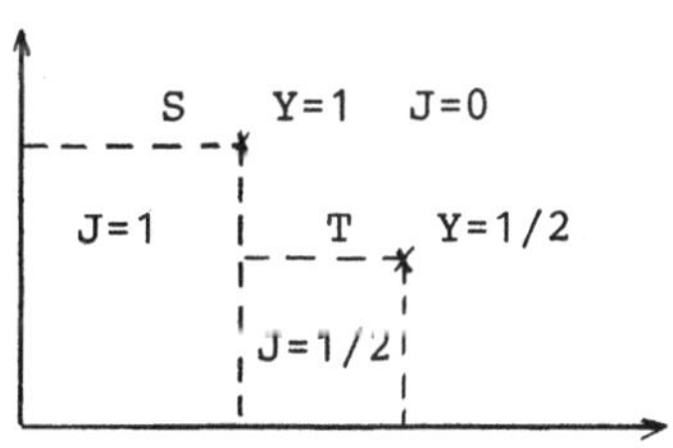

Debut $\{ Y = J \} = \{ S,T \}$
S is optimal, but T is not!

The existence of maximal s.p. can be deduced from Zorn's lemma under quite general assumptions on the Snell enveloppe. In the following we assume that J is left continuous in expectation on s.p. (that is to say: $\lim_n E(J_{T_n}) = E(J_{\lim_n T_n})$ for every increasing sequence $(T_n ; n \in \mathbb{N})$ of s.p.)

Then the main problem is to prove that a maximal s.p. is indeed an optimal s.p.. In the case of processes indexed by one-dimensional sets, $\mathbb{N}$ or $\mathbb{R}_+$, this could be deduced from the Mertens decomposition (6) of the supermartingale J. But for two-parameter supermartingales, we lack such a useful decomposition. For these reasons, we consider particular situations in which we can say more about the problem or the Snell envelope.

II- Optimal stopping on $\mathbb{N}^2$

The optimal stopping problem, for a process indexed by $\mathbb{N}^2$, presented here, generalizes in some ways that of (7) which is mainly concerned with sums of independent identically distributed r.v.. Moreover we solve by purely probabilistic methods the problem solved in (8) via dynamic programming. The main tool is due to (7) and (8). For that purpose we assume that the filtration satisfies the condition of conditional qualitative independence of (7) or more particularly, verifies the F4 property of (13).

To $t = (t_1,t_2)$ in $\mathbb{N}^2$, we associate the points t', t" and t^+, respectively defined by (t_1+1,t_2), (t_1,t_2+1) and (t_1+1,t_2+1).

A tactic is a random path in $\mathbb{N}^2$, increasing from 0 to ∞ : $Z = (Z_0 = 0;\ Z_{p+1} = Z'_p \text{ or } Z''_p,\ \forall p;\ Z_\infty = \infty)$, such that, for each p, Z_p is a s.p. and Z_{p+1} is measurable with respect to the σ-field $\underline{\underline{F}}_{Z_p}$.

The main advantage of tactics comes from the fact - proved in (7) and (12) - that, given two s.p. S and T such that $S \leq T$, there exists one tactic which goes through S and T, a.s..

To solve the optimal stopping problem, we define a tactic Z^*, and afterwards, a s.p. T^* on this tactic, by the following recurrence.

$$Z^*_0 = 0 \quad \text{and} \quad \forall p > 0 \quad \text{let}$$

$$Z^*_{p+1} = t'' \quad \text{on } \{Z^*_p = t\} \cap \{J_t = E(J_{t''}/\underline{\underline{F}}_t)\}$$

$$Z^*_{p+1} = t' \quad \text{on } \{Z^*_p = t\} \cap \{J_t > E(J_{t''}/\underline{\underline{F}}_t)\} ,$$

$$T^* = Z^*_p \quad \text{on } \{Z^*_p = t\} \cap \{J_t > \inf(E(J_{t''}/\underline{\underline{F}}_t), E(J_{t'}/\underline{\underline{F}}_t))\} , \ \forall\, t \in \mathbb{N}^2.$$

If we assume that Y is a.s. continuous up to infinity - i.e. $\lim_t Y_t$ exists and equals Y_∞ - then, T^* is a maximal s.p.. Moreover, we prove in (9) the following result.

Proposition: Every maximal stopping point is optimal.

III- Optimal stopping on $\mathbb{R}^2_+$

At present we cannot propose any general method to solve the optimal stopping problem of a process indexed by $\mathbb{R}^2_+$. The main reason is the lack of information about the behaviour of the supermartingale J, the Snell envelope. Consequently, we turn here towards particular cases in which J is better known, thanks to its functional features.

III-1 Optimal stopping of the Brownian sheet

Let $W = (W_t;\ t \in \mathbb{R}_+^2)$ be a Brownian sheet, defined on its canonical space $(\Omega, \underline{\underline{A}}, (\underline{\underline{F}}_t), \mathbb{P})$; let f be a positive, bounded continuous function and, α a positive constant. The pay-off process is given by:

$$Y_t = \exp(-\alpha t_1 t_2)\ f(W_t) \quad , \quad \forall\ t = (t_1, t_2) \in \mathbb{R}_+^2 .$$

The Snell envelope, J, is computed in (9), We obtain the following formula:

$$J_t = \exp(-\alpha t_1 t_2)\ q(W_t)$$

where q is the Snell reduite of f with respect to the classical Brownian semigroup on $\mathbb{R}_+$, see (11). In addition, when q is sufficiently differentiable, it can be defined as the solution of a classical system of variational inequations, (1).

The notion of optional increasing paths (o.i.p.), due to (12), generalizes that of tatics to $\mathbb{R}^2$. An o.i.p. is a family of s.p., $Z = (Z_u;\ u \in \mathbb{R}_+)$, increasing for the partial ordering (with $Z_0 = 0$ and $Z_\infty = \infty$) and such that the application $u \to Z_u$ is continuous from $\mathbb{R}_+$ into $\mathbb{R}_+^2$.

We prove in (9), the following result:

Proposition: On each optional increasing path, Z, not identically equal to the coordinate axes, there exists an optimal stopping point, T^*, given by:
$T^* = Z_{\tau^*}$, *with* $\tau^* = \inf\{\ u\ /\ q(W_{Z_u}) = f(W_{Z_u})\ \}$

III-2 Optimal stopping of a Bi-Markov Process

This example can be explained in the context of stochastic games theory. The evolutions of two independent players are modelized by two Markov processes defined on their canonical spaces; for $i = 1,2 : X^i = (\Omega^i, \underline{\underline{A}}^i, X^i_{t_i}, \underline{\underline{F}}^i_{t_i}, (\mathbb{P}^i_{x_i}; x_i \in E_i))$. They must stop at - possibly different- times T_1 and T_2 such that the average of common reward

$$E(\exp(-\alpha_1 T_1 - \alpha_2 T_2)\ f(X^1_{T_1}, X^2_{T_2}))$$

is maximum. In that formula, f is a given positive bounded function on $E_1 x E_2$ and α_1, α_2 are some positive constants. Times T_1 and T_2 have to be causally choosen; that is to say, knowing the sample paths of X^1 up to T_1 and X^2 up to T_2 only. In other words, for every given real t_1, t_2, the event ($T_1 \leq t_1$, $T_2 \leq t_2$) must depend only on the r.v. ($X^1_{s_1}$, $X^2_{s_2}$, $s_1 \leq t_1$, $s_2 \leq t_2$).

The two parameter model is defined on the following tensorial products:

$\Omega = \Omega^1 x \Omega^2$, $\underline{\underline{A}} = \underline{\underline{A}}^1 \otimes \underline{\underline{A}}^2$, $\mathbb{P}_x = \mathbb{P}^1_{x_1} \otimes \mathbb{P}^2_{x_2}$, $E = E^1 x E^2$, $\underline{\underline{F}}_t = \underline{\underline{F}}^1_{t_1} \otimes \underline{\underline{F}}^2_{t_2}$

$X_t = (X^1_{t_1}, X^2_{t_2})$, $Y_t = \exp(-\alpha.t)\ f(X_t)$ where $t = (t_1,t_2)$ and $\alpha.t = \alpha_1 t_1 + \alpha_2 t_2$.

The Snell envelope, J, for the optimal stopping of process Y defined on the filtered probability space $(\Omega,\underline{\underline{A}},(\underline{\underline{F}}_t;t \in \mathbb{R}^2_+),\mathbb{P}_x)$ is computed in (9). It satisfies

$$J_t = \exp(-\alpha.t)\ q(X_t)$$

where q is the smallest α-biexcessive function on E - i.e. for every fixed coordinate, the function q is α_i-excessive as in (2) with respect to the other coordinate.The function q is called the α-reduite of f. Moreover when q is sufficiently differentiable, it satisfies a system of variational inequalities. For example in case X^1 and X^2 are standard Brownian motions in some open sets and $\alpha_1 = \alpha_2 = 0$, this system is the following:

$$q \geq f$$

$$\Delta^1 q \leq 0\ ,\ \Delta^2 q \leq 0$$

$$\Delta^1\ \Delta^2 q = 0 \quad \text{on} \quad (A_1 x A_2)^c \quad \text{with} \quad A = \{q = f\}$$

where Δ^1(resp Δ^2) denotes the Laplacian on the first (resp second) variable and, for any $A \subset E$, A_1 (resp A_2) denotes its projection on E^1 (resp E^2).

This system leads to several remarks. First, it is not closed because nothing is said upon the state of q on $A_1 x A_2$ / A. Then we do not know under what appropriate conditions such system admits a unique solution which is is the Snell reduite of f. To our knowledge these problems are open.

Nevertheless, when the function $\Delta^1\Delta^2 q$ has a constant sign $\Delta^1\Delta^2 q \geq 0$ one can show, that there exists a unique maximal s.p. and this stopping point is optimal. However until now, we cannot say anything about the general case.

Bibliography

(1) A. BENSOUSSAN and J.L. LIONS : "Applications des inéquations variationnelles en contrôle stochastique" Dunod, 1978.

(2) J.M. BISMUT and B. SKALLI : "Temps d'arrêt optimal, théorie générale des processus et processus de Markov" Z.f.Wahr.V.Geb. 39,(1977) 301-313.

(3) R. CAIROLI : "Enveloppe de Snell d'un processus à paramètre bidimensionnel" Ann.Inst. H. Poincaré,VolXVIII N°1 (1982)47-53

(4) C. DELLACHERIE and P.A. MEYER : "Probabilités et potentiel" vol. 1 and 2, Hermann ,1975 and 1980.

(5) E.B. DYNKIN : "Markov processes" Springer Verlag, 1965.

(6) N. EL KAROUI : "Les aspects probabilistes du contrôle stochastique" Ecole d'été de St Flour 1979. Lect. Notes in Math. N°876, Springer-Verlag (1981) 74-239.

(7) U. KRENGEL and L. SUCHESTON : "Stopping rules and tactics for processes indexed by directed sets" J. of Mult. Anal.Vol 11 N°2 (1981) 199-229.

(8) A. MANDELBAUM and R.J. VANDERBEI : "Optimal stopping and supermartingales over partially ordered sets" Z.f.Wahr.V.Geb.57 (1981) 253-264.

(9) G. MAZZIOTTO and J. SZPIRGLAS : "Arrêt optimal sur le plan" C.R.Acad.Sc.Paris t.293 (1981) 87-90.

(10) P.A. MEYER : "Théorie élémentaire des processus à deux indices" Lect.Notes in Math.N°863, Springer-Verlag (1981) 1-39.

(11) A.N. SHIRYAYEV : "Optimal stopping rules" Springer-Verlag,1978.

(12) J.B. WALSH : "Optional increasing paths" Lect. Notes in Math. N°863, Springer-Verlag,(1981) 172-201.

(13) E. WONG and M. ZAKAI : "Martingales and stochastic integrals for processes with a multidimensional parameter" Z.f.Wahr.V. Geb.29 (1974) 109-122.

STOCHASTIC CONTROL PROBLEM FOR REFLECTED DIFFUSIONS IN A CONVEX BOUNDED DOMAIN

Jose Luis Menaldi(*)
Department of Mathematics
Wayne State University
Detroit, Michigan 48202, U.S.A.

1. INTRODUCTION

Let $\mathcal{O}$ be a convex bounded domain in $\mathbb{R}^n$. The state of our stochastic control problem is a normal reflected diffusion $(y(t),\ t \geq 0)$ on $\bar{\mathcal{O}}$, starting at x and with drift coefficient $g(y(t),v(t))$, diffusion term $\sigma(y(t),v(t))$. The functions g,σ are given and our control is the adapted process $v(t)$ taking values in a compact convex set V in $\mathbb{R}^m$.

The cost functional is defined by

$$(1.1) \qquad J_x(v) = E\int_0^\infty f(y(t),v(t))\exp\left(-\int_0^t c(y(s),v(s))ds\right)dt ,$$

where functions f,c are known.

Our control problem is to minimize the cost functional (1.1) over all admissible controls $v(\cdot)$,

$$(1.2) \qquad u(x) = \inf\{J_x(v) : v(\cdot)\} .$$

At least formally, by the argument of the dynamic programming, we can derive the following equation to be satisfied by u,

$$(1.3) \qquad \begin{aligned} &\sup\{A(v)u - f(\cdot,v) : v \in V\} = 0 \quad \text{in } \mathcal{O} \\ &\frac{\partial u}{\partial n} = 0 \quad \text{on } \partial\mathcal{O} , \end{aligned}$$

where $A(v) = -(1/2)\mathrm{tr}(\sigma\sigma^* \nabla^2) - g\cdot\nabla + c$, ∇ is the gradient operator n denotes the outward unit normal to $\mathcal{O}$, $\mathrm{tr}(\cdot)$ is the trace and $*$ denotes the transpose.

In this way, the initial stochastic control problem (1.2) is connected to the Hamilton-Jacobi-Bellman equation with Neumann boundary conditions (1.3).

We characterize the optimal cost function $u(x)$ as the unique stationary function of the corresponding nonlinear semigroup. Also, under some conditions, we construct an optimal Markovian control.

(*) This research has been supported in part by the University of Paris-Daphine and I.N.R.I.A., France.

On the other hand, we consider the diffusion $(y^\epsilon(t), t \geq 0)$ in $\mathbb{R}^n$, with drift coefficient $g(y^\epsilon(t),v(t)) - (1/\epsilon)\beta(y^\epsilon(t))$ and diffusion term $\sigma(y^\epsilon(t),v(t))$. The function β represents the penalization factor, which is exactly the gradient of the square of the distance to the set $\mathcal{O}$.

We denote by $J_x^\epsilon(v)$, $\epsilon > 0$ the cost functional (1.1) with $y^\epsilon(t)$ instead of $y(t)$. The optimal cost is

$$u_\epsilon(x) = \inf\{J_x^\epsilon(v) : v(\cdot)\}. \tag{1.4}$$

We show that the function $u_\epsilon(x)$ converges uniformly in $\bar{\mathcal{O}}$ to the initial optimal cost function $u(x)$. Also, under suitable assumptions, the diffusion process $(y^\epsilon(t), t \geq 0)$ converges to the reflected diffusion $(y(t), t \geq 0)$ in an appropriate sense.

The first rigorous results concerning the derivation of the HJB (1.3) in the whole space were obtained by Krylov [10]. The HJB equation (1.3) with Dirichlet boundary conditions was studied by Safonov [24] in a plane domain, by Evans and Friedman [5], Lions and Menaldi [14] in a n-dimensional domain but with constant coefficient σ, and by Lions [13], Evans and Lions [7], Evans and Lenhard [6], Evans [4], Krylov [12] in a general smooth domain. A weak formulation of HJB equation (1.3) in a general domain with Dirichlet boundary conditions was presented in Lions and Menaldi [14,15].

Let us mention that these kinds of problems were introduced in the book of Fleming and Rishel [8]. We also refer to the books of Bensoussan and Lions [1,2], Friedman [9], Krylov [11] for impulsive and continuous control problems. We remark that some results concerning impulsive control problems are given in [17,18] for degenerate diffusion processes, in [19,20] for degenerate reflected diffusion processes, and in Bensoussan and Menaldi [3] for diffusion processes with jumps.

In section 2, we give the main notations, assumptions and background to be used later on. Next, in section 3, we establish some results concerning the characterization and regularity of the optimal cost, and the approximation by the problem in the whole space.

2. PRELIMINARIES

Let $\mathcal{O}$ be an open, convex and bounded set in $\mathbb{R}^n$, and V be a compact convex set in $\mathbb{R}^m$. We suppose that

$$g : \mathbb{R}^n \times V \to \mathbb{R}^n, \quad \sigma : \mathbb{R}^n \times V \to \mathbb{R}^n \otimes \mathbb{R}^n,$$
$$f : \mathbb{R}^n \times V \to \mathbb{R}, \quad c : \mathbb{R}^n \times V \to \mathbb{R}^+$$

satisfy $g = (g_i, i = 1,\dots,n)$, $\sigma = (\sigma_{ij}, i,j = 1,\dots,n)$,

$$|\phi(x,v) - \phi(x',v')| \le C|x-x'| + \rho(|v-v'|) \ , \quad \forall \ x,x',v,v' \ ;$$

(2.1) $$|\phi(x,v)| \le C \ , \quad \forall \ x,v \ ; \quad \forall \ \phi = g_i \ , \ \sigma_{ij}, f, c \ ;$$

$$c(x,v) \ge c_0 > 0 \ , \quad \forall \ x,v \ ;$$

where ρ is a given continuous function such that $\rho(0)=0$ and C denotes generic constants.

We call an admissible system a set $\mathcal{A} = (\Omega,\mathcal{F},\mathcal{F}^t,P,w(t),v(t),y(t))$, where $(\Omega,\mathcal{F},P)$ is a probability space $(\mathcal{F}^t, t \ge 0)$ is a nondecreasing right continuous family of completed sub σ-fields of $\mathcal{F}$, $w(t)$ is a standard Wiener process in $\mathbb{R}^n$ with respect to $\mathcal{F}^t$, $v(t)$ is a measurable adapted process taking values in V , and $y(t)$ is the normal reflected diffusion defined by the following stochastic variational inequality (S.V.I.)

(2.2)

$y(t)$, $\eta(t)$ continuous, measurable and adapted processes in $\mathbb{R}^n$ such that

(i) $y(t)$ takes values in the closure $\bar{\mathcal{O}}$ and $\eta(t)$ has bounded variation on $[0,T]$, $0 < T < \infty$, $\eta(0)=0$.

(ii) $$y(t) + \eta(t) = x + \int_0^t g(y(s),v(s))ds + \int_0^t \sigma(y(s),v(s))dw(s) \ , \quad \forall t \ge 0$$

(iii) for any $z(t)$ continuous, measurable and adapted process taking values in the closure $\bar{\mathcal{O}}$, we have

$$\int_0^T (y(t) - z(t))d\eta(t) \ge 0 \ , \quad \forall T \ge 0 \ .$$

Notice that if the convex set $\mathcal{O}$ has smooth boundary, the normal reflected diffusion defined by the S.V.I. (2.2) becomes the classical one. The existence of such reflected diffusion in (2.2) is deduced from Tanaka [27], Lions, Menaldi and Sznitman [16], and [21].

Now, for any admissible system $\mathcal{A}$ and any real measurable bounded function $h(x)$, we define a functional

(2.3)
$$J(x,\mathcal{A},t,h) = E\{\int_0^t f(y(s),v(s))\varphi(s,\mathcal{A})ds + h(y(t))\varphi(t,\mathcal{A})\} \ ,$$

$$\varphi(t,\mathcal{A}) = \exp(-\int_0^t c(y(s),v(s))ds) \ ,$$

and and operator

(2.4) $$[Q(t)h](x) = \inf\{J(x,\mathcal{A},t,h) : \mathcal{A}\} \ .$$

Denote by $C_b(\bar{\mathcal{O}})$ the space of all real functions on $\bar{\mathcal{O}}$ which are uniformly continuous and bounded,[1] endowed with the supremum norm $\|\cdot\|$. Then as in Nisio [22,23], Bensoussan and Lions [1], Lions and Menaldi [14,15], we can prove that $(Q(t), t \geq 0)$ is a nonlinear semigroup acting on $C_b(\bar{\mathcal{O}})$, we have precisely the following properties:

(2.5) $$Q(t) : C_b(\bar{\mathcal{O}}) \to C_b(\bar{\mathcal{O}}) ,$$

(2.6) $$Q(t+s) = Q(t) \circ Q(s) , \quad Q(0) = \text{identity},$$

(2.7) $$\|Q(t)h - Q(s)h\| \to 0 \quad \text{as} \quad t \to s , \quad \forall h \in C_b(\bar{\mathcal{O}}) ,$$

(2.8) $$\|Q(t)h_1 - Q(t)h_2\| \leq \|h_1 - h_2\| , \quad \forall \; h_1, h_2 \in C_b(\bar{\mathcal{O}}) ,$$

(2.9) $$Q(t)h_1 \leq Q(t)h_2 \quad \text{if} \quad h_1 \leq h_2 , \qquad h_1, h_2 \in C_b(\bar{\mathcal{O}}) ,$$

(2.10) If h is twice continuously differentiable in a compact subset K of $\bar{\mathcal{O}}$ and the normal derivative to $\partial\mathcal{O}$ of h vanishes on $\partial K \cap \partial\mathcal{O}$, then we have

$$\frac{1}{t}[Q(t)h - h] \to -\sup\{A(v)h - f(\cdot,v) : v \in V\} \quad \text{as} \quad t \to 0 ,$$

uniformly on K,

where the operator $A(v)$ is defined by

(2.11) $$A(v) = -\frac{1}{2}\operatorname{tr}(\sigma\sigma^* \frac{\partial^2}{\partial x^2}) - g\frac{\partial}{\partial x} + c .$$

Under this formulation, we define a cost of each admissible system control $\mathcal{A}$,

(2.12) $$J_x(\mathcal{A}) = E\{\int_0^\infty f(y(t),v(t))\exp(-\int_0^t c(y(s),v(s))ds)dt\} ,$$

and the optimal cost

(2.13) $$u(x) = \inf\{J_x(\mathcal{A}) : \mathcal{A}\} .$$

On the other hand, we call $\mathbb{R}^n$-admissible system control a set $\mathcal{A}^\epsilon = (\Omega, \mathcal{F}, \mathcal{F}^t, P, w(t), y^\epsilon(t))$, $\epsilon > 0$ satisfying the same conditions as $\mathcal{A}$ but $y^\epsilon(t)$ is the diffusion process given by the following Ito equation

(2.14) $$dy^\epsilon(t) = g(y^\epsilon(t),v(t))dt + \sigma(y^\epsilon(t),v(t))dw(t) - \frac{1}{\epsilon}\beta(y^\epsilon(t))dt, \; t \geq 0$$
$$y^\epsilon(0) = x ,$$

where

(2.15) $$\beta(x) = (x - \pi(x))^* , \quad \pi : \text{orthogonal projection on } \bar{\mathcal{O}} .$$

---

[1] Since the domain is bounded, we just need continuous functions.

In a similar way to (2.3), (2.4) we define the operator

$$[Q_\varepsilon(t)h]x = \inf\{J(x,\mathcal{A}^\varepsilon,t,h) : \mathcal{A}^\varepsilon\} , \tag{2.16}$$

and as in (2.12), (2.13), we obtain the optimal cost

$$u_\varepsilon(x) = \inf\{J_x(\mathcal{A}^\varepsilon) : \mathcal{A}^\varepsilon\} . \tag{2.17}$$

Clearly, $(Q_\varepsilon(t) : t \geq 0)$ satisfies analogous properties (2.5), ..., (2.10).

We remark that the penalization (2.14) has been used in Shalaumov [25], in Lions, Menaldi and Sznitman [16], and in [21]. Also, similar kind of nonlinear semigroup have been treated in Stettner and Zabczyk [26], Zabczyk [28].

In order to construct an optimal Markovian admissible system with respect to (2.13), we will deal with the existence of a weak solution of the S.V.I. (2.2) for coefficients g,σ which are only bounded and measurable.

We assume that there is no degeneracy, i.e.

$$|\sigma(x,v)\xi| \geq \mu|\xi| , \quad \forall\, x,\xi \in \mathbb{R}^n , \ \forall\, v \in V , \tag{2.18}$$

for some constant $\mu > 0$, and we also suppose that we can not exert any control near to the boundary, i.e. ,

$$\begin{aligned} &\forall\, x \in \bar{\mathcal{O}} \text{ such that } \operatorname{dist}(x,\partial\mathcal{O}) \leq \delta \\ &\text{we have } \sigma(x,v) = \sigma(x) , \ \forall\, v \in V , \end{aligned} \tag{2.19}$$

where δ is some positive constant.

Lemma 2.1

Let $v(x)$ be a measurable function from $\bar{\mathcal{O}}$ into V . Then under assumptions (2.1), (2.18), (2.19)(1) there exists an admissible system $\mathcal{A} = (\Omega,\mathcal{F},\mathcal{F}^t,P,w(t), v(t), y(t))$ such that $v(t) = v(y(t))$ for all $t \geq 0$.

Outlines of the Proof.

First of all, combining Krylov's estimates for diffusion processes (cf. [11]) and classical estimates for reflected diffusions with smooth coefficients (cf. Bensoussan and Lions [2]), we obtain for any measurable function $h(x)$ the estimate

$$E\{\int_0^\infty |h(y(t))|e^{-c_0 t}dt\} \leq C(\int_{\mathcal{O}} |h(x)|^n dx)^{1/n} , \tag{2.20}$$

where $y(t)$ is the reflected diffusion associated to a smooth feed-back $v(x)$, and the constant C is independent of v,h .

(1) the boundary of $\mathcal{O}$ is supposed to be smooth.

Next, by a similar technique as in Krylov [11,p.86], the lemma follows.

3. CHARACTERIZATIONS

We have

Theorem 3.1

We assume (2.1). Then the function u defined by (2.13) is the unique solution of the problem

(3.1) $$u \in C_b(\bar{\mathcal{O}}) , \quad Q(t)u = u , \quad \forall t \geq 0 .$$

Moreover, the equation of the dynamic programming is satisfied

(3.2) $$u(x) = \inf\{J(x,\theta,\mathcal{A},u) : (\mathcal{A},\theta)\} ,$$

where J is the functional given by (2.3) and θ stands for any stopping time associated to the admissible system $\mathcal{A}$.

Outlines of the proof.

We just need to use properties (2.5),...,(2.9) in a similar way as in Nisio [22], Bensoussan and Lions [1], Lions and Menaldi [15].

We remark that if u is smooth, from Theorem 3.1 and property (2.10) we obtain equation (1.3). In this sense, problem (3.1) can be taken as an integral formulation of the HJB equation (1.3).

For the regular case, we have

Theorem 3.2

Let assumptions (2.1), (2.18), (2.19) hold. Suppose also that

(3.3) g_i, σ_{ij}, f, c have uniform Lipschitz continuous first derivative in x ,

(3.4) $\mathcal{O}$ has smooth boundary .

Then the optimal cost u defined by (2.13) is the unique solution of problem (1.3) in the Sobolev space $W^{2,\infty}(\mathcal{O})$. Moreover, there exists an optimal admissible system which is Markovian feed-back.

Outlines of the proof.

We just need to show the $W^{2,\infty}$ regularity for the optimal cost u . Since there is no control near the boundary $\partial\mathcal{O}$, the function u solves a linear equation near $\partial\mathcal{O}$. Hence u is smooth in a neighborhood of $\partial\mathcal{O}$ by using classical arguments. Next, similar to Evans and Lions [7], we deduce the $W^{2,\infty}$ regularity in the interior of $\mathcal{O}$.

The existence of an optimal control follows from the fact that u is $W^{2,\infty}$ and the Lemma 2.1.

Remark 3.1

Using the technique of Lions and Menaldi [15] and [19], the optimal cost u can be characterized as the maximum subsolution in an appropriate sense.

Remark 3.2

We can obtain the $C^{2,\beta}(\bar{\mathcal{O}})$ regularity of the solution u by using a recent result of Evans [4].

On the other hand, we have

Theorem 3.3

Under assumption (2.1) *we have the following convergence*

(3.5) $$u_\varepsilon(x) \to u(x) \quad \text{as} \quad \varepsilon \to 0 \;, \quad \text{uniformly in} \quad x \in \bar{\mathcal{O}} \;,$$

where u_ε, u *are given by* (2.17), (2.13) *respectively. Moreover, for any function* h *in* $C_b(\mathbb{R}^n)$ *we have, as* $\varepsilon \to 0$,

(3.6) $$Q_\varepsilon(t)h \to Q(t)h \quad \text{in} \quad C_b(\bar{\mathcal{O}}) \;, \quad \text{uniformly in} \quad t \geq 0 \;,$$

where $Q_\varepsilon(t)$, $Q(t)$ *are the nonlinear semigroup defined by* (2.16), (2.4).

Outlines of the proof.

We use the following fact (cf. [21]): $\forall\, T > 0$ we have

(3.7) $$E\{\sup|y^\varepsilon(t) - y(t)| \;:\; 0 \leq t \leq T\} \to 0 \quad \text{as} \quad \varepsilon \to 0 \;,$$

being the limit uniform in $v(\cdot)$ and x in $\bar{\mathcal{O}}$. The processes $y^\varepsilon(\cdot)$, $y(t)$ are given by the stochastic equations (2.14), (2.4) .

Theorem 3.4

Let ε *be any positive constant. Then under hypothesis* (2.1) *there exists a function* u_ε *such that*

$$u_\varepsilon \in C^{2,\beta}(\mathbb{R}^n) \quad \text{for some} \quad \beta > 0 \;,$$

(3.8) $$-\varepsilon \leq \sup\{A(v)u_\varepsilon - f(\cdot,v) \;:\; v \in V\} \leq \varepsilon \quad \text{in} \quad \bar{\mathcal{O}} \;,$$

$$-\varepsilon \leq u_\varepsilon(x) - u(x) \leq \varepsilon \;, \quad \forall\, x \in \bar{\mathcal{O}} \;,$$

where u *is the optimal cost* (2.13). *Moreover, we can construct an admissible system* $\mathcal{A}_\varepsilon$, *which is* ε*-optimal, i.e.*

(3.9) $$J_x(\mathcal{A}_\varepsilon) \leq u(x) + \varepsilon \;, \qquad \mathcal{A}_\varepsilon = \mathcal{A}_{\varepsilon,x} \;.$$

Outlines of the proof.

First of all, we note that without loss of generality we can assume that data g, σ, f, c satisfy (2.18), (3.3). Since the domain $\mathcal{O}$ is not supposed to be smooth, we can not apply directly Theorem 3.2. Then, we use (3.5) to deduce (3.8). We remark that in (3.8), the function $u_\epsilon = u_{\epsilon'}$, with $u_{\epsilon'}$ given by (2.17) for some small ϵ'.

Finally, we construct $\mathcal{A}_\epsilon$ by mean of a feed-back $v_\epsilon(x)$ which achieves the infimum in (2.17).

Remark 3.3

In almost all of this paper, assumptions (2.1), (2.18), (2.19) can be relaxed. We will have quite similar results.

Remark 3.4

We can extend all results to the parabolic case.

Remark 3.5

With analogous techniques, we can consider the case where we add an impulsive control in the system $\mathcal{A}$.

REFERENCES

[1] A. BENSOUSSAN and J.L. LIONS, Applications des Inéquations Variationnelles en Contrôle Stochastique, Dunod, Paris, 1978.

[2] A. BENSOUSSAN and J.L. LIONS, Contrôle Impulsionnel et Inéquations Quasi-Variationnelles, Dunod, Paris, 1981 (to appear).

[3] A. BENSOUSSAN and J.L. MENALDI, Optimal Stochastic Control of Diffusion Processes with Jumps Stopped at the Exit of a Domain, Advances in Probability Vol. 7, Stochastic Differential Equations, Ed. M.A. Pinsky, Marcel Dekker Inc. to appear.

[4] L.C. EVANS, Classical Solutions of the Hamilton-Jacobi-Bellman Equations for Uniformly Elliptic Operator, preprint.

[5] L.C. EVANS and A. FRIEDMAN, Optimal Stochastic Switching and the Dirichlet Problem for Bellman Equation, Trans. Am. Math. Soc., 253 (1979), pp.365-389.

[6] L.C. EVANS and S. LENHART, The Parabolic Bellman Equation, Nonlinear Analysis, 5 (1981), pp. 765-773.

[7] L.C. EVANS and P.L. LIONS, Résolution des Equations de Hamilton-Jacobi-Bellman pour des Opérateurs Uniformement Elliptiques, C. R. Acad. Sc. Paris, A-290 (1980), pp. 1049-1052.

[8] W.H. FLEMING and R. RISHEL, Optimal Deterministic and Stochastic Control, Springer-Verlag, New York, 1975.

[9] A. FRIEDMAN, Stochastic Differential Equations and Applications, Vol. I and II, Academic Press, New York, 1976.

[10] N.V. KRYLOV, Control of a Solution of a Stochastic Integral Equation, Theory Prob. Appl., 17 (1972), pp. 114-131.

[11] N.V. KRYLOV, Controlled Diffusion Processed, Springer-Verlag, New York, 1980.

[12] N.V. KRYLOV, Some New Results in the Theory of Controlled Diffusions Processes, Math. USSR Sbornik, 37 (1980), pp. 133-149.

[13] P.L. LIONS, Sur Quelques Classes d'Equations aux Dérivées Partielles Nonlinéaires et Leur Résolution Numérique, Thèse d'Etat, Université de Paris VI, 1979.

[14] P.L. LIONS and J.L. MENALDI, Problèmes de Bellman avec le Contrôle dans les Coefficients de Plus Haut Degre, C. R. Acad. Sc. Paris, A-287, (1978), pp. 409-412.

[15] P.L. LIONS and J.L. MENALDI, Control of Stochastic Integrals and Hamilton-Jacobi-Bellman Equation, Part I and II, SIAM J. Control Optim., 20 (1982), pp. 58-95. See also Proc. 20th IEEE CDC, San Diego, 1981, pp. 1340-1344.

[16] P.L. LIONS, J.L. MENALDI and A.S. SZNITMAN, Construction de Processus de Diffusion Réfléchis par Pénalisation du Domaine, C. R. Acad. Sc. Paris, I-292 (1981), pp. 559-562.

[17] J.L. MENALDI, On the Optimal Stopping Time Problem for Degenerate Diffusions, SIAM J. Control Optim., 18 (1980), pp. 697-721. See also C. R. Acad. Sc. Paris, A-284 (1977), pp. 1443-1446.

[18] J.L. MENALDI, On the Optimal Impulse Control Problem for Degenerate Diffusions, SIAM J. Control Optim., 18 (1980), pp. 722-739. See also C. R. Acad. Sc. Paris, A-284 (1977), pp. 1449-1502.

[19] J.L. MENALDI, Sur le Problème de Temps d'Arrêt Optimal pour les Diffusions Réfléchies Dégénérées, C. R. Acad. Sc. Paris, A-289 (1979), pp. 779-782. See also J. Optim. Theory Appl., 36 (1982), to appear.

[20] J.L, MENALDI, Sur le Problème de Contrôle Impulsionnel Optimal pour les Diffusions Réfléchies Dégénérées, C. R. Acad. Sc. Paris, A-290 (1980), pp. 5-8. See also Mathematicae Notae, 28 (1982), to appear.

[21] J.L. MENALDI, Stochastic Variational Inequality for Reflected Diffusion, to appear.

[22] M. NISIO, On Stochastic Optimal Controls and Envelope of Markovian Semi-Groups, Proc. of Intern. Symp. SDE, Kyoto 1976, pp. 297-325.

[23] M. NISIO, On a Non-Linear Semi-Group Attached to Stochastic Optimal Control, Publ. RIMS, Kyoto Univ., 13 (1976), pp. 513-537.

[24] M.V. SAFONOV, On the Dirichlet Problem for Bellman Equation in a Plane Domain, Math. USSR Sbornik, 31 (1977), pp. 231-284 and 34 (1978), pp. 521-526.

[25] V.A. SHALAUMOV, On the Behavior of a Diffusion Process with a Large Drift Coefficient in a Half Space, Theory Prob. Appl., 24 (1980), pp. 592-598.

[26] L. STETTNER and J. ZABCZYK, Strong Envelopes of Stochastic Processes and a Penalty Methods, Stochastics, 4 (1981), pp. 267-280.

[27] H. TANAKA, Stochastic Differential Equations with Reflecting Boundary Condition in Convex Regions, Hiroshima Math. J., 9 (1979), pp. 163-177.

[28] J. ZABCZYK, Semigroup Methods in Stochastic Control Theory, preprint 1978, University of Montreal.

Nonlinear Filtering of Diffusion Processes
A Guided Tour

by

Sanjoy K. Mitter

Department of Electrical Engineering and Computer Science and
Laboratory for Information and Decision Systems
Massachusetts Institute of Technology
Cambridge, MA 02139

1. Introduction

In this paper we give a guided tour through the development of nonlinear filtering of diffusion processes. The important topic of filtering of point processes is not discussed in this paper.

There are two essentially different approaches to the nonlinear filtering problem. The first is based on the important idea of innovations processes, originally introduced by Bode and Shannon (and Kolmogoroff) in the context of Wiener Filtering problems and later developed by Kailath and his students in the late sixties for nonlinear filtering problems. This approach reaches its culmination in the seminal paper of FUJISAKI-KALLIANPUR-KUNITA [1972]. A detailed account of this approach is now available in book form cf. LIPSTER-SHIRYAYEV [1977] and KALLIANPUR [1980]. The second approach can be traced back to the doctoral dissertation of MORTENSEN [1966], DUNCAN [1967] and the important paper of ZAKAI [1969]. In this approach attention is focused on the unnormalized conditional density equation, which is a bilinear stochastic partial differential equation, and it derives its inspiration from function space integration as originally introduced by KAC [1951] and RAY [1954]. Mathematically, this view is closely connected to the path integral formulation of Quantum Physics due to FEYNMAN [1965]. For an exposition of this analogy see MITTER [1980, 1981]. A detailed account of the second viewpoint can be found in the lectures given by Kunita, Pardoux and Mitter in the CIME lecture Notes on Nonlinear Filtering and Stochastic Control [1982] and in HAZEWINKEL-WILLEMS [1981].

2. Basic Problem Formulation

To simplify the exposition we consider the situation where all processes are scalar-valued.

Let $(\Omega, \mathscr{F}, P)$ be a complete probability space and let $\mathscr{F}_t$, $t \in [0, T]$ be an increasing family of sub σ-fields of $\mathscr{F}$. Let ξ_t be an $\mathscr{F}_t$-adapted process, considered to be the signal process and consider the observation process y_t given by

(2.1) $$y_t = \int_0^t h_s ds + n_t,$$

where n_t is an $\mathcal{F}_t$-Weiner process and it is assumed that $\sigma(n_t - n_s | t \geq s)$ is independent of the past of the joint signal-observation process $\sigma(y_u, h_u | u \leq s)$. Information about the ξ-process is contained in h which satisfies

$$E\int_0^t |h_s|^2 ds < \infty \quad \forall\ t \in [0,T].$$

Let $\mathcal{F}_t^Y = \sigma(y_s | s \leq t)$. Then the filtering problem consists of computing

$$E\left(\phi(\xi_t) | \mathcal{F}_t^Y\right) \stackrel{\Delta}{=} \hat{\xi}_t,$$

where ϕ, say is a bounded, continuous function (indeed any function such that the conditional expectation makes sense).

3. The Innovations Method

The fundamental paper of Fujisaki-Kallianpur-Kunita proceeds as follows:

Define the innovations process:

(3.1) $$\nu_t = y_t - \int_0^t \hat{h}_s ds \text{ , where}$$

(3.2) $$\hat{h}_s \stackrel{\Delta}{=} E(h_s | \mathcal{F}_s^Y).$$

Then it can be shown that

(i) ν_t is $\mathcal{F}_t^Y$-Wiener

(ii) $\sigma\left\{\nu_u - \nu_t | u \geq t\right\}$ is independent of $\mathcal{F}_t^Y$.

In general it is not true that:

(Innovations Property) $\mathcal{F}_t^Y = \mathcal{F}_t^\nu$ (Cirelson counter-example).

If is assumed that h_t and n_t are independent then innovaticns property holds (cf. ALLINGER-MITTER [1981]).

However even without the innovations property holding, it can be proved:

Every square-integrable $\mathcal{F}_t^Y$-martingale m_t can be represented as:

(3.3) $$m_t = m_0 + \int_0^t \phi_s d\nu_s \text{ , where}$$

ϕ_s is jointly measurable, adapted to

$\mathscr{F}_s^Y$ and $\int_0^T E|\phi_s|^2 ds < \infty$.

To proceed further, let us assume that ξ_t is a continuous semimartingale

(3.4) $\xi_t = \xi_0 + \int_0^t f_s ds + v_t$, where v_t is a square integrable $\mathscr{F}_t$-martingale. Then

(3.5) $m_t = \hat{\xi}_t - \xi_0 - \int_0^t \hat{f}_s ds$ [(1)] is a $\mathscr{F}_t^Y$- martingale and from the previous result

(3.6) $m_t = \int_0^t \phi_s d\nu_s$, where ϕ_s can be identified as

(3.7) $\phi_s = \widehat{\xi_s h_s} - \hat{\xi}_s \hat{h}_s + \hat{D}_s$, where D_s is a $\mathscr{F}_s$-predictable process and $\int_0^t D_s ds = \langle v,n\rangle_t$, where $\langle\ ,\ \rangle_t$ denotes the quadratic variation.

In case h_t and n_t are independent there is an essential simplification using the innovations property.

For example, one would have the representation

(3.8) $\hat{\xi}_t = \hat{\xi}_0 + \int_0^t \frac{\partial}{\partial s} E(\xi_t \nu_s) d\nu_s + \int_0^t \int_0^{s_1} \frac{\partial^2}{\partial s_1 \partial s_2} E(\xi_t \nu_{s_1} \nu_{s_2}) d\nu_{s_1} d\nu_{s_2} + \dots$

To proceed further let us assume that ξ_t is a Markov diffusion process satisfying the Ito equation

$$d\xi_t = f(\xi_t)\, dt + g(\xi_t)\, dB_t \text{ , and} \tag{3.9}$$

$$h_s = h(\xi_s) \ . \tag{3.10}$$

Let us also assume that B_t and n_t are independent (assumed throughout the rest of the paper). Then using the Ito differential rule for $\phi \in C_b^2$, one gets

---

(1) $\hat{}$ always denotes conditional expectation with respect to $\mathscr{F}_t^Y$

$$(3.11)\quad \hat{\phi}(\xi_t) = \hat{\phi}(\xi_0) + \int_0^t \widehat{L\,\phi(\xi_s)}\, ds + \int_0^t \left(\widehat{\phi(\xi_s)\, h_s} - \hat{\phi}(\xi_s)\, \hat{h}_s \right) d\nu_s ,$$

where L is the generator of the diffusion process.

4. The Innovations Method Continued: Kushner-Stratonovich Equation.(1)

Let $\Pi_t(dy,\omega)$ denote the conditional distribution of ξ_t given $\mathscr{F}_t^y$. Let $\phi \in C_b^2$ and denote by

$$(4.1)\quad \Pi_t(\phi) = \int \phi(y)\, \Pi_t(dy,\omega)$$

Then Π_t satisfies the Nonlinear Stochastic Partial Differential Equation (Kushner-Stratonovich Equation)

$$(4.2)\quad \begin{cases} \Pi_t(\phi) = \Pi_0(\phi) + \int_0^t \Pi_s(L\phi)\, ds \\ \qquad + \int_0^t \left[\Pi_s(\phi h_s) - \Pi_s(\phi)\Pi_s(h_s) \right] \left(dy_s - \Pi_s(h_s)\, ds \right). \end{cases}$$

We think of the filter as a dynamical system and we think of equation (4.2) as the input-output equation of the filter, the input being the observations y_s and the output the conditional distribution Π_t.

5. Zakai Equation

Let ρ_t be a continuous stochastic process with values in the set of finite positive measures on $\mathbb{R}$. Denote by

$$(5.1)\quad \rho_t(\phi) = \int \phi(y)\, \rho_t\, (dy) \, .$$

Consider the equation

$$(5.2)\quad \rho_t(\phi) = \rho_0(\phi) + \int_0^t \rho_s\, (L\phi)\, ds + \int_0^t \rho_s\, (h_s \phi)\, dy_s \, .$$

(weak form of Zakai equation)

Now it can be proved:

---

(1) The development here follows Kunita [1982].

If ρ_t is a solution of (5.2) then $\Pi_t(\phi) = \frac{\rho_t(\phi)}{\rho_t(1)}$ is a solution of equation (4.2).

Morever we have the Feynman-Kac formula:

$$(5.3)\quad \rho_t(\phi) = E\left[\phi(\xi_t)\exp\left(\int_0^t h_s\, dy_s - \frac{1}{2}\int_0^t h_s^2\, ds\right)\Big|\mathscr{F}_t^y\right]$$

For later use (and throughout the rest of the paper) we shall consider the Weak form of the Zakai equation in Stratonovich form:

$$(5.4)\quad \rho_t(\phi) = \rho_0(\phi) + \int_0^t \rho_s\left[(L - \frac{1}{2}h^2)\phi\right] ds + \int_0^t \rho_s(h\phi)\circ dy_s$$

If the solution $\rho_t(dy)$ has a smooth density $\rho_t(y)$ it satisfies the Zakai equation:

$$(5.5)\qquad d\rho_t = (L^* - \frac{1}{2}h^2)\quad \rho_t(y)\, dt + h\,\rho_t(y)\circ dy_t \quad ,$$

where * denotes formal adjoint.

$\rho_t(y)$ has the interpretation of an unnormalized density and is to be thought of as the "state" of the filter.

To compute conditional statistics, we need the state-output equation

$$(5.6)\qquad \hat{\phi}(x_t) = \frac{\int_{\mathbb{R}} \phi(x)\rho_t(x;y)\,dx}{\int_{\mathbb{R}} \rho_t(x;y)\,dx}$$

The fundamental problem of nonlinear filtering is the "invariant" study of equation (5.5). The analytic difficulty of this problem stems from the following:

(i) In most interesting situations the operator $x \mapsto h(x)$ is unbounded.

(ii) The paths of the y-process are only Hölder continuous of exponent $< \frac{1}{2}$.

6. Pathwise Nonlinear Filtering

The ideas of this section are due to CLARK [1978], DAVIS [1980], and MITTER [1980].

There is as yet no theory of nonlinear filtering where the

observations are:

(6.1) $$y_t = h(\xi_t) + \eta_t \ ,$$

where η_t is physical wide-band noise and hence smooth. Define $y_t = \dot{Y}_t$ and $\eta_t = \dot{N}_t$ where $\cdot$ denotes differentiation. Then (6.1) can be written as:

(6.2) $$dY_t = h(\xi_t)\, dt + dN_t \ , \text{ or}$$

(6.3) $$Y_t = \int_0^t h(\xi_s)\, ds + N_t \ .$$

Equation (6.3) is a mathematical model of the physical observation (6.1) where the wide band noise η_t has been approximated as "white noise" $\dot{N}_t$ and hence N_t is a Wiener process.

Now, if we wish to compute

$$E\left(\phi(\xi_t)\ |\mathscr{F}_t^Y\right) = \text{Functional of Y a.s. Wiener measure}$$

then this filter does not accept the physical observation y. The idea is to construct a suitable version of the conditional expectation so that the performance of the filter as measured by the mean-square error remains close to when the physical observation 'y' is replaced by the mathematical model of the observation.

This is most easily done by eliminating the stochastic integral in (5.5) by a suitable transformation (gauge transformation in the language of physicists).

Define $q_t(x;y)$ by

(6.4) $$\rho_t(x;y) = \exp\left(y_t h(x)\right) q_t(x;y) \ .$$

Then q_t satisfies the parabolic partial differential equation

(6.5) $$\frac{\partial q}{\partial t} = \frac{1}{2}\, a(x)\, \frac{\partial^2 q}{\partial x^2} + b^y(x,t)\, \frac{\partial q}{\partial t} + V^y(x,t) q$$

where $a(x) = g^2(x)\,, \quad b^y = -f + y_t\, a\, \frac{dh}{dx} + \frac{da}{dx}\,,$

$$V^y = \frac{1}{2} h^2 - y_t Lh + \frac{1}{2} y_t^2\, a\left(\frac{dh}{dx}\right)^2 - \frac{df}{dx} + y_t a \frac{dh}{dx} + \frac{1}{2}\frac{d^2 a}{dx^2}$$

Equation (6.5), the pathwise filter equation, is now needed to be solved for each y, the observation (which can be taken to be a physical observation).

7. Existence and Uniqueness Results for the Zakai Equation.

In case the observation h is bounded or linear, existence and uniqueness results for the Zakai equation has been given by PARDOUX [1982] by studying equation (5.5) directly. Existence, uniqueness and estimate of tail distributions for equation (5.5) including unbounded observations (in the scalar case all polynominal h) have been given by FLEMING-MITTER [1982], using stochastic control arguments. This approach studies equation (6.5) by transforming it into a Bellman-Hamilton-Jacobi equation (cf. paper of FLEMING this volume) using an exponential transformation and then showing that the measure constructed from q coincides with the measure given by formula (5.3). For other literature on this problem see the bibliography in FLEMING-MITTER (1982). See also PARDOUX [1981] and MITTER [1982] for an interpretation of the exponential transformation in the context of nonlinear filtering. For related variational considerations see MITTER [1980], BISMUT [1981], HIJAB [1980].

8. Geometrical Theory of Nonlinear Filtering.

How can one answer the question when two filtering problems have identical solutions? How can one decide whether the Zakai equation admits a finite-dimensional statistic?

The starting point of this analysis is the Zakai euqation (5.5) in Stratonovich form. Consider the two operators:

$$\mathscr{L}_0 = L^* - \frac{1}{2} h^2$$

$$\mathscr{L}_1 = h$$

where the two operators are considered as formal differential operators on $C_0^\infty(\mathbb{R})$. Denote by $\mathscr{F}$ the Lie algebra of operators generated by $\mathscr{L}_0$ and $\mathscr{L}_1$ under the standard bracket operation. This Lie algebra is invariant under (i) smooth change of coordinates $x \longmapsto \phi(x)$ and (ii) gauge transformations $\phi \longmapsto \phi\rho$ where ϕ is a C^∞-function which is invertible. Indeed, if the "invariance group" of the Zakai equation is suitably defined then the above constitutes the largest invariance group of the Lie algebra $\mathscr{F}$. The insight here is that two filtering problems with isomophic Lie algebras are likely to have the same filters. We say likely, since for a proof, analytic considerations such as the existence of a common dense set of analytic vectors must come into play. For a rigorous analysis in specific situations see OCONE [1980].

By a finite dimensional filter for the conditional statistic $\hat{\phi}(x_t)$ we mean a stochastic dynamical system.

(8.1) $$d\psi_t = \alpha(\psi_t)dt + \beta(\psi_t)\circ dy_t$$

where α and β are smooth vector fields on some finite-dimensional smooth manifold, and a state-output equation

(8.2) $\hat{\phi}(x_t) = \gamma(\psi_t)$, with γ a smooth real-valued function.

The idea of studying the Lie algebra is independently due to BROCKETT [1981] (cf. the bibliography of earlier Brockett papers cited there) and MITTER [1980] (cf. the bibliography of earlier paper of Mitter cited there) and the idea of implementing the Lie algebra $\mathscr{F}$ as a Lie Algebra of vector fields is due to Brockett. The first examples of finite-dimensional filters for nonlinear filtering problems were constructed by BENES [1981] using functional integral methods. For a generalisation of Benes results see OCONE-BARAS-MARCUS [1982].

In most situations the Lie algebra F is infinite dimensional (cf. IGUSA [1981]) and in many situations simple. If the Lie algebra F is infinite dimensional it does not necessarily mean that a finite-dimensional filter does not exist. For a precise result in this direction see HAZEWINKEL-MARCUS [1982], SUSSMANN [1981].

9. Final Remarks.

There are other topics of interest and the theory represented by the second point of view is far from complete.

(i) Asymptotic Expansions: see BLANKENSHIP-LIU-MARCUS [1982]. Much work remains to be done here.

(ii) Lower Bounds on Nonlinear Filtering: see article of BOBROVSKY-ZAKAI in HAZEWINKEL-MARCUS [1981], (and the bibliography cited there). The stochastic control interpretation of FLEMING-MITTER [1982] should be important here.

(iii) Filtering on Manifolds: see DUNCAN [1977] for a beginning.

(iv) Smoothness of densities: see KUNITA [1982].

(v) For a partial solution to the vector "cubic-sensor" problem, see DELFOUR-MITTER [1982].

REFERENCES

1. Allinger D., and Mitter, S.K. (1981) New Results on the Innovations Problem of Nonlinear Filtering, Stochastics, vol 4, pp. 339-348.

2. Beneš, V.E., (1981) Exact Finite Dimensional Filters for Cetain Diffusions with Nonlinear Drift, Stochastics, vol 5, pp. 65-92.

3. Bismut, J.M., (1981) Mecanique Aléatoire, Springer Lecture Notes in Mathematics, vol. 866.

4. Blankenship, G.L., Liu, C., and Marcus, S., (1982) Asymptotic Expansions and Lie Algebras for Some Nonlinear Filtering Problems, to appear IEEE Trans on Auto. Control.

5. Brockett, R.W., (1981) Paper in Hazewinkel-Willems, loc. cit.

6. Clark, J.M.C., (1978) The Design of Robust Approximations to the Stochastic Differential Equations of Nonlinear Filtering, in Communication Systems and Random Process Theory: ed. J.K. Skwirzynski, Sithoff and Noordhoff.

7. CIME Lectures on Nonlinear Filtering and Stochastic Control (1982), to be published as Springer Lecture Notes, eds. S.K. Mitter & A. Moro.

8. CIME Lectures on Nonlinear Filtering and Stochastic Control [1982], to be published as Springer Lecture Notes, ed. A. Moro.

9. Davis, M.H.A., [1980] On a Multiplicative Functional Transformation Arising in Nonlinear Filtering Theory, Z. Wahr. verw Gebiete 54 pp. 125-139.

10. Delfour, M.C., and Mitter, S.K., to appear.

11. Duncan, T.E., [1967] Doctoral Dissertation, Department of Electrical Engineering, Stanford University

12. Duncan, T.E., [1977] Some Filtering Results on Riemannian Manifolds, Information and Control, 35, pp. 182-195.

13. Feynman, R.P., and Hibbs, A.R., [1965] Quantum Mechanics and Path Integrals, McGraw Hill, New York.

14. Fleming, W.H., and Mitter, S.K., [1982] Optimal Control and Nonlinear Filtering for Nondegenerate Diffusion Processes, to appear Stochastics.

15. Fujisaki, M., Kallianpur, G., and Kunita, H., [1972] Stochastic Differential Equations for the Nonlinear Filtering Problem, Osaka J. of Mathematics, Vol. 9, pp. 19-40.

16. Hazewinkel, M. and Marcus, S., [1982] On Lie Algebras and Finite Dimensional Filtering, to appear in Stochastics.

17. Hazewinkel, M., and Willems, J. C., [1982] Stochastic Systems: The Mathematics of Filtering and Identification and Applications: eds. M. Hazewinkel and J. C. Willems, pp. 479-503, D. Reidel Publishing Company.

18. Hijab, O, [1980] Minimum Energy Estimation, Doctoral Dissertation, University of Calfornia, Berkeley.

19. Igusa, J.I., [1981] On Lie Albegras Generated by Two Differential Operators in Lie Groups and Manifolds, pp. 187-195. Birkhauser Verlag Boston.

20. Kac, M., [1951] On Some Connections Between Probability Theory and Differential and Integral Equations, Proc. 2nd Berkeley Symposium Math. Stat. and Prob., pp. 189-215.

21. Kallianpur, G., [1980] Stochastic Filtering Theory, Application of Math. Springer Verlag, vol. 13.

22. Kunita, H., [1982] Stochastic Partial Differential Equations Connected with Nonlinear Filtering, to appear in Proceedings on CIME School on Nonlinear Filtering and Stochastic Control.

23. Liptser R.S., and Shiryayev, A.N., [1977] Statistics of Random Processes, Springer-Verlag, New York.

24. Mitter, S.K., [1980] On the Analogy Between Mathematical Problems of Nonlinear Filtering and Quantum Physics, Ricerche di Automatica, vol. 10, No. 2, pp. 163-216.

25. Mitter, S.K., [1981] Non-linear Filtering and Stochastic Mechanics in Stochastic Systems: The Mathematics of Filtering and Identifcation and Applications: eds. M. Hazewinkel and J.C. Willems, pp. 479-503. D. Reidel Publishing Company.

26. Mitter, S.K.,[1982] Lectures Given at CIME School on Nonlinear Filtering and Stochastic Control, Cortona, Italy, July 1981. Proceedings to be published by Springer-Verlag.

27. Mortensen, R.E., [1966] Doctoral Dissertation, Department of Electrical Engineering, University of California, Berkeley, Calif.

28. Ocone, D., [1980] Topics in Nonlinear Filtering Theory, Doctoral Dissertation, M.I.T.

29. Ocone, D., Baras, J.S., and Marcus, S., [1982] Explicit Filters for Certain Diffusions with Nonlinear Drift, to appear Stochastics.

30. Pardoux, E. [1981] The Solution of the Nonlinear Filtering Equation as a Likelihood Function, Proceedings of 20th CDC Conference, San Diego, California.

31. Pardoux, E. [1982] Equations du Filtrage non lineaire de la Prediction et du Lissage, to appear in Stochastics.

32. Ray. D.B., [1949] On Spectra of Second Order Differential Operators, Trans. Am. Math. Soc., 77, pp. 299-321.

33. Sussmann, H., [1981] Rigorous Results on the Cubic Sensor Problem in WILLEMS-HAZEWINKEL, loc. cit.

Note on uniqueness of semigroup associated with Bellman operator

Makiko Nisio

Department of Mathematics, Kobe University
Kobe, 657, Japan

1. Introduction. As a mathematical formulation of Bellman principle, we define a nonlinear semigroup, using the value function of stochastic optimal control. [2, 4, 7] The generator of this semigroup is the Bellman operator. The purpose of this note is to consider the uniqueness of semigroup with generator of Bellman operator, appealing to results of integral solution (Benilan solution)[1] of Cauchy problem.

First we recall our nonlinear semigroup. Let Γ be a compact convex subset of R^k, called a control region. $B(t)$, $t \geq 0$, denotes an n-dimensional Brownian motion on a probability space (Ω, F, P). Any Γ-valued $\sigma_t(B)$-progressible measurable process is called an admissible control. $\mathfrak{A}$ denotes the totality of admissible controls. For $U \in \mathfrak{A}$, we consider the following controlled stochastic differential equation

(1. 1) $$dX(t) = \alpha(X(t), U(t))dB(t) + \gamma(X(t), U(t))dt$$

$$X(0) = x(\in R^n).$$

we assume that $n \times n$ symmetric non-negative definite matrix $\alpha(x, u)$ and n-vector $\gamma(x, u)$ satisfy the following conditions (A1) and (A2).

(A1) $|h(x, u)| \leq b$ for any x, u

(A2) $|h(x, u) - h(y, v)| \leq K|x - y| + \rho(|u - v|)$

where b and K are constants and ρ continuous on $[0, \infty)$ with $\rho(0) = 0$. Then there exists a unique solution $X(t) = X(t; x, U)$, called the response for U. The value function is defined as follows

(1. 2) $$S(t, x, \phi) = \sup_{U \in \mathfrak{A}} E_x \int_0^t e^{-\int_0^s c(X(\theta), U(\theta))d\theta} f(X(s), U(s))ds$$

$$+ e^{-\int_0^t c(X(s),\ U(s))ds} \phi(X(t)),$$

where $X(t)$ is the response for U and x its starting point. Now we assume (A3),

(A3) $c \geq 0$ and f satisfy (A1) and (A2).

Let C be a Banach lattice of bounded and uniformly continuous functions on R^n, with the supremum norm and the usual order. Then we can easily see that $S(t, \cdot, \phi) \in C$ whenever $\phi \in C$. Moreover the operator $S(t)\colon C \to C$, defined by

(1. 3) $\quad S(t)\phi(x) = S(t, x, \phi)$

has the following properties

(i) order preserving, $\phi \leq \psi \Rightarrow S(t)\phi \leq S(t)\psi$

(ii) contraction, $\|S(t)\phi - S(t)\psi\| \leq \|\phi - \psi\|$

(iii) continuous, $\|S(t)\phi - S(\theta)\phi\| \to 0$ as $t - \theta \to 0$

(iv) semigroup, $S(t + \theta) = S(t)S(\theta)$.

(v) generator $\mathcal{G}$, $D(\mathcal{G}) = \{\phi \in C\colon \frac{1}{t}(S(t)\phi - \phi)$

converges in norm, as $t \downarrow 0\} \supset C^2$, where $C^2 = \{\phi \in C;\ \frac{\partial \phi}{\partial x_i},$

$\frac{\partial^2 \phi}{\partial x_i \partial x_j} \in C,\ i, j = 1, \ldots n\}$, and

$$\mathcal{G}\phi = \sup_{u \in \Gamma} L^u \phi + f(x, u), \text{ for } \phi \in C^2$$

where $L^u = \sum_{i,j} a_{ij}(x, u) \frac{\partial^2}{\partial x_i \partial x_j} + \sum_i \gamma_i(x, u) \frac{\partial}{\partial x_i} - c(x, u)I$ and

$a = \frac{1}{2}\alpha^2$, I = identity.

(vi) Define G^u by $G^u\phi = L^u\phi + f(x, u)$. Let $T^u(t)$, $t \geq 0$, be a semigroup on C with generator G^u. Then

(1. 4) $\quad T^u(t)\phi \leq S(t)\phi$ for any t, ϕ, u.

(Hereafter semigroup means strongly continuous one.)

(vii) If semigroup $\Lambda(t)$, $t \geq 0$, on C satisfy (1. 4), then

$$S(t)\phi \leq \Lambda(t)\phi \text{ for any } t, \phi.$$

In §2 we investigate some properties of integral solution, and in §3 we apply these results to $S(t)$ and show that under some conditions $S(t)$, $t \geq 0$, is a unique contraction semigroup with generator of Bellman operator [Theorem 3].

2. Integral solution. In this section we state some properties of integral solution, which are useful in §3.

Let A be a densely defined single valued strict dissipative operator in C, namely $\overline{\mathcal{D}(A)} = C$ and $A\colon \mathcal{D}(A) \to C$ satisfies

(2. 1) $\tau(A\phi - A\psi, \phi - \psi) \leq 0$ for any $\phi, \psi \in \mathcal{D}(A)$

where τ is the tangential functional, that is

(2. 2) $$\tau(g, h) = \lim_{\lambda \downarrow 0} \frac{1}{\lambda}(\|h + \lambda g\| - \|h\|) = \inf_{\lambda > 0} \frac{1}{\lambda}(\|h + \lambda g\| - \|h\|)$$

Hereafter dissipative means strict dissipative.

Definition. $V\colon [0, \infty) \to C$ is called an integral solution of Cauchy problem,

(2. 3) $$\frac{dV(t)}{dt} = AV(t), \quad V(0) = \phi \ (\in C)$$

if, $V(t)$ is continuous in t, $V(0) = \phi$ and satisfies the following inequality

(2. 4) $$\|V(t) - \psi\| - \|V(s) - \psi\| \leq \int_s^t \tau(A\psi, V(\theta) - \psi)\,d\theta$$

for any t, s and $\psi \in \mathcal{D}(A)$.

Hence an integral solution may not be differentiable. An integral solution is weaker than a solution (= V is Lipschitz continuous, strongly differentiable at $a.\ a.\ t$ and satisfies (2. 3) at $a.\ a.\ t$).

Proposition 1. Let $\Lambda(t)$, $t \geq 0$, be a contraction semigroup on C, whose generator $\mathcal{G}$ is an extension of A, that is, $\mathcal{D}(\mathcal{G}) \supset \mathcal{D}(A)$ and $\mathcal{G} = A$ on $\mathcal{D}(A)$. Then $\Lambda(t)\phi$ is an integral solution of (2. 3).

Proof. Putting $V(t) = \Lambda(t)\phi$, we will show (2. 4) according to [6]. Define A_h by

(2. 5) $\quad A_h = \frac{1}{h}(\Lambda(h) - I),\ \mathcal{D}(A_h) = C.$

Since $\Lambda(h)$ is contraction, A_h is dissipative. Moreover there exists a unique semigroup $\Lambda_h(t)$ on C with generator A_h. Furthermore $\Lambda_h(\cdot)\phi$ is continuously differentiable on $(0, \infty)$ and $\Lambda_h(t)\phi$ is a unique solution of Cauchy problem

(2. 6) $\quad \frac{dW(t)}{dt} = A_h W(t),\ W(0) = \phi.$

Since $\Lambda_h(t)\phi$ turns out an integral solution of (2. 6), we have,

(2. 7) $$\|\Lambda_h(t)\phi - \psi\| - \|\Lambda_h(s)\phi - \psi\| \le \int_s^t \tau(A_h\psi, \Lambda_h(\theta)\phi - \psi)d\theta$$

$$\le \int_s^t \frac{1}{\lambda}(\|\Lambda_h(\theta)\phi - \psi + \lambda A_h\psi\| - \|\Lambda_h(\theta)\phi - \psi\|)d\theta, \text{ for } \lambda > 0$$

For $\psi \in \mathcal{D}(A)$,

(2. 8) $\quad \|A_h\psi - A\psi\| \to 0$, as $h \downarrow 0$.

Hence we see

(2. 9) $$|\text{integrand of right side of (2. 7)}| \le \|A_h\psi\|$$

$$\le \|A\psi\| + 1 \qquad \text{for small } h$$

On the other hand, as $h \downarrow 0$

(2. 10) $\quad \|\Lambda_h(\theta)\phi - \Lambda(\theta)\phi\| \to 0$

uniformly on any bounded time interval. Therefore, tending h to 0, we have, from (2. 7) $\sim$ (2. 10),

(2. 11) $$\|\Lambda(t)\phi-\psi\| - \|\Lambda(s)\phi-\psi\| \le \int_s^t \frac{1}{\lambda}(\|\Lambda(\theta)\phi-\psi+\lambda A\psi\| - \|\Lambda(\theta)\phi-\psi\|)d\theta$$

Tending λ to 0, the bounded convergence theorem derives (2. 4). This completes the proof.

Next we recall the following proposition,

Proposition 2 [3. 6]. Suppose the condition (2. 12).

(2. 12) $$\lim_{\lambda\downarrow 0} \frac{1}{\lambda} \text{dist. } (R(I - \lambda A), \phi) = 0, \text{ for any } \phi \in C$$

where R means the range. Then we have

(i) for any $\phi \in C$, an integral solution $V(t; \phi)$ of (2. 3) exists uniquely.

(ii) the operator $W(t)$, defined by $W(t)\phi = V(t; \phi)$ becomes a unique semigroup which provides an integral solution of (2. 3).

From Proposition 1 and 2, we see

Proposition 3. Under condition (2. 12), a semigroup, whose generator is an extension of A, is unique, if exists.

3. Semigroup $S(t)$. Put $G\phi = \sup_{u\in\Gamma} L^u\phi + f(x, u)$ and

(3. 1) $$D = \{\phi \in W^2_\infty \cap C;\ G\phi \text{ has a version in } C\}.$$

Since coefficients α, γ and c are continuous in u, $G\phi$ can be defined for $\phi \in W^2_\infty$, if versions of $\frac{\partial\phi}{\partial x_i}$ and $\frac{\partial^2\phi}{\partial x_i \partial x_j}$ are fixed independent of u. Since $D \supset C^2$ and C^2 is dense in C, D is dense in C. $\tilde{G} = G_{/D}$ can be regarded as a mapping from D into C.

Theorem 1. Under (A1) $\sim$ (A3), the generator $\mathcal{G}$ of $S(t)$ satisfies the following conditions.

(3. 2) $$\mathcal{D}(\mathcal{G}) \supset D \text{ and } \mathcal{G} = G \text{ on } D.$$

namely $\mathcal{G}$ is an extension of $\tilde{G}$.

Lemma. Suppose that (A1) $\sim$ (A3) and (A4) hold.

(A4) there exists $\mu > 0$ such that

(3. 3) $$\sum_{ij} \alpha_{ij}(x, u)\theta_i\theta_j \geq \mu|\theta|^2 \quad \text{for any } x, \theta, u.$$

Let $\psi \in W^2_\infty \cap C$. If $G\psi = g + h$ with $g \in C$ and $h \in L_\infty$, then, for $\lambda > 0$, there exists $t_0 = t_0(\lambda, b, k, g) > 0$, independent of μ and h, such that

(3. 4) $$\left\| \frac{1}{t}(S(t)\psi - \psi) - g \right\| \leq \|h\|_\infty + \lambda, \text{ whenever } t < t_0$$

Proof. By (A4) Ito's formula [4] tells us

(3. 5) $$S(t)\psi(x) - \psi(x) = \sup_{U\in\mathfrak{U}} E_x \int_0^t e^{-\int_0^s c(X(\theta),\, U(\theta))d\theta} (f(X(s), U(s)) + L^{U(s)}\psi(X(s)))ds$$

$$\leq \sup_{U\in\mathfrak{U}} E_x \int_0^t e^{-\int_0^s c(X(\theta),\, U(\theta))d\theta} G\psi(X(s))ds$$

$$\leq \sup_{U\in\mathfrak{U}} E_x \int_0^t e^{-\int_0^s c(X(\theta),\, U(\theta))d\theta} g(X(s)ds + t\|h\|_\infty$$

Since $g \in C$, we see that, as $t \downarrow 0$, $\frac{1}{t} E_x \int_0^t e^{-\int_0^s c(X(\theta), U(\theta))d\theta} g(X(s))ds$ converges to $g(x)$ uniformly in x and U. Moreover, by routine method, we can take $t_1 = t_1(\lambda, b, K, g) > 0$ which is independent of μ and h, so that

$$(3.\ 6) \qquad \frac{1}{t}(S(t)\psi(x) - \psi(x)) - g(x) \le \|h\|_\infty + \lambda, \text{ for } x \in R^n,\ t < t_1$$

On the other hand, by measurable selection theorem we can take a Γ-valued Borel function $u(\cdot)$ such that

$$(3.\ 7) \qquad G\psi(x) = L^{u(x)}\psi(x) + f(x, u(x)) \qquad \text{for any } x.$$

The stochastic differential equation

$$d\xi(t) = \alpha(\xi(t), u(\xi(t))dB(t) + \gamma(\xi(t), u(\xi(t))dt$$

$$\xi(0) = x$$

has a weak solution by (A4), [4]. Although $u(\xi(t))$ may not be $\sigma_t(B)$-measurable, we can see [7],

$$S(t)\psi(x) \ge E_x \int_0^t e^{-\int_0^s c(\xi(\theta), u(\xi(\theta))d\theta} f(\xi(s), u(\xi(s))ds + e^{-\int_0^t c(\xi(\theta), u(\xi(\theta))d\theta} \psi(\xi(t))\ .$$

Hence Ito's formula implies

$$(3.\ 8) \qquad S(t)\psi(x) - \psi(x) \ge E_x \int_0^t e^{-\int^s c(\xi(\theta), u(\xi(\theta))d\theta} (f(\xi(s), u(\xi(s)) + L^{u(\xi(s))} \psi(\xi(s)))ds$$

$$= E_x \int_0^t e^{-\int_0^s c(\xi(\theta), u(\xi(\theta))d\theta} G\psi(\xi(s))ds$$

$$\ge E_x \int_0^t e^{-\int_0^s c(\xi(\theta), u(\xi(\theta))d\theta} g(\xi(s))ds - t\|h\|_\infty\ .$$

Therefore there exists $t_2 = t_2(\lambda, b, K, g)$, independent of μ and h, such that

$$(3.\ 9) \quad \frac{1}{t}(S(t)\psi(x) - \psi(x)) - g(x) \ge -\|h\|_\infty - \lambda, \text{ for } x \in R^n,\ t < t_2.$$

Combining (3. 6) and (3. 9), we conclude Lemma.

Proof of Theorem 1. $\alpha^\varepsilon = \alpha + \varepsilon I$ satisfies (A4). When we use α^ε instead of α, we put superscript ε. Consider controlled stochastic differential equations

$$(3.10)\qquad \begin{cases} dX(t) = \alpha(X(t), U(t))dB(t) + \gamma(X(t), U(t))dt \\ X(0) = x \end{cases}$$

and

$$(3.11)\qquad \begin{cases} dX^{\varepsilon}(t) = \alpha^{\varepsilon}(X^{\varepsilon}(t), U(t))dB(t) + \gamma(X^{\varepsilon}(t), U(t))dt \\ X^{\varepsilon}(0) = x \end{cases}$$

Applying the routine method, we can see that, for $T > 0$, there exists $K_1 = K_1(T)$, independent of ε, x and U, such that

$$E|X^{\varepsilon}(t) - X(t)|^2 \le \varepsilon K_1 t \qquad \text{for } t < T.$$

Therefore we have, as $\varepsilon \to 0$,

$$(3.12)\qquad ||S^{\varepsilon}(t)\phi - S(t)\phi|| \to 0, \text{ uniformly on } [0, T], (\forall T > 0).$$

For $\psi \in D$, $G\psi$ has a version in C and

$$(3.13)\qquad |G^{\varepsilon}\psi(x) - G\psi(x)| \le \varepsilon \sup_{u\in\Gamma} |\sum \alpha_{ij}(x, u) \frac{\partial^2\psi}{\partial x_i \partial x_j} + \frac{\varepsilon}{2}\Delta\psi(x)|, \text{ a.e.}$$

Hence, with some constant $d = d(\alpha, \psi) > 0$,

$$(3.14)\qquad ||G^{\varepsilon}\psi - G\psi||_{\infty} < \varepsilon d$$

Now we apply Lemma to $G^{\varepsilon}\psi$. α^{ε} satisfies (A4) and we can suppose that $G\psi \in C$. Hence, for $\varepsilon' > 0$, there exists $t_0 = t_0(\varepsilon', b, k, G\psi)$ such that

$$(3.15)\qquad ||\frac{1}{t}(S^{\varepsilon}(t)\psi - \psi) - G\psi|| < d\varepsilon + \varepsilon', \text{ for } t < t_0 \text{ and } \varepsilon < 1.$$

From (3.12) and (3.15), we get

$$(3.16)\qquad ||\frac{1}{t}(S(t)\psi - \psi) - G\psi|| \le \varepsilon', \text{ for } t < t_0.$$

This means that $\psi \in \mathcal{D}(\mathcal{G})$ and $\mathcal{G}\psi = G\psi$.

For the condition (2.12) we introduce (A5) and (A6) according to [5].

(A5) $\alpha_{i,j}$, γ_i, c and f are twice continuously differentiable and, putting $h = \alpha_{k,\ell}, \gamma_k, c, f$,

$$\sup_{x,u} \left(\left| \frac{\partial^2 h(x,u)}{\partial x_i \partial x_j} \right| + \left| \frac{\partial h(x,u)}{\partial x_i} \right| \right) < \infty, \qquad i, j = 1,\dots,n,$$

(A6) there exists a constant $\nu > 0$, such that for all $x \in R^n$ there are an integer m, and

$u_1, \ldots, u_m \in \Gamma$, $\theta_1, \ldots \theta_m \in [0, 1]$, $\sum_{i=1}^{m} \theta_i = 1$, such that

$$\sum_{ij} \sum_{k} \theta_k \, a_{ij}(x, u_k)\xi_i\xi_j \geq \nu|\xi|^2, \text{ for } \forall\xi \in R^n.$$

Theorem 2. Under (A1)(A2)(A3)(A5) and (A6), $R(I - \lambda\tilde{G})$ is dense in C, for small $\lambda > 0$.

Proof. Consider Bellman equation for $\psi \in C^2$

$$\text{(3. 17)} \qquad \sup_{u \in \Gamma} L^u W(x) + f(x, u) - \frac{1}{\lambda} W(x) + \frac{1}{\lambda} \psi(x) = 0, \text{ a.e.}$$

For small λ (depending to α and γ), (3. 17) has a solution $W \in W^2_\infty \cap C$. [5]. Therefore $GW = \frac{1}{\lambda}(W - \psi)$ a.e. Since $W - \psi \in C$, GW has a version in C. So $W \in D$. This means $\psi \in R(I - \lambda\tilde{G})$. Since C^2 is dense, this completes the proof.

Appealing to Propositions 1 ∿ 3, we have

Theorem 3. (i) Under (A1) ∿ (A3), $S(t)\phi$ is an integral solution of Cauchy problem in C,

$$\frac{dV(t)}{dt} = \tilde{G}V(t), \quad V(0) = \phi.$$

(ii) Under (A1)(A2)(A3)(A5) and (A6), $S(t)$, $t \geq 0$ is a unique contraction semigroup on C, whose generator is an extension of $\tilde{G}(= G_{/D})$.

4. Optimal stopping. Put $m(t) = \{B\text{-stopping time} \leq t\}$. For $U \in \mathfrak{A}$ and $\tau \in m(t)$ we have the motion X of (1. 1) until stopping time τ. So, the value function S^* is defined as follows,

$$\text{(4. 1)} \qquad S^*(t, x, \phi) = \sup_{\tau \in m(t), U \in \mathfrak{A}} E_x \int_0^\tau e^{-\int_0^s c(X(\theta), U(\theta))d\theta} f(X(s), U(s))ds + e^{-\int_0^\tau c(X(s), U(s))ds} \phi(X(\tau)).$$

We can easily see that $S^*(t, \cdot, \phi) \in C$ whenever $\phi \in C$. Moreover the operator $S^*(t); C \to C$, defined by

$$\text{(4. 2)} \qquad S^*(t)\phi(x) = S^*(t, x, \phi)$$

is an order preserving, contraction semigroup on C, whose generator

$\mathcal{G}^*$ is given by

(4. 3) $\mathcal{G}^*\phi = \max\{0, \sup_{u\in\Gamma} L^u\phi + f(x, u)\}$, for $\phi \in C^2$.

Using the same method as in §3, we can get similar results. That is, define $\tilde{G}^*$ by

(4. 4) $\tilde{G}^*\phi = \max(0, \tilde{G}\phi)$ for $\phi \in D$

Then, $\mathcal{G}^*$ is an extension of $\tilde{G}^*$ under the conditions (A1) ∿ (A3).

Theorem 4. (i) Under (A1) ∿ (A3), $S^*(t)\phi$ is an integral solution of Cauchy problem in C,

(4. 5) $\frac{dV(t)}{dt} = \tilde{G}^*V(t)$, $V(0) = \phi$.

(ii) Under (A1)(A2)(A3)(A5) and (A6), $S^*(t)$ is a unique contraction semigroup on C, whose generator is an extension of $\tilde{G}^*$.

References

1. Ph. Benilan, Equation d'evolution dans un espace de Banach quelconque et applications, Thése Orsay, 1972.
2. A. Bensoussan and J. L. Lions, Applications des Inéquations Variationneles en Contrôle Stochastique, Dunod, Paris, 1978.
3. Y. Kobayashi, Difference approximation of evolution equation and generation of nonlinear semigroups, Proc. Japan Acad., 51(1975), 406-410.
4. N. V. Krylov, Controlled Diffusion Processes, Appl. Math. 14, Springer Verlag, 1980. (Orig. Russian 1977).
5. P. L. Lions, Control of diffusion processes in R^n , Comm. Pure & Appl. Math. 34(1981) 121-147.
6. I. Miyadera, Hisenkei Hangun (Nonlinear semigroups), Kinokuniya, 1977 (Japanese).
7. M. Nisio, Stochastic Control Theory, ISI Lect. Notes 9, Macmillan India, 1981.

P D E with random coefficients :

asymptotic expansion for the moments

R. BOUC[+] - E. PARDOUX[++]

Introduction

Suppose Z_t is a diffusion process with values in $\mathbb{R}^d$, and $\forall z \in \mathbb{R}^d$, we are given an unbounded linear operator A(z) on some Hilbert space H (in pratical examples, A(z) will be a PDE operator), such that the following Cauchy problem is well - posed :

$$(0.1) \quad \frac{d\,u_t}{d\,t} + A(Z_t)u_t = o, \; u_o \text{ given in H}$$

In [2] it is shown, under appropriate conditions, that if q(t,z) denotes the solution of:

$$(0.2) \quad \begin{aligned} &\frac{\partial q}{\partial t}(t,z) + A(z)q(t,z) = L^* q(t,z) \\ &q(o,z) = u_o \, p_o(z) \end{aligned}$$

where L^* is the adjoint of the infinitesimal generator of Z_t, p_o the density of the law of Z_o, then $E[\,u_t\,] = \int q(t,z)dz$

Thus the computation of the first moment of u_t is reduced to the computation of the solution of (0.2). Similar results apply for higher-order moments.

+ L M A - C N R S 31, chemin Joseph Aiguier, 13274 Marseille Cedex 2

++ UER de Mathématique, Université de Provence,

3 place V. Hugo, 13331 Marseille Cedex 3.

As noted in [2], this result is not completely satisfactory, since the dimension of the space variable in equation (0.2) will often be too large for computing a numerical approximation. Indeed, if the space variable x implicit in equation (0.1) is of dimension n, the space variable $\binom{x}{z}$ of equation (0.2) is of dimension $n+d$. This is the price one has to pay for the fact that (0.1) is non-linear in $\binom{u_t}{Z_t}$. The above result exploits the linearity with respect to u_t . Unfortunately, one is typically interested in cases where d is large - the limiting case $d = +\infty$ would correspond to a random field input $Z(t,x)$.

On the other hand, if u_t is the solution of a bilinear Ito equation :

$$(0.3) \quad du_t + \mathcal{A}\, u_t\, dt = \mathcal{B}\, u_t\, dw_t$$

then the first moment $\bar{u}_t = E[u_t]$ solves the deterministic evolution equation :

$$(0.4) \quad \frac{d\bar{u}_t}{dt} + \mathcal{A}\, \bar{u}_t = o, \quad \bar{u}_o = E[u_o]$$

whose space variable is of the same dimension as that of (0.3).

The result of this paper is that if in some sense the noise entering (0.1) is "wide-band", then u_t is "close in law" to the solution of an equation similar to (0.3). This is a natural generalisation of known results in finite dimension-see e.g.BLANKENSHIP-PAPANICOLAOU [1], KUSHNER [5]- with the other difficulty that Z_t takes values in a (non compact) euclidean space. Our main concern here is to show how arbitrary accurate (1) approximations to $E[u_t]$ can be computed, solving only PDEs with space variable of the same dimension as that of (0.1). Similar results would hold for higher order moments. The details of proofs of the results announced here will be published elsewhere.

----

(1) at least in principle, since the coefficients in the equation become more and more complicated, when one wants to increase the accuracy.

I. The PDE with coefficients perturbed by wide-band noise

We replace (0.1) by :

$$(1.1)\quad \begin{cases} \dfrac{d u_t^{\varepsilon}}{dt} + A u_t^{\varepsilon} + \dfrac{1}{\varepsilon} B(Z_{t/\varepsilon^2}) u_t^{\varepsilon} = o \\ u_o^{\varepsilon} = u_o \end{cases}$$

where ε is a "small parameter", and we now make precise assumptions on Z_t, A and B.

I. a The process Z_t

We suppose that Z_t is a Markov diffusion process with values in $\mathbb{R}^d$, whose infinitesimal generator is given by :

$$L = \frac{1}{2} \sum_{i,j} a_{ij}(z) \frac{\partial^2}{\partial z_i \partial z_j} + \sum_i b_i(z) \frac{\partial}{\partial z_i}$$

we suppose that the coefficients $a_{ij}(.)$ and $b_i(.)$ are measurable and bounded[(1)], and that the following ellipticity condition is satisfied :

$$(1.2)\quad \exists \beta > o \text{ s.t. } \sum_{i,j} a_{ij}(z)\xi_i\xi_j \geqslant \beta |\xi|^2, \ \forall z, \xi \in \mathbb{R}^d$$

We suppose moreover that the process Z_t has an invariant probability measure, which has a density p, which then satisfies :

$$L^* p = o$$

where L^* is the formal adjoint of L. We suppose moreover that :

$$(1.3)\quad p^{-1} \sum_j \frac{\partial}{\partial z_j} (a_{ij} p) \in L^{\infty}(\mathbb{R}^d), \ i = 1...d$$

We add the following hypothesis (which is not necessary, but does simplify the exposition):

$$(1.4)\quad p^{-1} L^*(p.) = L$$

We choose p as initial density. Z. is then a stationary process. (1.4) implies that Z. is time-reversible, in other words : $\forall t > o$, $\forall \varphi$ bounded measurable from $\mathbb{R}^d$ into $\mathbb{R}$,

---

(1) All the results that follow would be true with minor changes, in the case where the b_i's are allowed to grow at most linearly in z.

$$E\,[\varphi(Z_o)/Z_t = z] = E\,[\varphi(Z_t)/Z_o = z]$$

In the sequel, we will write E_z for $E[\,./Z_o = z\,]$.

I. b The operators A and B

We suppose that A and B are PDE operators acting on functions of x, where $x \in \mathbb{R}^n$ (we consider (1.1) as an equation in the whole space, without boundary conditions). We denote by $|.|$ the norm of $L^2(\mathbb{R}^n)$, $\|.\|$ the norm of $H^1(\mathbb{R}^n)$, $(.,.)$ and $<.,.>$ the scalar product in $L^2(\mathbb{R}^n)$ and pairing between $H^1(\mathbb{R}^n)$ and $H^{-1}(\mathbb{R}^n)$.

Let us assume that A is a second order elliptic operator, which belongs to $\mathcal{L}(H^1(\mathbb{R}^n), H^{-1}(\mathbb{R}^n))$, and s.t. $\exists \gamma > o$ and λ with :

$$(1.5) \quad < Au,u > + \lambda\, |u|^2 \geq \gamma\, \|u\|^2, \ \forall u \in H^1(\mathbb{R}^n)$$

Moreover, $\forall z \in \mathbb{R}^d$, B(z) is a first order differential operator, and

$$B(.) \in L^\infty(\mathbb{R}^d\, ;\, \mathcal{L}(H^1(\mathbb{R}^n)\, ;\, L^2(\mathbb{R}^n)))$$

We assume that :

$$B(z) + B^*(z) = \ell(.,z)$$

i.e. $\forall z \in \mathbb{R}^d$, $B(z) + B^*(z)$ is the multiplication by $\ell(x,z)$ - this is the case as soon as the coefficients of B have some regularity in x - and that :

$$(1.6) \quad \ell \in L^\infty(\mathbb{R}^{n+d})$$

Now the crucial hypothesis on B is :

$$(1.7) \quad E\,[B(Z_t)] = o, \quad \text{i.e.} \quad \int B(z)p(z)dz = o$$ (1)

We finally suppose that each coefficient of A and B is C^∞ in x, all derivatives being bounded functions of x - resp - of (x,z).

II. Convergence in law

Under some additional hypothesis, which are essentially those that we will introduce in the next sections, one can show that u_ε converges in law to u, the

----

(1) This means in fact that the same equality holds for each coefficient of the PDE operator B, a.e. in x.

solution of the following Ito - type stochastic PDE :

$$du_t + [A - \int_0^\infty E(B(Z_\theta)B(Z_o))d\theta]u_t dt = dM_t(u)$$

where $M_t(u)$ is a continuous $L^2(\mathbb{R}^n)$ - valued martingale, such that $\forall \varphi \in L^2(\mathbb{R}^n)$, the quadratic variation process of the real - valued martingale $(M_t(u),\varphi)$ is :

$$<(M.(u),\varphi)>_t = 2\int_0^t ds \int_0^\infty E[(B(Z_o)u_s, \varphi)(B(Z_\theta)u_s,\varphi)]d\theta$$

This result is analogous to known results in finite - dimension, see e.g. BLANKENSHIP - PAPANICOLAOU [1]. The difficulty due to the PDE can be taken care of by using techniques from VIOT [8]. We also need here the results in § IV and V below.

III Expansion of the first moment in powers of ε : formal derivation

It follows from the result quoted in the introduction that :

$$E[u_t^\varepsilon] = \int q^\varepsilon(t,z)dz$$

where q^ε solves :

$$\frac{\partial q^\varepsilon}{\partial t} + Aq^\varepsilon + \frac{1}{\varepsilon} B q^\varepsilon = \frac{1}{\varepsilon^2} L^* q^\varepsilon$$

$$q^\varepsilon(o) = p\, u_o$$

It is convenient to divide the above equation by p, i.e. to consider $\overline{v}^\varepsilon(t,z)$ s.t.

$$q^\varepsilon(t,z) = p(z)\, \overline{v}^\varepsilon(t,z)$$

Using (1.4), we get that $\overline{v}^\varepsilon$ solves :

$$(3.1)\qquad \begin{cases} \dfrac{\partial \overline{v}^\varepsilon}{\partial t} + A\,\overline{v}^\varepsilon + \dfrac{1}{\varepsilon} B\,\overline{v}^\varepsilon = \dfrac{1}{\varepsilon^2} L\,\overline{v}^\varepsilon \\ \overline{v}^\varepsilon(o) = u_o \end{cases}$$

$\overline{v}^\varepsilon(t,z)$ is also given by :

$$\overline{v}^\varepsilon(t,z) = E[u_t^\varepsilon / Z_t^\varepsilon = z]$$

At that point, we need to introduce one complication. There are two time scales in our problem : t and t/ε^2 . We now have to introduce the "fast time variable" explicitly, in order to treat the initial layer. That is, we introduce a new function $v^{\varepsilon}(t,\tau,z)$, such that :

$$\bar{v}^{\varepsilon}(t,z) = v^{\varepsilon}(t,t/_{\varepsilon^2},z)$$

From (3.1), v^{ε} must satisfy :

$$(3.2)\quad \begin{aligned} &\frac{\partial v^{\varepsilon}}{\partial t} + A\, v^{\varepsilon} + \frac{1}{\varepsilon} B\, v^{\varepsilon} + \frac{1}{\varepsilon^2}\left(\frac{\partial v^{\varepsilon}}{\partial \tau} - L\, v^{\varepsilon}\right) = o \\ &v^{\varepsilon}(o,o,z) = u_o \end{aligned}$$

We do not claim that v^{ε} is caracterised by (3.2) - part of the boundary conditions are missing - but this does not matter, since (3.2) will be used only as a guide to guess the correct expansion. We will then go back to (3.1) in order to justify it.

Let us now write v^{ε} as :

$$v^{\varepsilon}(t,\tau,z) = \bar{u}^{\varepsilon}(t,\tau) + \hat{v}^{\varepsilon}(t,\tau,z)$$

with the constraint :

$$\int \hat{v}^{\varepsilon}(t,\tau,z)p(z)dz = o \quad , \text{ so that :}$$

$$E[u_t^{\varepsilon}] = \int v^{\varepsilon}(t,t/_{\varepsilon^2},z)p(z)dz$$

$$= \bar{u}^{\varepsilon}(t,t/_{\varepsilon^2})$$

$\bar{u}^{\varepsilon}$ is the quantity of interest. Our problem comes from the fact that we cannot compute $\bar{u}^{\varepsilon}$ without computing the whole function $v^{\varepsilon}(t,\tau,z)$.

We are now going to look for an expansion of the form :

$$(3.3)\quad v^{\varepsilon}(t,\tau,z)= v^{o}(t,\tau,z)+\varepsilon\, v^{1}(t,\tau,z)+ \varepsilon^2 v^2(t,\tau,z)+ \dots$$

and we write each v^i as :

$$(3.4)\quad v^i(t,\tau,z)= \bar{u}^i(t,\tau)+ \hat{v}^i(t,\tau,z)$$

with the constraint ;

$$\int \hat{v}^i(t,\tau,z)p(z)dz = o$$

so that :

$$\bar{u}^{\varepsilon}(t,\tau)= \bar{u}^{o}(t,\tau)+\varepsilon\, \bar{u}^{1}(t,\tau)+ \varepsilon^2\, \bar{u}^{2}(t,\tau)+ \dots$$

The reason why this expansion is useful is that we will get "explicit" expressions for the $\hat{v}^i$'s, and the $\bar{u}^i$'s will be solutions of PDEs with state variable x.

We now replace in (3.2) v^ε by the right hand side of (3.3), and equate to zero the coefficient of each power of ε. Considering the coefficients of ε^{-2}, ε^{-1}, ε^o and ε^1, we get the equations :

(3.4) $$\frac{\partial v^o}{\partial \tau} - L\, v^o = o$$

(3.5) $$\frac{\partial v^1}{\partial \tau} - L\, v^1 + B\, v^o = o$$

(3.6) $$\frac{\partial v^2}{\partial \tau} - L\, v^2 + B\, v^1 + \frac{\partial v^o}{\partial t} + A\, v^o = o$$

(3.7) $$\frac{\partial v^3}{\partial \tau} - L\, v^3 + B\, v^2 + \frac{\partial v^1}{\partial t} + A\, v^1 = o$$

These equations will be enough for deriving the equations satisfied by $\bar{u}^o$ and $\bar{u}^1$. Let us recall that what follows is not a rigorous derivation. We are going to "guess" from (3.4)...(3.7) the equations that should satisfy the $\bar{u}^i$'s and $\hat{v}^i$'s. The justification will be made later on.

The following condition implies that (3.4) is satisfied :

(3.8) $v^o(t,\tau,z) = \bar{u}^o(t)$

which implies in particular that $\hat{v}^o = o$.

Using (3.8), (3.5) becomes :

$$\frac{\partial v^1}{\partial \tau} - L\, v^1 + B\, \bar{u}^o(t) = o$$

we multiply the above equation by p, and integrate with respect to z. We then get, using $L^* p = o$ and (1.7):

$$\frac{\partial \bar{u}^1}{\partial \tau} = o, \text{ i.e. } \bar{u}^1(t,\tau) = \bar{u}^1(t)$$

we then deduce :

$$\frac{\partial \hat{v}^1}{\partial \tau} - L\, \hat{v}^1 + B\, \bar{u}^o(t) = o$$

to which we associate the boundary condition :

$$\hat{v}^1(t,o,z) = o$$

Then $\hat{v}^1$ is given in term of $\bar{u}^o$ by :

$$(3.9)\quad \hat{v}^1(t,\ ,z) = -\left(\int_o^\tau E_z[B(Z_s)]ds\right)\bar{u}^o(t)$$

We now consider (3.6) :

$$\frac{\partial v^2}{\partial \tau} - L\, v^2 + B\, v^1 + \frac{d\bar{u}^o}{dt} + A\,\bar{u}^o = o$$

Again, multiplying by p and integrating against dz, we get :

$$(3.10)\quad \frac{\partial \bar{u}^2}{\partial \tau} + \int B\,\hat{v}^1\, p\, dz + \frac{d\bar{u}^o}{dt} + A\,\bar{u}^o = o$$

we expect that all quantities should have limits for $\tau \to +\infty$, and that :

$$\frac{\partial \bar{u}^2}{\partial \tau}(t,\infty,z) = o$$

we then get from (3.10), using (3.9) :

$$(3.11)\quad \begin{aligned} &\frac{d\bar{u}^o}{dt}(t) + \left[A - \int_o^\infty E(B(Z_s)B(Z_o))ds\right]\bar{u}^o(t) = o \\ &\bar{u}^o(o) = u_o \end{aligned}$$

(3.11) is the equation for the limit of $E[u_t^\varepsilon]$, which is also the expectation of the limit in law of u_t^ε . The difference between A and the operator in (3.11) can be interpreted as an Ito - Stratonovitch correcting term.

It follows from (3.10),(3.9) and (3.11):

$$\frac{\partial \bar{u}^2}{\partial \tau} + \left[\int_\tau^\infty E(B(Z_o)B(Z_s))ds\right]\bar{u}^o = o$$

we now choose the boundary condition :

$$\bar{u}^2(t,o) = o$$

we can make this choice because we do not want really to compute $\bar{u}^2$, but only $\bar{u}^o$ and $\bar{u}^1$. By direct integration, we get :

$$\bar{u}^2(t,\tau) = -\left[\int_o^\tau d\theta \int_\theta^\infty E(B(Z_s)B(Z_o))ds\right]\bar{u}^o(t)$$

Subtracting (3.10) from (3.6), we get :

$$\frac{\partial \hat{v}^2}{\partial \tau} - L\,\hat{v}^2 + B\,\bar{u}^1 + B\,\hat{v}^1 - \int B\,\hat{v}^1\, p\, dz = o$$

$$\hat{v}^2(t,o,z) = o$$

Using (3.9), we get the following expression for $\hat{v}^2$:

$$\hat{v}^2(t,\tau,z) = -\left[\int_o^\tau E_z(B(Z_s))ds\right]\bar{u}^1(t) + \left[\int_o^\tau\int_\theta^\tau (E_z - E)(B(Z_\theta)B(Z_s))ds\,d\theta\right]\bar{u}^o(t)$$

By similar argument, using the above expressions for $\hat{v}^1$, $\bar{u}^2$ and $\hat{v}^2$, we get from (3.7):

$$\frac{d\bar{u}^1}{dt}(t)+[A-\int_o^\infty E(B(Z_s)B(Z_o))ds]\bar{u}^1(t)+$$

$$(3.12)\qquad +[\int_o^\infty\int_\theta^\infty E(B(Z_s)B(Z_\theta)B(Z_o))dsd\theta]\bar{u}^0(t)=o$$

$$\bar{u}^1(o)=o$$

and one can express $\bar{u}^3$ and $\hat{v}^3$ in terms of $\bar{u}^o$ and $\bar{u}^1$. Again, we do not compute the "true" value of $\bar{u}^3$, but just a quantity which will be necessary for estimating $\bar{u}^\varepsilon(t) - \bar{u}^o(t) - \varepsilon\bar{u}^1(t)$.

IV Ergodic properties of $\{Z_t\}$

We need some conditions to guarantee that the integrals appearing in (3.11) and (3.12) do converge.

What we need is to make sense of

$$\int_o^\infty E_z[\varphi(Z_t)]dt$$

for φ bounded mesurable from $\mathbb{R}^d$ into $\mathbb{R}$, such that :

(4.1) $\int\varphi(z)p(z)dz = o$

and to show that it is the limit, as $\tau\to+\infty$, of :

$$\int_o^\tau E_z[\varphi(Z_t)]dt$$

This will be true if, for some norm $\|\cdot\|$, (4.1) implies that $\exists\alpha>o$ s.t.:

(4.2) $\|E.\ \varphi(Z_t)\| \leqslant \|\varphi\| e^{-\alpha t}$

Doeblin's condition (see e.g. PAPANICOLAOU [6]) would imply (4.2) with $\|\cdot\|$ equal to the sup norm.

But Doeblin's condition cannot be true here, essentially because Z_t takes values in the non compact set $\mathbb{R}^d$.

Following CAUGHEY - PAYNE [3], we now define the Hilbert space :

$\hat{L}^2 \triangleq L^2(\mathbb{R}^d ; p(z)dz)$. We denote by $|\cdot|_\wedge$ and $(.,.)_\wedge$ the norm and scalar product on $\hat{L}^2$. Consider L as an unbounded operator on $\hat{L}^2$, with domain D(L). For $u\in D(L)$, one gets :

$$(4.3)\quad (L u ,u)_\Lambda = - \sum_{i,j} (a_{ij} \frac{\partial u}{\partial z_i}, \frac{\partial u}{\partial z_j})_\Lambda \leq o$$

It then follows from (1.4) that L is a selfadjoint operator on $\hat{L}^2$. o is an eigenvalue of L, the corresponding eigenvectors being the constants.

Define $\hat{L}^2_o = \{u \in \hat{L}^2; \int u(z)p(z)dz = o\}$, and L_o to be the operator L considered as an unbounded operator on $\hat{L}^2_o$.

From (4.3), the spectrum of L, Sp(L) is contained in $\mathbb{R}-$. Suppose that o is an isolated point in Sp(L). Then $\exists \alpha > o$ such that $Sp(L_o) \subset]-\infty, -\alpha]$, and (4.2) follows from (4.1), with $\| . \|$ replaced by $| . |_\Lambda$. We now state :

<u>Theorem 4.1</u> Suppose in addition to (1.2) that $\exists M$ and $c > o$ s.t.

$$\forall z \in \mathbb{R}^d, \ |z| \geq M,$$

$$(4.4)\quad b(z). \frac{z}{|z|} \leq - c$$

Then o is an isolated point of Sp(L).

□

The proof of the Theorem uses a criterion due to H. Weyl (see RIESZ-NAGY [7]), which says that the conclusion of the theorem is equivalent to the non existence of sequences $\{u_n\} \subset D(L)$ s.t.

$$|u_n|_\Lambda = 1, \ u_n \to o \text{ in } \hat{L}^2 \text{ weakly}$$

$$Lu_n \to o \text{ in } \hat{L}^2 \text{ strongly}$$

The same criterion shows that (4.4) is close to be a necessary condition for o to be an isolated point of Sp(L).

<u>Remark</u> : Suppose for simplicity that d=1, and define the following unbounded operator on $L^2(\mathbb{R})$:

$$A v = \sqrt{p}\ L(\frac{v}{\sqrt{p}})$$

A has the same spectrum as -L on $\hat{L}^2$.

In the particular case where a (z)=2, A is the Schrodinger operator $\Delta + V$, where

$$V(z) = - \frac{1}{4}(b^2 + 2 b')$$

one can check that $Sp(A)=\mathbb{R}_-$, when $b^2(z)+2b'(z) \to 0$ as $|z| \to +\infty$.

A similar result in dimension $d=3$ can be found in KATO [4] p.304.

V The Poisson equation

Under the hypothesis (4.4), one easily shows that if $f \in \hat{L}^2_o$, then the Poisson equation

(5.1) $LW + f = o$

has a unique solution $W \in \hat{L}^2_o$, which is precisely given by :

$$W(z) = \int_o^\infty E_z[f(Z_t)]dt$$

It will prove crucial for us to have conditions on f under which the solution W is bounded.

Again, Doeblin's condition would insure that f bounded implies W bounded. This is not true here, as the following trivial example shows :

$d = 1$, $f(z)=-b(z)$; then $W(z) = z$

We now state a result, which is due to P.L. Lions :

Theorem 5.1 Suppose that (4.4) holds and in addition there exist $p > \frac{d}{2}$ and $N > o$ s.t :

(i) $f \in L^p_{loc}(\mathbb{R}^d)$

(ii) $\int_N^{+\infty} \operatorname{ess\,sup}_{|z|\geq t} |f(z)|dt < \infty$

Then $W \in L^\infty(\mathbb{R}^d)$, where W is the unique solution of (5.1) which belongs to $\hat{L}^2_o$.

□

It is worth noting that a condition like $f \in L^p(\mathbb{R}^d)$ for some large p would not imply the theorem. Indeed, a smooth function equal to $\log|z|$ for $|z|\geq 1$ would satisfy (5.1), with $|f(z)| \leq \frac{c}{|z|}$ for $|z|$ large.

VI Jusfification of the expansion

We assume that (4.4) holds. Fix $T > o$.

It is then easy to prove that equations (3.11) and (3.12) have unique solutions belonging to:

$$L^2(o,T,H^1(\mathbb{R}^n)) \cap C([o,T] \; ; \; L^2(\mathbb{R}^n))$$

And from the regularity assumptions on the coefficients of A and B, one can check that any partial derivative of any order with respect to the z_i's has the same regularity.

We now want to show :

$$(6.1) \quad |E[\, u_t^\varepsilon \,] - \bar{u}^o(t) - \varepsilon\bar{u}^1(t)| \leq C(t)\varepsilon^2$$

where again $|\,.\,|$ stands for the norm in $L^2(\mathbb{R}^n)$.

We will use the following notation : if H is a Hilbert space, $\hat{L}^2(H) \triangleq L^2(\mathbb{R}^d, p(z)dz; H)$.

Define :

$$\rho^\varepsilon(t) = v^\varepsilon(t, t/\varepsilon^2) - \bar{u}^o(t) - \varepsilon\, v^1(t, t/\varepsilon^2) - \varepsilon^2 v^2(t, t/\varepsilon^2) - \varepsilon^3 v^3(t, t/\varepsilon^2)$$

where, for $i = 1,2,3$, $v^i(t,\tau) = \bar{u}^i(t,\tau) + \hat{v}^i(t,\tau)$, and these are the quantities whe have computed in § III.

Lemma 6.1 A sufficient condition for (6.1) to hold is that $\forall t > o$, $\exists\, \bar{C}(t) \in \mathbb{R}_+$ s.t :

$$(6.2) \quad \|\rho^\varepsilon(t)\|_{\hat{L}^2(L^2(\mathbb{R}^d))} \leq \bar{C}(t)\varepsilon^2$$

Proof : From Cauchy-Schwarz inequality,

$$|\int \rho^\varepsilon(t,z)p(z)dz| \leq \|\rho^\varepsilon(t)\|_{\hat{L}^2(H)}$$

so that (6.2) implies :

$$|E\, u_t^\varepsilon - \bar{u}^o(t) - \varepsilon\,\bar{u}^1(t) - \varepsilon^2\bar{u}^2(t, t/\varepsilon^2) - \varepsilon^3\bar{u}^3(t, t/\varepsilon^2)| \leq \bar{C}(t)\varepsilon^2$$

But one can show that $\forall t > o$, $\exists\, C'$ s.t. :

$$\sup_{\tau > o} |\bar{u}^2(t,\tau)| \leq C', \quad \sup_{\tau > o} |\bar{u}^3(t,\tau)| \leq C'$$

(6.1) then follows.

□

Now ρ^ε satisfies :

$$\left(\frac{\partial}{\partial t} + A + \frac{1}{\varepsilon} B - \frac{1}{\varepsilon^2} L\right)\rho^\varepsilon(t) = \varepsilon^2 g(t,\varepsilon); \rho^\varepsilon(o) = o$$

where $g(t,\varepsilon)$ is an expression involving v^2 and v^3.

One can show, under the above hypothesis :

$\forall$ T > o , $\exists$ C'' s.t :

(6.3) $$\int_0^T \|g(t,\varepsilon)\|^2_{\hat{L}^2(H^{-1}(\mathbb{R}^d))} dt \leq C''$$

Define $y^{\varepsilon}(t) = \frac{1}{\varepsilon^2}\rho^{\varepsilon}(t)$. y^{ε} solves :

(6.4) $$\left(\frac{\partial}{\partial t} + A + \frac{1}{\varepsilon} B - \frac{1}{\varepsilon^2} L\right) y^{\varepsilon}(t) = g(t,\varepsilon);\ y^{\varepsilon}(o) = o$$

(6.2) is a consequence of (6.3) and :

<u>Theorem 6.2</u> Suppose that the following holds for $h = \ell, \frac{\partial \ell}{\partial x_i}, i = 1\ldots n$:

$\exists$ N and K > o s.t.:

(6.5) $$\int_N^{+\infty} \operatorname*{ess\,sup}_{|z|\geq t} |h(x,z)|\,dt \leq K,\ \forall x \in \mathbb{R}^n$$

and that $\frac{\partial \ell}{\partial x_i} \in L^{\infty}(\mathbb{R}^{n+d})$, $i = 1\ldots n$.

then $\forall t > o$, $\exists k(t)$ s.t.

$$\|y^{\varepsilon}(t)\|^2_{\hat{L}^2(L^2(\mathbb{R}^d))} \leq k(t)\int_0^t \|g(s,\varepsilon)\|^2_{\hat{L}^2(H^{-1}(\mathbb{R}^d))} ds$$

<u>Sketch of proof</u> : It follows from the hypotheses of the Theorem, using Theorem 5.1, that χ, unique solution of :

$$L\,\chi(x,z) = \ell(x,z)$$

$$\chi(x,.) \in L^2_o,\ \forall x \in \mathbb{R}^n$$

satisfies:

(6.6) $$\chi \in L^{\infty}(\mathbb{R}^{n+d})\ ;\ \frac{\partial \chi}{\partial x_i} \in L^{\infty}(\mathbb{R}^{n+d}),\ i = 1\ldots n$$

One can then get, using (4.3) and standard PDE techniques :

$$\|y^{\varepsilon}(t)\|^2_{\hat{L}^2(L^2(\mathbb{R}^d))} + \varepsilon\,(y^{\varepsilon}(t),\chi y^{\varepsilon}(t))_{\hat{L}^2(L^2(\mathbb{R}^d))} \leq \bar{k}(t)\int_0^t \|g(s,\varepsilon)\|^2_{\hat{L}^2(H^{-1}(\mathbb{R}^d))} ds$$

which yields the desired result, from (6.6). □

<u>Remark</u> We have supposed that B(z) is of the form :

$$B(z) = \beta_1(x,z).\nabla + \beta_o(x,z)$$

If $\beta_o = o$, then one can get estimates for $y^{\varepsilon}(t)$ uniformly in ε, using the maximum principle and then avoiding the restrictive assumption of Theorem 6.2. □

Bibliography

[1] G. BLANKENSHIP. G. PAPANICOLAOU.- Stability and Control of Stochastic Systems with wide - band noise disturbances. *Siam J. Appl - Math.* 34, 3, pp. 437-476 (1978)

[2] R. BOUC. E. PARDOUX.- Moments of semilinear random evolutions, *Siam J. Appl - Math.* 41, 2, pp. 370-399 (1981)

[3] T. CAUGHEY. H. PAYNE.- On the response of a class of self-excited oscillators to stochastic excitation. *Int.J. of Nonlinear Mechanics* 2, pp. 125-151 (1967)

[4] T. KATO.- *Perturbation theory for linear operators.* Springer (1976)

[5] H. KUSHNER.- article in this volume

[6] G. PAPANICOLAOU.- Asymptoptic analysis of stochastic equations. in *Studies in Probability Theory* M. Rosenblatt, ed. MAA Studies in Applied Mathematics, vol. 18

[7] F. RIESZ. B. SZ.-NAGY.- *Leçons d'Analyse Fonctionnelle* Gauthier-Villars (1972)

[8] M. VIOT.- Solutions faibles d'équations aux dérivées partielles stochastiques non linéaires. Thèse, Univ. Paris VI (1976)

A DISCRETE TIME STOCHASTIC DECISION MODEL

Stanley R. Pliska
Northwestern University
Evanston, IL 60201/USA

Discrete time Markov decision chains are usually defined in terms of a Markov transition matrix. A less common approach, but one that is more useful for applications, is to formulate the model in terms of a state transition function, where the next state is a function of the current state, the current action, and an exogenous random variable.

For most applications these exogenous random variables (one for each period) have an explicit, physical interpretation. Moreover, in the case of Markov decision chains, they are independent and identically distributed. A natural and important generalization, therefore, and the subject of this paper, is the stochastic decision process that results when these exogenous random variables are not independent and identically distributed, but rather they comprise a general stochastic process.

Upon making this generalization, the underlying process being controlled may become non-Markovian. A few authors have studied non-Markovian stochastic control models. Davis [2] and Rishel [6] studied very general, continuous-time models. Discrete-time models were presented by Dynkin [3] and Gihman and Skorohod [4]. The stochastic decision model studied here is considerably more structured and less abstract than any of these. Indeed, as mentioned above, it is only a modest, yet significant, generalization of the state-transition-function kind of Markov decision chain model.

After formulating the stochastic decision model in Section 1, its potential usefulness as a practical tool is illustrated with the brief presentation of five different applications. Section 3 provides a martingale type of optimality condition and explains how to use dynamic programming to solve the problem. An alternative solution technique that is sometimes more efficient than dynamic programming is sketched out in Section 4; this method involves stochastic calculus and convex optimization theory. Sections 5, 6, and 7 solve a fairly general example problem with both dynamic programming and the alternative solution technique. The paper ends with some concluding remarks.

1. Formulation of the Model

The basic elements of the model are a filtered probability space $(\Omega, \mathcal{F}, \mathbb{F}, P)$ and a time horizon $T < \infty$. For technical reasons it is assumed the sample space Ω is discrete. However, most of what is done here is also true when the sample space is uncountable. Thus in the case of general sample spaces the reader should regard the results here as being formal but not rigorous.

The filtration $\mathbb{F} = \{\mathcal{F}_t;\ t = 0,1,..,T\}$, where each $\mathcal{F}_t$ is a σ-algebra of subsets of Ω and $F_0 \subseteq \cdot \cdot \subseteq F_T$. Without any real loss of generality, it is assumed that $F_0 = \{\emptyset, \Omega\}$ and $F_T = F$.

A stochastic process $Z = \{Z_t;\ t = 1,2,..,T\}$ is specified and fixed. This will be called the environmental process. It is assumed that Z is real-valued and adapted, that is, the function $\omega \to Z_t(\omega)$ is measurable with respect to F_t (written $Z_t \in F_t$) for each t.

The admissible controls are defined in terms of a predictable set-valued process $A = \{A_t;\ t = 1,2,..,T\}$ called the constraint process. Here $\emptyset \neq A_t(\omega) \subseteq \mathbb{R}$ for each t and ω; for example, $A_t(\omega)$ is an interval. One should think of A_t as defining the admissible actions or decisions. Predictable means $A_t \in F_{t-1}$ for each t.

Throughout this paper controls and policies will be called decision processes. An admissible decision process will be any predictable, real-valued stochastic process $D = \{D_t;\ t = 1,2,..,T\}$ satisfying $D_t(\omega) \in A_t(\omega)$ for all ω and t. Let $\underline{\underline{D}}$ denote the set of all such decision processes. Viewing the sequence of control actions as a predictable stochastic process is a crucial feature of this decision model. It will be seen that this approach is not really different from that taken with Markov decision chains, say, where the control is taken to be a function of the current state.

The process to be controlled is denoted $X = \{X_t;\ t = 1,2,..,T\}$ and called the controlled process. It evolves according to a specified state transition function $f\colon\ \mathbb{R}^3 \to \mathbb{R}$. The way this works is very simple. The initial state X_1 is specified, that is, $X_1 \in F_0$, and then for any particular decision process D one has

$$X_{t+1} = f(X_t,\ Z_t,\ D_t), \quad t = 1,2,..,T-1.$$

Note that X is predictable.

The decision model generates rewards according to a specified reward function $r\colon\ \mathbb{R}^3 \to \mathbb{R}$. The reward

$$R_t = r(X_t,\ Z_t,\ D_t)$$

is generated at time t, and this sequence of rewards defines a reward process $R = \{R_t;\ t = 1,2,..,T\}$. Note that R is adapted.

Corresponding to the reward process is another adapted stochastic process $W = \{W_t;\ t = 0,1,..,T\}$ called the wealth process. The initial wealth $W_0 \in F_0$ is specified, and then one has

$$W_t = R_t + W_{t-1}, \quad t = 1,2,..,T.$$

To understand how the decision model operates it is useful to think of the time parameter t as the index for a sequence of periods. At the beginning of period t the decision maker observes the information F_{t-1} which includes $X_1, X_2, .., X_t$; $D_1, D_2, .., D_{t-1}$; $Z_0, Z_1, .., Z_{t-1}$; $A_1, A_2, .., A_t$; and $W_0, W_1, ..,$ and W_{t-1}. In particular, one should think of X_t as the current state and W_{t-1} as the current wealth. The decision maker then uses this information to choose the action D_t, after which the next state Z_t of the environmental process is observed, the reward R_t for the period is generated, and the new wealth W_t is realized. This sequence is repeated period-by-period until the terminal wealth W_T is realized. The applications in the next section will give further insight into how this decision model functions.

The decision maker's objective may be to maximize the expected terminal wealth W_T. However, it will be useful for purposes of economic modeling to be more general than this. Let u be a specified real-valued function measuring the utility of the decision maker's terminal wealth. Then the problem is to choose a decision process D so as to maximize the expected utility $E[u(W_T)]$. Later sections will explain how to solve this problem.

It is important to recognize that if the random variables in the sequence $\{Z_t\}$ are independent and identically distributed and if the rest of the decision model is suitably defined, then the decision model becomes an ordinary Markov decision chain. Indeed, it becomes identical to the kind of Markov decision chain treated by Bertsekas [1] which, in turn, is equivalent to the conventional kind of Markov decision chain that is formulated in terms of a Markov transition matrix.

2. Some Applications

A primary reason for the importance of the decision model is its suitability and usefulness for many different applications. The following table presents five possible applications. These applications are meant to be suggestive, not definitive. The columns indicate the applications, while the rows specify the various elements of the model. Note that the constraint process A is sometimes specified in terms of the controlled process X; this is allowed, since X is predictable.

All of the applications involve an environmental process Z that has an explicit, physical interpretation. In the special case where Z is a sequence of independent and identically distributed random variables, all of these problems specialize to standard applications of Markov decision chains. But for all of these problems it is both natural and meaningful to allow the environmental process to be more general.

The first three applications are simple generalizations of classical problems from the operations research literature. For all three of these problems it may be important to take the environmental process Z to be more general than a sequence of independent and identically distributed random variables. For the controlled queueing problem the term r_1 of the reward function is meant to be the service cost, while r_2 is the waiting cost. In the production-inventory problem r_1 is the ordering cost, r_2 is the holding cost, r_3 is the shortage cost, and there is complete backlogging. In the replacement-maintenance problem r_1 is the cost of maintaining an item that has received quantity X_t of shocks and now receives shock Z_t, while the scalar c is the replacement cost.

The fourth application is one example of an optimal portfolio problem, an important and well-studied problem in finance. The investor can buy a stock, with $1

	controlled queues	production-inventory	replacement-maintenance	optimal portfolio	consumption investment
Z_t	arrivals during period t	demand during period t	shocks & wear during period t	one period rate-of-return or stock	one period rate-of-return of investment
X_t	number waiting	beginning inventory level	cumulative wear	current wealth	current wealth
D_t	number served	inv. level after ordering	service action 0 = replace 1 = keep	fraction of wealth in stock versus bank at rate $1+\alpha$	consumption during period t
A_t	$\{0,1,..,X_t\}$	$\{X_t,X_t+1,..\}$	$\{0,1\}$	$[0,1]$	$[0,X_t]$
R_t	$-r_1(D_t)$ $-r_2(X_t-D_t)$	$-r_1(D_t-X_t)$ $-r_2((D_t-Z_t)\vee 0)$ $-r_3((Z_t-D_t)\vee 0)$	$-r(X_t,Z_t)D_t$ $-c(1-D_t)$	$X_t(D_tZ_t+(1-D_t)\alpha)$	$r(D_t)$
X_{t+1}	$X_t+Z_t-D_t$	D_t-Z_t	$D_tX_t+Z_t$	X_t+R_t	$(1+Z_t)(X_t-D_t)$

TABLE. Some Applications of the Decision Model

invested at time t becoming worth $\$(Z_t + 1)$ at time $t+1$, and/or put money in a bank at a fixed interest rate α. The problem is to optimally divide his money between the two investments. Note that $X = W$.

The last application is a consumption-investment problem. Consumption-investment problems, as well as variations such as optimal capital accumulation under uncertainty and resource allocation under uncertainty, have been extensively studied in the economics literature. As with the optimal portfolio problem, the environmental process Z is the rate of return of an investment, and X is current wealth available for investment. However, now $W \neq X$. Each period the decision maker must consume the portion D_t of his wealth and invest the balance $X_t - D_t$. The consumption generates immediate utility $r(D_t)$, while the investment yields wealth $(1+Z_t)(X_t - D_t)$ next period. Now W should be thought of as the cumulative utility, so one should take $u(w) = w$. Incidentally, thinking of how one might model the prime interest rate, it may be appropriate for the environmental process Z of these last two applications to be a Markov chain.

3. Dynamic Programming and Martingale Optimality Conditions

Just as dynamic programming is used to solve Markov decision problems, so can it be used to solve the stochastic decision problem. This will be explained in this section, as will be a martingale type of necessary and sufficient condition for a decision process to be optimal.

Let W^D denote a wealth process under decision process D, and similarly for X^D. One says "a" wealth process rather than "the" wealth process because there is no restriction on its initial value W_0^D. Similarly for X^D. For each $t = 0,1,..,T$ and $D \in \underline{\underline{D}}$, let V_t^D be the real-valued function on $\mathbb{R}^2 \times \Omega$ defined by

$$V_t^D(w,x,\cdot) = E[u(W_T^D) \mid W_t^D = w, X_{t+1}^D = x, F_t] - u(w).$$

In other words, $V_t^D(w,x,\cdot)$ is the conditional expected change in utility from the end of period t, that is, from the beginning of period $t+1$, given the wealth then is w, the state then is x, and the information corresponding to F_t has been observed. The problem, of course, is to choose $D \in \underline{\underline{D}}$ so as to maximize $V_0^D(W_0,X_1,\cdot)$, where W_0 and X_1 are the specified initial values.

For each $t = 0,1,..,T$, let V_t be the real-valued function on $\mathbb{R}^2 \times \Omega$ defined by

$$V_t(w,x,\omega) = \sup_{D \in \underline{\underline{D}}} V_t^D(w,x,\omega).$$

Thus V_t is the maximum expected change in utility from the end of period t. If $D \in \underline{\underline{D}}$ is such that $V_0^D(W_0,X_1,\cdot) = V_0(W_0,X_1,\cdot)$, then D will be called <u>optimal</u>, for clearly this D maximizes $E[u(W_T^D)]$ over $\underline{\underline{D}}$ subject to $W_0^D = W_0$ and $X_1^D = X_1$. Note that $V_T = 0$.

The function V will be called the <u>value function</u>. In order to avoid annoying technicalities, it will be assumed that $V_t^D(w,x,\omega)$ and $V_t(w,x,\omega)$ are well-defined and finite for every D, t, w, x, and ω. The main result of this section is that V can be computed by solving a dynamic programming functional equation.

(1) <u>Theorem</u>. Suppose there exist real-valued functions $v_0, v_1, \ldots, v_T$, each with domain $\mathbb{R}^2 \times \Omega$, satisfying $v_T = 0$ and

$$(2) \quad v_{t-1}(w,x,\cdot) = \sup_{D_t \in A_t} \{E[u(w+r(x,Z_t,D_t)) \mid F_{t-1}] - u(w) + E[v_t(w+r(x,Z_t,D_t),\ f(x,Z_t,D_t),\cdot) \mid F_{t-1}]\}$$

for $t = T, T-1, \ldots, 1$. Then $V_t = v_t$ for each t.

<u>Remark</u>. This dynamic programming equation says the maximum expected change in utility equals the maximum of the sum of the expected change in utility over the current period plus the expected remaining change. The notation here is somewhat confusing, so it deserves an explanation. For each fixed w, x, and D_t the expression on the right hand side being taken a supremum of, that is, the expression within the parentheses {, }, is an F_{t-1} measurable function on Ω. Thus, for each fixed $\omega \in \Omega$, $v_{t-1}(w,x,\omega)$ equals the supremum of this expression as the scalar $D_t(\omega)$ varies over the set $A_t(\omega)$. Since Ω is discrete, it follows that $v_{t-1}(w,x,\cdot)$ is F_{t-1} measurable and, moreover, if $D_t(\omega)$ attains the supremum for every $\omega \in \Omega$, then D_t is F_{t-1} measurable.

<u>Proof</u>. This induction proof is similar to that for conventional dynamic programming problems, so it will only be sketched. After easily showing $V_{T-1} = v_{T-1}$, one assumes $V_t = v_t$ and shows $V_{t-1} = v_{t-1}$ by carrying out the following computation:

$$\begin{aligned}
V_{t-1}(w,x,\cdot) &= \sup_{D \in \underline{\underline{D}}} \{V^D_{t-1}(w,x,\cdot) + E[u(W^D_t) - u(W^D_t) \mid W^D_{t-1} = w,\ X^D_t = x,\ F_{t-1}]\} \\
&= \sup_{D \in \underline{\underline{D}}} \{E[u(W^D_t) \mid W^D_{t-1} = w,\ X^D_t = x,\ F_{t-1}] - u(w) \\
&\qquad + E[u(W^D_T) - u(W^D_t) \mid W^D_{t-1} = w,\ X^D_t = x,\ F_{t-1}]\} \\
&= \sup_{D \in \underline{\underline{D}}} \{E[u(w + r(x,Z_t,D_t)) \mid F_{t-1}] - u(w) \\
&\qquad + E[V^D_t(w+r(x,Z_t,D_t),\ f(x,Z_t,D_t),\cdot) \mid F_{t-1}]\} \\
&= \sup_{D_t \in A_t} \{E[u(w+r(x,Z_t,D_t)) \mid F_{t-1}] - u(w) \\
&\qquad + E[\sup_{D_{t+1},\ldots} V^D_t(w+r(x,Z_t,D_t),\ f(x,Z_t,D_t),\cdot) \mid F_{t-1}]\} \\
&= \sup_{D_t \in A_t} \{E[u(w+r(x,Z_t,D_t)) \mid F_{t-1}] - u(w) \\
&\qquad + E[V_t(w+r(x,Z_t,D_t),\ f(x,Z_t,D_t),\cdot) \mid F_{t-1}]\} \\
&= v_{t-1}.
\end{aligned}$$

□

Just as with conventional dynamic programming problems, if the supremum is attained in (2) then the corresponding decision process is optimal (the fact that Ω is discrete makes it easy to show this process is predictable). In other words, one has the following.

(3) Corollary. For $D \in \underline{\underline{D}}$ to be optimal it is necessary and sufficient that, for $t = 1,2,..,T$,

$$(4)\quad V_{t-1}(W^D_{t-1},X^D_t,\cdot) = E[u(W^D_{t-1}+r(X^D_t,Z_t,D_t))|F_{t-1}] - u(W^D_{t-1}) + E[V_t(W^D_{t-1}+r(X^D_t,Z_t,D_t),\ f(X^D_t,Z_t,D_t),\cdot)|F_{t-1}]\}.$$

To use this Corollary to determine an optimal decision process, first set $t = 1$, $W^D_0 = W_0$, and $X^D_1 = X_1$. Then choose D_1 so as to satisfy (4). Next, for $t = 2$, set $W^D_1 = W^D_0 + r(X^D_1,Z_1,D_1)$ and $X^D_2 = f(X^D_1,Z_1,D_1)$, and then choose D_2 so as to satisfy (4). Continue in this way until D_T has been determined. This procedure will be illustrated in Section 6.

The recent literature on stochastic control (see, e.g., Davis [2]) contains optimality conditions stated in terms of certain martingale properties. It turns out Theorem (1) and Corollary (3) can be transformed into such a statement.

For each $D \in \underline{\underline{D}}$ let $M^D = \{M^D_t;\ t=0,1,..,T\}$ be the stochastic process defined by setting

$$M^D_t = u(W^D_t) - u(W^D_0) + V_t(W^D_t,\ X^D_{t+1},\cdot),$$

where $W^D_0 = W_0$ and $X^D_1 = X_1$. Thus M^D_t equals the expected change in utility over all periods if D is used through the end of period t and then optimal decisions are made. Recall that a process such as M^D is a supermartingale if $M^D_t \geq E[M^D_{t+1}|F_t]$ for all t and that M^D is a martingale if both M^D and $-M^D$ are supermartingales. This leads to the following result.

(5) Theorem. The process M^D is a supermartingale for every $D \in \underline{\underline{D}}$. The decision process D is optimal if and only if M^D is a martingale.

Proof. Let $D \in \underline{\underline{D}}$ and t be arbitrary. Note that

$$M^D_{t+1} - M^D_t = u(W^D_{t+1}) - u(W^D_t) + V_{t+1}(W^D_{t+1},X^D_{t+2},\cdot) - V_t(W^D_t,X^D_{t+1},\cdot).$$

Also, by the dynamic programming results,

$$V_t(W^D_t,X^D_{t+1},\cdot) \geq E[u(W^D_{t+1})|F_t] - u(W^D_t) + E[V_{t+1}(W^D_{t+1},X^D_{t+2},\cdot)|F_t],$$

so

$$M^D_{t+1} - M^D_t \leq u(W^D_{t+1}) + V_{t+1}(W^D_{t+1},X^D_{t+2},\cdot) - E[u(W^D_{t+1})|F_t] - E[V_{t+1}(W^D_{t+1},X^D_{t+2},\cdot)|F_t].$$

Hence $E[M^D_{t+1} - M^D_t \mid F_t] \leq 0$, which means M^D is a supermartingale.

For the second part of this theorem, one sees by Corollary (3) that D is optimal if and only if all of these inequalities are actually equalities. But, of course, $E[M^D_{t+1} - M^D_t \mid F_t] = 0$ for all t if and only if M^D is a martingale. □

4. An Alternative Solution Technique

The dynamic programming approach amounts to maximizing the real-valued function $D \to E[u(W_T^D)]$ over the space $\underline{\underline{D}}$ of admissible decision processes. Even with a finite sample space, however, the "curse of dimensionality" may make this a formidable undertaking from the computational standpoint.

An alternative approach would be to proceed with the following three steps. First, identify in the space of random variables the set that is the range of the function $D \to W_T^D$. In other words, if $\underline{\underline{W}}$ denotes this set of attainable wealths, then the random variable $Y \in \underline{\underline{W}}$ if and only if $Y = W_T^D$ for some $D \in \underline{\underline{D}}$.

Second, solve the optimization problem of maximizing $E[u(Y)]$ subject to $Y \in \underline{\underline{W}}$. This problem may be easy to solve, particularly if the utility function u is concave, for then this is a conventional convex optimization problem for which there is a large literature.

Finally, having determined an optimal $Y \in \underline{\underline{W}}$ for the preceding optimization problem, determine the decision process that generates this terminal wealth, that is, the $D \in \underline{\underline{D}}$ satisfying $W_T^D = Y$. Depending on the circumstances, this third step is usually the easiest of the three to carry out.

The first step is the crucial one; it may not be feasible to identify the range $\underline{\underline{W}}$. But the contention is that if $\underline{\underline{W}}$ can be readily identified, then this alternative solution procedure may be computationally more efficient than dynamic programming. Unfortunately, there are no general results to support this statement. However, one general class of stochastic decision models has been identified for which this alternative solution procedure works very well. This is where the wealth process W can be expressed as a stochastic integral of the decision process with respect to a martingale. In this case stochastic calculus theory can be applied to identify $\underline{\underline{W}}$. Then if u is concave the second step can be carried out with convex analysis. Finally, the optimal decision process is readily determined by solving a martingale representation problem. This all will be illustrated in Section 7 where an example of this kind of problem will be solved in detail.

5. An Example

This section presents an example of a stochastic decision model that will be solved with dynamic programming in Section 6 and with martingale methods and convex analysis in Section 7.

Let the sample space Ω be all the T-dimensional vectors whose components are either 1 or -1. Thus Ω has 2^T elements.

Let the probability measure P be arbitrary, subject only to the requirement that $P(\omega) > 0$ for all $\omega \in \Omega$.

Let the environmental process Z be defined by setting $Z_t(\omega) = \omega_t$, the $t^{\underline{th}}$ component of ω.

Let the filtration $\mathbb{F} = \{F_t;\ t=0,1,..,T\}$ be the one generated by Z with $F_0 = \{\emptyset,\Omega\}$. Note that $F_T = F$, the set of all subsets of Ω.

For the constraint process A, set $A_t = \mathbb{R}$ for all t. Thus a decision process $D \in \underline{\underline{D}}$ can be any scalar-valued, predictable stochastic process.

There is no controlled process X with this example. The reward associated with period t is simply $R_t = D_t Z_t$. With initial wealth $W_0 = 0$, the wealth at the end of period t is $W_t = R_1 + R_2 + .. + R_t$.

The utility function u is initially taken to be strictly concave and increasing with either $u'(w) \to 0$ as $w \to \infty$ or $u'(w) \to -\infty$ as $w \to -\infty$. Moreover, u has a continuous second derivative. Later this will be specialized even further.

6. Solving the Example with Dynamic Programming

To solve the example problem of Section 5, one begins by computing the value function V. Since there is no controlled process X with this example, the notation will be modified accordingly. In particular, the dynamic programming equation (2) becomes

$$(6)\quad V_{t-1}(w,\cdot) = \sup_{D_t \in F_{t-1}} E\left[u(w+D_tZ_t) - u(w) + V_t(w+D_tZ_t,\cdot)\,|\,F_{t-1}\right].$$

With V_t' denoting the partial derivative of V_t with respect to its first argument, one has

(7) Proposition. There exists an optimal control. For each fixed $\omega \in \Omega$ and each t the function $w \to V_t(w,\omega) + u(w)$ is strictly concave and increasing with a continuous second derivative and either $V_t'(w,\omega) + u'(w) \to 0$ as $w \to \infty$ or $V_t'(w,\omega) + u'(w) \to -\infty$ as $w \to -\infty$.

Proof. The proof is by induction. Since $V_T = 0$, the function $V_T + u$ clearly has the specified properties, so for the induction step it will be assumed that so does the function h, which is defined by

$$h(w,\omega) = V_t(w,\omega) + u(w).$$

Substituting h into (6) yields

$$(8)\quad V_{t-1}(w,\cdot) = \sup_{D_t \in F_{t-1}} E\left[h(w+D_tZ_t,\cdot)\,|\,F_{t-1}\right] - u(w).$$

To analyze this, consider how the σ-algebra F_{t-1} is equivalent to a partition of Ω, and then focus on an arbitrary cell of this partition, say $B \subset \Omega$. (thus $B \in F_{t-1}$, but no proper subset of B is in F_{t-1}). The cell B will subdivide into two cells, say B_1 and B_{-1}, according to whether $Z_t = 1$ or $Z_t = -1$. Let $\omega_1 \in B_1$ and $\omega_{-1} \in B_{-1}$ be arbitrary, so that $h(w+d,\omega) = h(w+d,\omega_1)$ for all $\omega \in B_1$ and $h(w-d,\omega) = h(w-d,\omega_{-1})$ for all $\omega \in B_{-1}$. Then with $p = P(Z_t = 1|B)$, (8) becomes, for all $\omega \in B$,

$$V_{t-1}(w,\omega) = \sup_{d \in \mathbb{R}} \{p\,h(w+d,\omega_1) + (1-p)h(w-d,\omega_{-1})\} - u(w).$$

Note that $\omega \to V_{t-1}(w,\omega)$ is constant on B. By the induction assumption about the asymptotic properties of h and some elementary calculus, there exists a continuous real-valued function f on the real line such that, for each $w \in \mathbb{R}$, f(w) attains the supremum. In particular, f satisfies the first-order optimality condition

(9) $p\, h'(w+f(w),\omega_1) = (1-p)\, h'(w-f(w),\omega_{-1}), \quad \omega \in B,$

where h' denotes the partial derivative of h with respect to its first argument.

Thus f(w) is the optimal decision for all $\omega \in B$ when the wealth is w, and

$$V_{t-1}(w,\omega) + u(w) = p\, h(w+f(w),\omega_1) + (1-p)h(w-f(w),\omega_{-1}).$$

Differentiating this last expression with respect to w one obtains, after using (9),

(10) $V'_{t-1}(w,\omega) + u'(w) = p\, h'(w+f(w),\omega_1) + (1-p)h'(w-f(w),\omega_{-1}).$

Meanwhile, differentiating (9) yields, for all $\omega \in B$,

$$f'(w) = \frac{(1-p)h''(w-f(w),\omega_{-1}) - p\, h''(w+f(w),\omega_1)}{(1-p)h''(w-f(w),\omega_{-1}) + p\, h''(w+f(w),\omega_1)},$$

where h'' denotes the second partial derivative of h with respect to its first argument. Thus f' is continuous with $-1 < f'(w) < 1$. Hence, by (10), not only is $V_{t-1}(w,\omega) + u(w)$ increasing with respect to w, but $V'_{t-1}(w,\omega) + u'(w)$ either converges to 0 as $w \to \infty$ or converges to $-\infty$ as $w \to -\infty$.

Finally, differentiating (10) with respect to w yields

$$\begin{aligned} V''_{t-1}(w,\omega) + u''(w) &= p\, h''(w+f(w),\omega_1)[1+f'(w)] \\ &\quad + (1-p)h''(w-f(w),\omega_{-1})[1-f'(w)], \end{aligned}$$

where V''_{t-1} denotes the second partial derivative of V_{t-1} with respect to its first argument. Hence $w \to V_{t-1}(w,\omega)$ is strictly concave and has a continuous second derivative. □

What can be said about the computational effort that is required to obtain the value function V? For each cell in the partition corresponding to each F_t one needs to solve for the function f that was defined in the preceding proof. This amounts to solving the first-order optimality condition (9). For example, if

$$u(w) = a - (b/c)\exp(-c\, w)$$

for arbitrary scalars a, $b > 0$, and $c > 0$, then $u'(w) = b\exp(-c\, w)$. For the last period, $h = u$ and each cell in the partition corresponding to F_{T-1} consists of two elements, say ω_1 and ω_{-1}. Hence (9) becomes $p/(1-p) = \exp(2c\, f(w))$, where the conditional probability $p = P(\omega_1)/(P(\omega_1) + P(\omega_{-1}))$. For $\omega \in \{\omega_1, \omega_{-1}\}$, the optimal value of $D_T(\omega)$ is then given by

$$f(w) = \frac{1}{2c}\log(P(\omega_1)/P(\omega_{-1})).$$

Moreover, for the same ω,

$$V_{T-1}(w,\omega) = a - 2(b/c)\sqrt{P(\omega_1)P(\omega_{-1})}\exp(-c\, w)/(P(\omega_1) + P(\omega_{-1})) - u(w).$$

Note that $f(w)$ is independent of w. This is a manifestation of a well known property of the exponential utility function: the risk adjusted value of a lottery is not a function of the wealth. Indeed, it is not difficult to show the optimal decision D_t is independent of current wealth for earlier time periods as well. For example, with $h(w,\omega) = V_{T-1}(w,\omega) + u(w)$ and V_{T-1} as above one gets $h'(w,\omega) = \hat{b}\exp(-c\,w)$, where $\hat{b}$ is a function that may depend upon ω but not w. Substituting this into the first-order optimality condition (9) yields $f(w) = (1/2\,c)\log(g)$, where g is a function that may depend upon ω but not w. Finally, substituting this into the expression for V_{T-2} shows that the only dependence of this function on the wealth w is through the factor $\exp(-c\,w)$.

7. Solving the Example with Convex Analysis

To carry out the alternative solution technique described in Section 4, first define a stochastic process $M = \{M_t;\ t=0,1,..,T\}$ by putting $M_0 = 0$ and

$$M_t = Z_1 + Z_2 + .. + Z_t, \qquad t=1,2,..,T.$$

Thus M is a random walk on the integers, and under the probability measure $P'(\omega) = (1/2)^T$ M is a martingale with respect to $\mathbb{F}$. Let E' denote the expectation operator corresponding to P'.

Since $Z = \Delta M$, it is clear that the wealth process W^D under any decision process D can be represented as the stochastic integral of D with respect to the martingale M, that is,

$$W_t^D = \sum_{s=1}^{t} D_s\,\Delta M_s, \qquad t=1,2,..,T.$$

By standard results, each such wealth process will be a martingale under P'. Furthermore, since P' is the unique probability measure equivalent to P under which M is a martingale, it follows (see, e.g., Jacod [5, Ch. XI]) that every martingale (under P') can be represented as a stochastic integral of a decision process with respect to M.

The implications of this are as follows. Let $\underline{\underline{Y}}$ denote the space of random variables Y on Ω, and let $\underline{\underline{W}}$ be as in Section 5, that is, $\underline{\underline{W}}$ consists of all $Y \in \underline{\underline{Y}}$ such that $Y = W_T^D$ for some $D \in \underline{\underline{D}}$. Since W^D is a martingale under P' null at zero, it follows that $E'[Y] = 0$ for all $Y \in \underline{\underline{W}}$. Conversely, if $Y \in \underline{\underline{Y}}$ satisfies $E'[Y] = 0$, then upon considering the martingale N defined by $N_t = E'[Y|F_t]$ it follows from the martingale representation property described above that there exists some decision policy $D \in \underline{\underline{D}}$ such that $W_T^D = Y$. Hence

$$\underline{\underline{W}} = \{Y \in \underline{\underline{Y}}:\ E'[Y] = 0\}.$$

This completes the first step in the alternative solution technique. The second step is to find an optimal terminal wealth, that is, some $\hat{Y} \in \underline{\underline{W}}$ such that $E[u(\hat{Y})] \geq E[u(Y)]$ for all $Y \in \underline{\underline{W}}$. This will be done with some convex optimization theory (see, e.g., Rockafellar[7]).

Let $\underline{\underline{Y}}^*$ denote the space that is dual to $\underline{\underline{Y}}$ under the linear functional $E'[Y\,Y^*]$. Let $\underline{\underline{W}}^*$ denote the <u>orthogonal</u> <u>complement</u> of $\underline{\underline{W}}$, that is

$$\underline{\underline{W}}^* = \{Y^* \in \underline{\underline{Y}}^*:\ \ E'[Y\,Y^*] = 0,\ \text{all } Y \in \underline{\underline{W}}\}.$$

Since $E'[Y] = 0$ for all $Y \in \underline{\underline{W}}$, it is clear that $\underline{\underline{W}}^*$ contains all the constant functions in $\underline{\underline{Y}}^*$. If $Y^* \in \underline{\underline{Y}}^*$ is not constant, then one can readily find some $Y \in \underline{\underline{W}}$ such that $E'[Y\,Y^*] \neq 0$, so actually

$$\underline{\underline{W}}^* = \{Y^* \in \underline{\underline{Y}}^*:\ \ Y^* \text{ is constant}\}.$$

Denoting $U(Y) = E[u(Y)]$, the fact that u is concave means U is a concave functional on $\underline{\underline{Y}}$. Hence step 2 amounts to solving the concave optimization problem

(11) maximize $U(Y)$

subject to $Y \in \underline{\underline{W}}$

Let U^* denote the concave conjugate functional of U, that is, for each $Y^* \in \underline{\underline{Y}}^*$,

$$U^*(Y^*) = \inf_{Y \in \underline{\underline{Y}}}\{E'[Y\,Y^*] - U(Y)\}.$$

(12) <u>Proposition</u>. $\hat{Y}$ solves (11) if and only if there exists some $\hat{Y}^* \in \underline{\underline{W}}^*$ such that

(13) $$U^*(\hat{Y}^*) = E'[\hat{Y}\,\hat{Y}^*] - U(\hat{Y}).$$

<u>Proof</u>. To show sufficiency, since $E'[\hat{Y}\,\hat{Y}^*] = 0$ one has $U^*(\hat{Y}^*) = -U(\hat{Y})$. But the definition of U^* means $U^*(\hat{Y}^*) \leq E'[Y\,\hat{Y}^*] - U(Y)$ for all $Y \in \underline{\underline{Y}}$, so in particular $U^*(\hat{Y}^*) = -U(\hat{Y}) \leq E'[Y\,\hat{Y}^*] - U(Y) = -U(Y)$ for all $Y \in \underline{\underline{W}}$, that is, $\hat{Y}$ is optimal.

Conversely, by a version of the Fenchel duality theorem there exists some $\hat{Y}^* \in \underline{\underline{W}}^*$ such that $U^*(\hat{Y}^*) = -U(\hat{Y})$. Since $E'[\hat{Y}\,\hat{Y}^*] = 0$, this means (13) holds. ☐

With $g(\omega) \equiv \dfrac{dP(\omega)}{dP'}$, the Radon-Nykodym derivative, let $u^*\colon\ \Omega \times \mathbb{R} \to \mathbb{R}$ be the concave conjugate functional

$$u^*(\omega, y) = \inf_{w \in \mathbb{R}}\{wy - g(\omega)u(w)\}.$$

Since $U(Y) = \int u(Y(\omega))dP(\omega) = \int u(Y(\omega))g(\omega)dP'(\omega)$, by Rockafellar's [7] results on integral functionals one has

(14) $$U^*(Y^*) = \int u^*(\omega, Y^*(\omega))dP'(\omega).$$

This leads to the following.

(15) <u>Proposition</u>. $\hat{Y}$ solves (11) if and only if there exists a positive, constant function $\hat{Y}^* \in \underline{\underline{W}}^*$ such that

(16) $$u^*(\omega, \hat{Y}^*(\omega)) = \hat{Y}(\omega)\hat{Y}^*(\omega) - g(\omega)u(\hat{Y}(\omega)), \qquad \text{all } \omega \in \Omega.$$

Moreover,

(17) $$U(\hat{Y}) = -\sup_{y \in \mathbb{R}_+} \int u^*(\omega, y)\,dP'(\omega).$$

<u>Proof</u>. By Rockafellar [7], (13) holds for $\hat{Y} \in \underline{\underline{W}}$ and constant $\hat{Y}^*$ if and only if (16) holds. Equation (14) and the fact that $U(\hat{Y}) = -\sup_{Y^* \in \underline{\underline{W}}^*} U^*(Y^*)$ imply

$U(Y) = -\sup_{y \in \mathbb{R}} \int u^*(\omega,y)dP'(\omega)$. This supremum is attained by $y = \hat{Y}^*$,

and $\hat{Y}^* \geq 0$ because for $y < 0$ $u^*(\omega,y) = -\infty$. □

Proposition (15) can readily be used to solve for the optimal $\hat{Y}$, as will shortly be illustrated. First, however, mention should be made of how to carry out the third and final step of the alternative solution technique, namely, finding the $D \in \underline{\underline{D}}$ such that $W_T^D = \hat{Y}$.

The idea is very simple. Let $\hat{W}$ denote the wealth process under the optimal decision process $\hat{D}$. It has just been determined that $\hat{W}_T = \hat{Y}$. As was stated previously $\hat{W}$ is a martingale under P', so

(18) $$\hat{W}_t = E'[\hat{Y} \mid F_t], \qquad t = 0,1,..,T.$$

It is also known that $\hat{W}$ can be represented as a stochastic integral with respect to M, that is,

$$\hat{W}_t = \sum_{s=1}^{t} D_s \, \Delta M_s$$

for some $D \in \underline{\underline{D}}$. Hence to determine $\hat{D}$ it remains to solve this easy representation problem.

Returning to step 2, the use of Proposition (15) will be illustrated with the specific utility function already studied in Section 6, namely $u(w) = a - (b/c)\exp(-cw)$. One begins by computing

$$u^*(\omega,y) = \begin{cases} -\infty, & y < 0 \\ -ag(\omega), & y = 0 \\ (y/c)\log(bg(\omega)/y) - ag(\omega) + y/c, & y > 0. \end{cases}$$

During this computation one notes that for $y > 0$ the argument in the definition of u^* is minimized by

(19) $$w = \frac{1}{c}\log\,(bg(\omega)/y).$$

Next, for any $y > 0$ one computes

$$E'[u^*(\omega,y)] = \frac{y}{c}[\log b + E'[\log g] - \log y + 1] - a,$$

since $E'[g] = 1$. This expression is maximized by

$$\hat{y} = b\exp(E'[\log g]),$$

so $\hat{Y}^*(\omega) = \hat{y}$ for all ω. Substituting this back into $E'[u^*(\omega,y)]$ and using (17) gives

$$U(\hat{Y}) = a - \hat{y}/c = a - (b/c)\exp(E'[\log g]).$$

for the optimal value of the objective function.

To compute $\hat{Y}$ one can solve equation (16). It is apparent the solution is given by (19) with $y = \hat{y}$, that is

(20) $$\hat{Y}(\omega) = \frac{1}{c}(\log(g(\omega)) - E'[\log g]).$$

This completes step 2.

For step 3 one could use (18) to solve for the optimal wealth process $\hat{W}$ and then use

$$\begin{aligned}(21)\qquad \hat{W}_t(\omega) &= \hat{W}_{t-1}(\omega) + \hat{D}_t(\omega)\,\Delta M_t(\omega)\\ &= \hat{W}_{t-1}(\omega) + \hat{D}_t(\omega)Z_t(\omega)\end{aligned}$$

to solve for $\hat{D}$. Alternatively, (21) can be used in a recursive manner to compute $\hat{W}$ and $\hat{D}$ simultaneously if one keeps track of the various partitions (corresponding to the σ-algebras F_t, as explained in Section 6). For example, suppose $\{\omega_1,\omega_{-1}\}$ is one cell in the partition corresponding to F_{T-1} with $Z_T(\omega_1) = 1$ and $Z_T(\omega_{-1}) = -1$. Substituting into (21) gives

$$\begin{aligned}(22)\qquad \hat{W}_T(\omega_1) &= \hat{W}_{T-1}(\omega_1) + \hat{D}_T(\omega_1)\\ \hat{W}_T(\omega_{-1}) &= \hat{W}_{T-1}(\omega_{-1}) - \hat{D}_T(\omega_{-1}).\end{aligned}$$

But $\hat{W}_{T-1}$ and $\hat{D}_T$, being F_{T-1} measurable, are constant over the cell $\{\omega_1,\omega_{-1}\}$, so these two equations suffice to solve for the two constant values. Indeed, $\hat{W}_{T-1}$ and $\hat{D}_T$ can be determined by solving $2^T = |\Omega|$ such equations, after which $\hat{W}_{T-2}$ and $\hat{D}_{T-1}$ can be determined by solving 2^{T-1} equations, and so forth. Overall, to solve for $\hat{W}$ and $\hat{D}$ one needs to solve $2^{T+1} - 2$ equations of the form (22). Note that with $\hat{W}_T = \hat{Y}$ given by (20), the equations in (22) imply $\hat{D}_T(\omega_1) = (1/2c)\log(P(\omega_1)/P(\omega_{-1}))$, the same as the answer computed by dynamic programming in Section 6.

8. Concluding Remarks

It should be apparent the alternative solution technique can be applied to considerably more general stochastic decision models. The essential feature is that each wealth process can be represented as a stochastic integral of the decision process with respect to an environmental process that, with a change of measure, is a martingale. The example assumed this change of measure was unique, but this requirement can be relaxed without too much difficulty.

The alternative solution technique is successful when one is able to give the environmental process special structure, namely, the martingale property. A research topic deserving attention is to see what other kinds of simplifications result when the environmental process is given other kinds of structure, such as the Markov property.

Another topic, duality theory for the stochastic decision problem, will be presented in a forthcoming paper.

References

1. Bertsekas, D. P., Dynamic Programming and Stochastic Control, Academic Press, New York, 1976.

2. Davis, M. H. A., "Martingale Methods in Stochastic Control," Lecture Notes in Control and Information Sciences 16, Springer-Verlag, New York - Berlin, 1979.

3. Dynkin, E. B., "Controlled Random Sequences," Theory of Probability and Its Applications 10(1966), 1-14.

4. Gihman, I. I., and A. V. Skorohod, Controlled Stochastic Processes, Springer-Verlag, New York - Berlin, 1979.

5. Jacod, J., Calcul Stochastique et Problèmes de Martingales, Lecture Notes in Mathematics 74, Springer-Verlag, New York - Berlin, 1979.

6. Rishel, R., "Necessary and Sufficient Dynamic Programming Conditions for Continuous Time Stochastic Optimal Control," Siam J. Control 8(1970), 559-571.

7. Rockafellar, R. T., "Conjugate Duality and Optimization," Regional Conference Series in Applied Mathematics 16, Society for Industrial and Applied Mathematics, Philadelphia, 1974.

ON THE APPROXIMATION OF CONTROLLED JUMP DIFFUSION PROCESSES

H. Pragarauskas

Institute of Mathematics and Cybernetics
Academy of Sciences of the Lithuanian SSR
Vilnius, K. Požėlos 54, USSR

Let R^d be a d-dimensional Euclidean space, $T \in (0,\infty)$, $H_T = [0,T] \times R^d$, $S_R = \{x \in R^d : |x| < R\}$, $(A, \mathcal{B}(A))$ be a separable metric space with Borel σ-algebra, an integer $d_1 \geq 1$ and $L_2(R^d, \Pi)$ be the space of functions $u : R^d \to S_1$ such that

$$\|u\|_{2,\Pi} = \{\int |u(z)|^2 \Pi(dz)\}^{1/2} < \infty \quad \text{where} \quad \Pi(dz) = dz/|z|^{d+1}.$$

Suppose that for all $\alpha \in A$, $(t,x) \in H_T$ we are given: a matrix $\sigma(\alpha,t,x)$ of dimension $d \times d_1$, a d-dimensional vector $b(\alpha,t,x)$, an element $c(\alpha,t,x,\cdot)$ of $L_2(R^d,\Pi)$ and a number $g(x)$.

1. Condition. a) σ, b are continuous in α, continuous in t uniformly in α; c is Borel in (α,t,x,z) and in the sense of the norm $\|\cdot\|_{2,\Pi}$: continuous in α, continuous in t uniformly in α; g is continuous. For all $(t,x) \in H_T$

$$\sup_{\alpha \in A} \int_{|z| \leq \varepsilon} |c(\alpha,t,x,z)|^2 \Pi(dz) \to 0$$

as $\varepsilon \to 0$.

b) For some constants $m, K \geq 0$ and all $\alpha \in A$, $(t,x) \in H_T$, $y \in R^d$

$$\|\sigma(\alpha,t,x)\| + |b(\alpha,t,x)| + \|c(\alpha,t,x,\cdot)\|_{2,\Pi} \leq K(1 + |x|),$$
$$\|\sigma(\alpha,t,x) - \sigma(\alpha,t,y)\| + |b(\alpha,t,x) - b(\alpha,t,y)| +$$
$$\|c(\alpha,t,x,\cdot) - c(\alpha,t,y,\cdot)\|_{2,\Pi} \leq K|x - y|,$$
$$|g(x)| \leq K(1 + |x|)^m.$$

Let (Ω, F, P) be a complete probability space with an increasing family $(F_t,\ t \geq 0)$ of complete σ-algebras $F_t \subset F$, (w_t, F_t) be a d_1-dimensional Wiener process, (z_t, F_t) be a d-dimensional Cauchy process independent of w_t with Levy measure Π, $q(dtdz)$ be the Poisson martingale measure, constructed from the jumps of z_t.

Let $\mathfrak{A}$ be the set of all processes $\alpha_t(\omega)$ progressively measurable w.r. to (F_t) having values in A. To each strategy $\alpha \in \mathfrak{A}$, $(s,x) \in H_T$ we set into correspondence a solution $x_t^{\alpha,s,x}$ of Ito's equation

$$x_t = x + \int_s^t \sigma(\alpha_u,u,x_u)dw_u + \int_s^t b(\alpha_u,u,x_u)du + \int_s^t \int c(\alpha_u,u,x_u,z) q(dudz).$$

For $(s,x) \in H_T$ let

$$v(s,x) = \sup_{\alpha \in \mathfrak{A}} Eg(x_T^{\alpha,s,x}).$$

Fix an arbitrary $s \in [0,T)$. Let $\{I_n\}$ be a sequence of embedded subdivisions $s = t_0^n < t_1^n < \dots < t_n^n = T$ of the interval $[s,T]$ and $\operatorname{diam} I_n \to 0$ as $n \to \infty$. Suppose that for all $\alpha \in A$, $n = 1,2, \dots$ we are given one-step Markov transition functions $p_n^\alpha(t_i^n,x,t_{i+1}^n,\Gamma) \equiv p_{ni}^\alpha(x,\Gamma)$, $i = 0,1, \dots ,n-1$, $x \in R^d$, $\Gamma \in \mathcal{B}(R^d)$ which are Borel in (α,x). Let $\mathcal{B}^n$ be the class of all families $\beta(n) = (q_0^n, \dots q_{n-1}^n)$ of functions $q_0^n(d\alpha_0|x_0)$, $q_i^n(d\alpha_i|x_0,\alpha_0, \dots ,\alpha_{i-1},x_i)$, $1 \le i \le n-1$, which are probability measures on $\mathcal{B}(A)$ in the first argument and Borel in the other arguments. An initial point (s,x), a strategy $\beta(n) \in \mathcal{B}^n$ and a family of transition functions define a controlled process $\xi^n(s) = x$, $\alpha^n(s)$, $\dots$ $\alpha^n(t_{n-1}^n)$, $\xi^n(T)$ on the probability space $(\Omega_n, \mathcal{B}(\Omega_n), Q_{s,x}^{\beta(n)})$, where $\Omega_n = (R^d)^{n+1} \times A^n$ (see §6, ch. I [1]).

2. Condition. For some constant $m' > m$ and every $x \in R^d$

$$\sup_n \sup_{\beta(n)\in\mathcal{B}^n} E_{s,x}^{\beta(n)} \sup_{t\in I_n} |\xi^n(t)|^{m'\vee 2} < \infty.$$

For $x \in R^d$, $n = 1,2, \dots$ let

$$V_n(s,x) = \sup_{\beta(n)\in\mathcal{B}^n} E_{s,x}^{\beta(n)} g(\xi^n(T)).$$

Let $D[s,T]$ be the space of all right continuous and having left hand limits functions x_t with values in R^d defined on $[s,T]$ with Skorochod topology. Set $\mathcal{D}[s,T] = \sigma\{x_t \in \Gamma;\ s \le t \le T,\ \Gamma \in \mathcal{B}(R^d)\}$.

Let us define the process ξ_t^n setting $\xi_t^n = \xi^n(t_i^n)$ if $t \in [t_i^n, t_{i+1}^n)$. Denote by $P_{s,x}^{\beta(n)}$ the measure on $(D[s,T], \mathcal{D}[s,T])$ induced by ξ^n.

3. Condition. For an arbitrary sequence $\{\beta(n)\}$ of strategies $\beta(n) \in \mathcal{B}^n$ the sequence of measures $\{P_{s,x}^{\beta(n)}\}$ is tight on $(D[s,T], \mathcal{D}[s,T])$.

For all $\alpha \in A$, $(t,x) \in H_T$ define a measure $\pi(\alpha,t,x,\cdot)$ on $\mathcal{B}(R^d)$ by the formula: $\pi(\alpha,t,x,dy) = \Pi(z : x + c(\alpha,t,x,z) \in dy\setminus\{x\})$, $\pi(\alpha,t,x,\{x\}) = 0$. Set $\Delta_i^n = t_{i+1}^n - t_i^n$, $a = \sigma\sigma^*$ (σ^* is the transpose matrix of σ).

4. Condition. For some number $\rho > 1$ every $R > 0$ and every continuous bounded function ψ on R^d which vanishes in some neighborhood of the origin

$$\sum_{i=0}^{n-1} \sup_{|x|\le R} \sup_{\alpha\in A} \sum_{j=1}^{d} \Big| \int_{|y-x|\le\rho} (y-x)_j p_{ni}^\alpha(x,dy) - \Delta_i^n b_j(\alpha,t_i^n,x) \Big| \to 0,$$

$$\sum_{i=0}^{n-1} \sup_{|x|\le R} \sup_{\alpha\in A} \sum_{j,k=1}^{d} \Big| \int_{|y-x|\le\rho} (y-x)_j (y-x)_k\, p_{ni}^\alpha(x,dy) - \Delta_i^n [a_{jk}(\alpha,t_i^n,x) + \int (y-x)_j (y-x)_k \pi(\alpha,t_i^n,x,dy)] \Big| \to 0,$$

$$\sum_{i=0}^{n-1} \sup_{|x|\le R} \sup_{\alpha\in A} \Big| \int \psi(y-x) p_{ni}^\alpha(x,dy) - \Delta_i^n \int \psi(y-x) \pi(\alpha,t_i^n,x,dy) \Big| \to 0$$

as $n \to \infty$.

5. Theorem. Let the conditions 1-4 hold. Then $V_n(s,x) \to v(s,x)$ as $n \to \infty$ for all $x \in R^d$.

The particular cases of this problem were considered by H. J. Kushner [2], H. J. Kushner and G. DiMasi [3] and J. P. Quadrat [7]. These authors did not use the fact that under suitable assumptions the payoff function v satisfies the Bellman equation.

Denote by $C^{1,2}$ the set of functions $u = u(t,x)$ on R^{d+1} twice continuously differentiable in x and once continuously differentiable in t. Let

$$L^{\alpha}(t,x)u(r,x) = u_r(r,x) + \frac{1}{2}\sum_{i,j=1}^{d} a_{ij}(\alpha,t,x)u_{x_i x_j}(r,x) +$$

$$+ \sum_{i=1}^{d} b_i(\alpha,t,x)u_{x_i}(r,x) - \int[u(r,y) - u(r,x) - \sum_{i=1}^{d} u_{x_i}(r,x)(y-x)_i]\pi(\alpha,t,x,dy),$$

$$L^{\alpha}(t,x)u(t,x) \equiv L^{\alpha}u(t,x),$$

$$Fu(t,x) = \sup_{\alpha\in A} L^{\alpha}u(t,x), \tag{1}$$

Fix $x \in R^d$. Let $\{\beta(n)\}$ be an arbitrary sequence of stategies $\beta(n) \in \mathcal{B}^n$, $(\xi^n(s),\alpha^n(s), \ldots ,\alpha^n(t^n_{n-1}),\xi^n(T))$ the process controlled by strategy $\beta(n)$, $P^n_{s,x}$ the measure on $(D[s,T], \mathcal{D}[s,T])$ induced by the process $\xi^n(t) \equiv \xi^n(t^n_i)$, $t \in [t^n_i,t^n_{i+1})$, $i = 0,1, \ldots ,n-1$.

By assumption 3 there exist a subsequence $\{m\} \subset \{n\}$ and a measure $P_{s,x}$ on $(D[s,T], \mathcal{D}[s,T])$ such that the measures $P^m_{s,x}$ converge weakly to $P_{s,x}$ as $m \to \infty$. By lemma from §6, ch. 1 [5] there exist a probability space $(\tilde{\Omega},\tilde{F},\tilde{P})$ and processes $\tilde{\xi}^m(\cdot)$, $\tilde{\xi}(\cdot)$ defined on this space such that the processes $(x(\cdot),P^m_{s,x})$ and $(\tilde{\xi}^m(\cdot),\tilde{P})$ as well as the processes $(x(\cdot),P_{s,x})$ and $(\tilde{\xi}(\cdot),\tilde{P})$ are equivalent and $\tilde{\xi}^m(t) \to \tilde{\xi}(t)$ in probability as $m \to \infty$ for every $t \in [s,T]$.

Denote by $\tilde{F}_t$ the completion (with respect to $\tilde{P}$) of $\sigma\{\tilde{\xi}(r),\ s < r < t\}$.

<u>6. Lemma</u>. Let the assumptions 1-4 hold, $u \in C^{1,2}$, $R > 0$, τ is the time of first exit of $(t,\tilde{\xi}(t))$ from $[s,T)\times S_R$.

Then

$$u(t\wedge\tau,\ \tilde{\xi}(t\wedge\tau)) - u(s,x) - \int_s^{t\wedge\tau} Fu(r,\tilde{\xi}(r))dr$$

is a $(\tilde{F}_t,\tilde{P})$ - supermartingale.

<u>Proof</u>. Without loss of generality we can suppose that $u(t,x) = 0$ if $x \notin S_{R+2}$. Then Fu is a finite continuous function.

Denote by F^n_k the completion (with respect to $Q^{\beta(n)}_{s,x}$) of $\sigma\{\xi^n(s),\alpha^n(s), \ldots ,\xi^n(t^n_k),\alpha^n(t^n_k)\}$. The sequence

$$u(t^n_k,\xi^n(t^n_k)) - u(s,x) - \sum_{i=0}^{k-1} E^{\beta(n)}_{s,x}[u(t^n_{i+1},\xi^n(t^n_{i+1})) - u(t^n_i,\xi^n(t^n_i))\,|\,F^n_i] =$$

$$= u(t^n_k,\xi^n(t^n_k)) - u(s,x) - \sum_{i=0}^{k-1} \int[u(t^n_{i+1},y) - u(t^n_i,\xi^n(t^n_i))]p_{ni}^{\alpha^n(t^n_i)}(\xi^n(t^n_i),dy)$$

is a $(F^n_k,Q^{\beta(n)}_{s,x})$ - martingale. From this it follows that

$$\eta_t^n \equiv u(t,x(t)) - u(s,x) - \sum_{i:t_{i+1}^n \le t} \sup_{\alpha\in A} \int [u(t_{i+1}^n,y) - u(t_i^n,x(t_i^n))] p_{ni}^{\alpha}(x(t_i^n),dy)$$

is a $(\mathcal{D}[s,t],\ P_{s,x}^n)$ - supermartingale.

Let

$$\tilde{\eta}_t^m \equiv u(t,\tilde{\xi}^m(t)) - u(s,x) - \sum_{i:t_{i+1}^n \le t} \sup_{\alpha\in A} \int [u(t_{i+1}^m,y) - u(t_i^m,\tilde{\xi}^m(t_i^m))] p_{mi}^{\alpha}(\tilde{\xi}^m(t_i^m),dy).$$

Suppose that

$$\tilde{\eta}_t^m \to u(t,\tilde{\xi}(t)) - u(s,x) - \int_s^t Fu(r,\tilde{\xi}(r))dr \equiv \tilde{\eta}_t \tag{2}$$

in probability as $m \to \infty$ for every $t \in [s,T]$. Then it is easy to prove that $\tilde{\eta}_t$ is a $(\tilde{F}_t,\tilde{P})$ - supermartingale. This implies the required result.

Therefore it suffices to prove (2). To this end it suffices to prove that for every $R > 0$

$$\sum_{i=0}^{m-1} \sup_{|x|\le R} \sup_{\alpha\in A} \left| \int [u(t_{i+1}^m,y) - u(t_i^m,x)] p_{mi}^{\alpha}(x,dy) - \Delta_i^m L^{\alpha} u(t_i^m,x) \right| \to 0 \tag{3}$$

as $m \to \infty$.

Let ρ be the number from assumption 4, $\varepsilon \in (0,1)$, ψ_ε, ψ_ρ be continuous functions on R^d with values in $[0,1]$ such that $\psi_\varepsilon(x) = 1$ if $|x| \le \varepsilon/2$, $\psi_\varepsilon(x) = 0$ if $|x| \ge \varepsilon$, $\psi_\rho(x)=1$ if $|x| \le \rho$ and $\psi_\rho(x) = 0$ if $|x| \ge \rho + \varepsilon$.

Using Taylor's expansion, we obtain

$$\begin{aligned}
&\int [u(t,y) - u(r,x)] p^{\alpha}(x,dy) - \Delta L^{\alpha} u(r,x) = u(t,x) - u(r,x) - \\
&- \Delta u_t(t,x) + \Delta L^{\alpha}(r,x)[u(t,x) - u(r,x)] + \\
&+ \sum_{k=1}^{d} u_{x_k}(t,x)\Big[\int_{|y-x|\le\rho} (y-x)_k p^{\alpha}(x,dy) - \Delta b_k(\alpha,r,x)\Big] + \\
&+ \frac{1}{2} \sum_{k,\ell=1}^{d} u_{x_k x_\ell}(t,x)\Big[\int_{|y-x|\le\rho} (y-x)_k (y-x)_\ell p^{\alpha}(x,dy) - \Delta a_{k\ell}(\alpha,r,x) - \\
&- \Delta \int (y-x)_k (y-x)_\ell \pi(\alpha,r,x,dy)\Big] - \\
&- \int [\psi_\rho - \psi_\varepsilon](y-x) R_1(t,x,y)[p^{\alpha}(x,dy) - \Delta\pi(\alpha,r,x,dy)] + \\
&+ \int [1 - \psi_\varepsilon](y-x)[u(t,y) - u(t,x)][p^{\alpha}(x,dy) - \Delta\pi(\alpha,r,x,dy)] + \\
&+ \int_{|y-x|>\rho} \psi_\rho(y-x) R_1(t,x,y) p^{\alpha}(x,dy) + \\
&+ \int \psi_\varepsilon(y-x) R(t,x,y)[p^{\alpha}(x,dy) - \Delta\pi(\alpha,r,x,dy)],
\end{aligned}$$

where $R_1(t,x,y) = \sum_{k=1}^{d} u_{x_k}(t,x)(y-x)_k + \frac{1}{2} \sum_{k,\ell=1}^{d} u_{x_k x_\ell}(t,x)(y-x)_k(y-x)_\ell$,

$$R(t,x,y) = u(t,y) - u(t,x) - R_1(t,x,y)$$

and for convenience we use the notation: $t_i^m = r$, $t_{i+1}^m = t$, $\Delta_i^m = \Delta$, $p_{mi}^{\alpha}(x,dy) = p^{\alpha}(x,dy)$.

Substituting these expressions in the left hand side of (3) and using the assumptions of the lemma it is easy to prove (3). The lemma is proved.

Let $\varepsilon > 0$. Denote by A^ε the set of all matrices a of dimension $d \times d_1$ with elements $a_{ij} \in [-\varepsilon,\varepsilon]$, B^ε the set of all d-dimensional vectors with components $b_i \in [-\varepsilon,\varepsilon]$, C^ε the set of all elements $c \in L_2(R^d,\Pi)$ such that $\|c\|^2_{2,\Pi} \le \varepsilon$.

Let $\tilde{A}^\varepsilon = A^\varepsilon \times B^\varepsilon \times C^\varepsilon$, $\bar{A}^\varepsilon = A \times \tilde{A}^\varepsilon$. We denote the elements of $\bar{A}^\varepsilon$ by $\bar{\alpha} = (\alpha,\tilde{\alpha})$, where $\alpha \in A$, $\tilde{\alpha} \in \tilde{A}^\varepsilon$.

Let $\zeta_1(x), \zeta_2(t)$ be nonnegative infinitely differentiable functions of arguments $x \in R^d$, $t \in R^1$ equal to zero for $|x| \ge 1$, $|t| \ge 1$ and such that $\int \zeta_1(x)dx = \int \zeta_2(t)dt = 1$. Let $\zeta^n(t,x) = n^{d+1}\zeta_1(nx)\zeta_2(nt)$, $n = 1,2, \ldots$.

Denote by $\sigma^{(n)}, b^{(n)}, c^{(n)}, g^{(n)}$ the convolutions of functions σ,b,c,g with function ζ^n with respect to (t,x) (in computing these convolutions we assume that $\sigma(\alpha,t,x) = \sigma(\alpha,0,x)$ for $t \le 0$, $\sigma(\alpha,t,x) = \sigma(\alpha,T,x)$ for $t \ge T$, etc.).

Let $\bar{\alpha} \in \bar{A}^\varepsilon$. Furthermore, let

$$\sigma^n_\varepsilon(\bar{\alpha},t,x) = \sigma^{(n)}(\alpha,t,x) + a^{\tilde{\alpha}}, \quad a^{\tilde{\alpha}} \in A^\varepsilon,$$

$$b^n_\varepsilon(\bar{\alpha},t,x) = b^{(n)}(\alpha,t,x) + b^{\tilde{\alpha}}, \quad b^{\tilde{\alpha}} \in B^\varepsilon,$$

$$c^n_\varepsilon(\bar{\alpha},t,x,z) = c^{(n)}(\alpha,t,x,z) + c^{\tilde{\alpha}}(z), \quad c^{\tilde{\alpha}}(\cdot) \in C^\varepsilon.$$

Replacing here $\sigma^{(n)}, b^{(n)}, c^{(n)}$ by σ,b,c we construct functions σ_ε, b_ε, c_ε. Using the collections $(\bar{A}^\varepsilon, \sigma^n_\varepsilon, b^n_\varepsilon, c^n_\varepsilon, g^{(n)})$, $(\bar{A}^\varepsilon, \sigma_\varepsilon, b_\varepsilon, c_\varepsilon, g)$ we construct controlled processes $x_t^{\bar{\alpha},s,x}(n,\varepsilon)$, $x_t^{\bar{\alpha},s,x}(\varepsilon)$ and payoff functions $v_{n\varepsilon}$, v_ε in the same way as we constructed the above controlled process $x_t^{\alpha,s,x}$ and the payoff function v on the basis of the collection (A,σ,b,c,g).

7. Lemma. Let assumption 1 hold. Then $v_{n\varepsilon} \to v_\varepsilon$ as $n \to \infty$, $v_\varepsilon \to v$ as $\varepsilon \to 0$ uniformly on every bounded subset H_T.

The proof of this lemma is analogous to the proof of Theorem III 1.12 [6].

8. Proof of Theorem 5. First we shall prove the inequality

$$\overline{\lim_{k\to\infty}}\, V_k(s,x) \le v(s,x).$$

Let the subsequence $\{n\} \subset \{k\}$ be such that $\overline{\lim\limits_{k\to\infty}}\, V_k(s,x) = \lim\limits_{n\to\infty} V_n(s,x)$. If $\{\beta(n)\}$ is a sequence of ε-optimal strategies, then for some subsequence $\{n'\} \subset \{n\}$ we have

$$V_{n'}(s,x) - \varepsilon \le E^{\beta(n')}_{s,x}\, g(\xi^{n'}(T)) = \tilde{E}\, g(\tilde{\xi}^{n'}(T)) \to \tilde{E}\, g(\tilde{\xi}(T))$$

as $n' \to \infty$. Here $\varepsilon > 0$ is an arbitrary positive number. Therefore, for (4) it suffices to show that $\tilde{E}\, g(\tilde{\xi}(T)) \le v(s,x)$.

Fix $\varepsilon > 0$, $R > 0$. Let

$\Gamma(t,x) = \{a(\alpha,t,x), b(\alpha,t,x), c(\alpha,t,x,\cdot);\ \alpha\in A\}$,

$\Gamma_{n\varepsilon}(t,x) = \{a^n_\varepsilon(\bar\alpha,t,x), b^n_\varepsilon(\bar\alpha,t,x), c^n_\varepsilon(\bar\alpha,t,x,\cdot);\ \bar\alpha\in \bar A^\varepsilon\}$,

where $a^n_\varepsilon = \sigma^n_\varepsilon[\sigma^n_\varepsilon]^*$. Using assumption 1 we derive that $\|a^n - a\| \to 0$, $|b^{(n)} - b| \to 0$, $\|c^{(n)} - c\|_{2,\Pi} \to 0$ as $n\to\infty$ uniformly over $A\times[0,T]\times S_R$, where $a^n = \sigma^{(n)}[\sigma^{(n)}]^*$. Therefore, for sufficiently large n

$$\Gamma(t,x) \subset \Gamma_{n\varepsilon}(t,x) \tag{5}$$

for all $(t,x)\in[0,T]\times S_R$.

The functions σ^n_ε, b^n_ε, c^n_ε, $g^{(n)}$ are smooth in (t,x) and for all $(t,x)\in H_T$

$$\inf_{|y|=1}\ \sup_{\bar\alpha\in\bar A^\varepsilon} (a^n_\varepsilon(\bar\alpha,t,x)y,y) > 0.$$

By Theorem 1.4 [4], there exist locally bounded partial derivatives $v_{n\varepsilon t}$, $v_{n\varepsilon x_i}$, $v_{n\varepsilon x_i x_j}$ in Sobolev sense such that $F_{n\varepsilon} v_{n\varepsilon} = 0$ (a.e. H_T) and $v_{n\varepsilon}(T,\cdot) = g^{(n)}(\cdot)$, where $F_{n\varepsilon}$ is the operator defined by the formula (1) if we replace in this formula A,a,b,c by $\bar A^\varepsilon$, a^n_ε, b^n_ε, c^n_ε. Therefore, using (5) we obtain that for sufficiently large n

$$Fv_{n\varepsilon} \le 0 \quad \text{a.e. on } [0,T]\times S_R. \tag{6}$$

Denote by $v^{(m)}_{n\varepsilon}$, $(L^\alpha v_{n\varepsilon})^{(m)}$ the convolutions with respect to (t,x) of functions $v_{n\varepsilon}$, $L^\alpha v_{n\varepsilon}$ with the function ζ^m. Fix a number h such that $s+h < T-h$. From Lemma 6 and (6) we obtain that for $m > 1/h$ and sufficiently large n

$$\begin{aligned}&\tilde E\, v^{(m)}_{n\varepsilon}(\gamma_2,\tilde\xi(\gamma_2)) - \tilde E\, v^{(m)}_{n\varepsilon}(\gamma_1,\tilde\xi(\gamma_1)) \le \\ &\le \tilde E \int_{\gamma_2}^{\gamma_1} Fv^{(m)}_{n\varepsilon}(r,\tilde\xi(r))\,dr \le \int_{s+h}^{T-h} \delta_m(t)\,dt,\end{aligned} \tag{7}$$

where $\gamma_1 = (s+h)\wedge\tau_R$, $\gamma_2 = (T-h)\wedge\tau_R$, τ_R is the time of first exit of $\tilde\xi(t)$ from S_R and

$$\delta_m(t) = \sup_{|x|\le R}\ \sup_{\alpha\in A} |L^\alpha v^{(m)}_{n\varepsilon} - (L^\alpha v_{n\varepsilon})^{(m)}|(t,x).$$

Using assumption 1 it is easy to derive that the functions δ_m are bounded on $(s+h, T-h)$ uniformly with respect to m and $\delta_m(t)\to 0$ as $m\to\infty$ for every $t\in(s+h, T-h)$. Letting $m\to\infty$ in (7) we obtain

$$\tilde E\, v_{n\varepsilon}(\gamma_2,\tilde\xi(\gamma_2)) \le \tilde E\, v_{n\varepsilon}(\gamma_1,\tilde\xi(\gamma_1)).$$

From this inequality in view of arbitrariness h and stochastic continuity $\tilde\xi(\cdot)$ we obtain

$$\tilde E\, v_{n\varepsilon}(\gamma_3,\tilde\xi(\gamma_3)) \le v_{n\varepsilon}(s,x),$$

where $\gamma_3 = T\wedge\tau_R$. Letting $n\to\infty$, $\varepsilon\to 0$, $R\to\infty$ in this inequality and using lemma

7 we obtain (4).

Now for the proof of the theorem it suffices to show that

$$\varliminf_{k\to\infty} V_k(s,x) \geq v(s,x).$$

Fix $\varepsilon > 0$. Using the same arguments as in Corollary III 2.9 [6] we conclude that there exist a subdivision I_k of an interval $[s,T]$, a finite subset $A_N = \{\alpha^1, \ldots, \alpha^N\} \subset A$ and a strategy $\alpha \in \mathfrak{A}$ having values in A_N, $\alpha(t) = \alpha(t_i^k)$ if $t \in [t_i^k, t_{i+1}^k)$ such that

$$E\, g(x_T^{\alpha,s,x}) \geq v(s,x) - \varepsilon.$$

Denote by $\xi^n(t)$ the process controlled by the strategy $\beta(n)=(q_0^n,\ldots,q_{n-1}^n)\in\mathcal{B}^n$, where

$$q_0^n(\alpha_0 = \beta | x_0) = P(\alpha(s) = \beta),$$

$$q_j^n(\alpha_j = \beta | x_0, \alpha_0, \ldots, \alpha_{j-1}, x_j) = P(\alpha(t_j^n) = \beta | \alpha(s) = \alpha_0, \ldots$$
$$\alpha(t_{j-1}^n) = \alpha_{j-1}\}, \; j = 1,2, \ldots, n, \; n \geq k, \; \beta \in A_N.$$

By the assumption 3, for some subsequence $\{n'\} \subset \{n\}$ the measures $P_{s,x}^{n'}$ induced by $\xi^{n'}(\cdot)$ on $(D[s,T], \mathcal{D}[s,T])$ converge weakly to some measure $P_{s,x}$. It is not difficult to show that the measure on $(D[s,T],\mathcal{D}[s,T])$ induced by $x^{\alpha,s,x}$ coincides with $P_{s,x}$. From this in view of arbitrariness $\varepsilon > 0$ (8) follows. (8) together with (4) proves the theorem.

9. Remark. The complete proof of the theorem and related results will be published in Lith. Math. J., vol. XXIII (1983).

REFERENCES

1. I. I. Gichman, A. V. Skorochod, "Controlled stochastic processes", Kiev, Naukova dumka, 1977 (in Russian).

2. H. J. Kushner, "Probability Methods for Approximations in Stochastic Control and for Elliptic Equations", Academic Press, New York, 1977.

3. H. J. Kushner, G. DiMasi, Approximations for Functionals and Optimal Control Problems on Jump Diffusion Processes, J. Math. Anal. Appl. 63 (1978), 772-800.

4. H. Pragarauskas, On Bellman equation for weakly nondegenerated general stochastic processes, Liet. matem. rink., 20 (1980), 129-136 (in Russian).

5. A. V. Skorochod, "Studies in the Theory of Random Processes", Kiev, Naukova dumka, 1961 (in Russian).

6. N. V. Krylov, "Controlled Diffusion Processes", Moscow, Nauka, 1977 (in Russian).

7. J. P. Quadrat, Existence de solution et algorithme de resolution numerique de probleme de controle optimal de diffusion stochastique degeneree ou non, SIAM J. Cont. Opt., vol. 18, N2, 1980, 199-226.

ON OPTIMAL STOCHASTIC CONTROL PROBLEM OF LARGE SYSTEMS

J.P. QUADRAT

Domaine de Voluceau - B.P. 105
78153 LE CHESNAY Cédex

I - INTRODUCTION.

We discuss three different approaches, leading to numerical methods, for the solution of optimal stochastic control problem of large dimension :

- Optimization in the class of local feedbacks,
- Monte Carlo and stochastic gradient techniques,
- Perturbation methods in the small intensity noise case.

We consider the stochastic control problem of diffusion processes in the complete observation case

$$(1) \qquad \begin{cases} dX_t = b\,(X_t,U_t)dt + dW_t \quad , \quad X_t \in \mathbb{R}^n,\ U_t \in \mathbb{R}^m \\ V(o,y) = \underset{u}{\text{Min}}\ \mathbb{E}\ \{\int_o^{+\infty} e^{-\lambda t}\ C(X_t,U_t)dt\ |X(o) = y\} \end{cases}$$

The solution of the Hamilton Jacobi equation

$$(2) \qquad \underset{u}{\text{Min}}\ \{b(x,u)\ \text{grad}\ V + C(x,u)\} + \Delta V - \lambda V = o$$

gives the optimal cost and the optimal strategies of (1).

The numerical solution of (2) is almost impossible in the general situation when n is large. The difficulty is not a problem of numerical analysis but an irreducible difficulty. To see that consider the simpler problem

$$(3) \qquad \begin{cases} \Delta V - \lambda V = C \qquad x \in \mathcal{O} = [0,1]^n \\ V_{\partial\mathcal{O}} = 0 \end{cases}$$

where $\partial\mathcal{O}$ denotes the boundary of $\mathcal{O}$.

For such problem it is easy to show that the number of eigen vectors associated to an eigen value smaller than a fixed value, increases exponentially with the dimension. But we need to have a good representation of the eigen vector associated to eigen value of small modulus, in any good finite dimensional approximation of (2). And thus, whatever could be the approximation, the obtention of a given precision will be obtain at a cost which increases exponentially with the dimension.

In the three following points of view we avoid this difficulty but we have a loss of optimality.

II - OPTIMIZATION IN THE CLASS OF LOCAL FEEDBACKS.

In this paragraph we give the optimality conditions in the class of local feedbacks, and show that it is more difficult to solve these conditions than to compute the solution of the Hamilton-Jacobi equation. Then we study two particular cases :

- the case of the uncoupled dynamics,
- the case of systems having the product form property.

In these cases only it is possible to compute the optimal local feedbacks for large systems. Finally we discuss briefly the decoupling point of view.

II-1. The general situation.

Given I the indexes of the subsystems $I = \{1,2,..,k\}$ n_i,[resp-m_i] denotes the dimension of the states[resp.the controls] of the subsystem $i \in I$. The local feedback S_i is a mapping of $\mathbb{R}^+ \times \mathbb{R}^{n_i}$ in $\mathcal{V}_i \subset \mathbb{R}^{m_i}$ the set of the admissible values of the control i. $\mathcal{S}_L$ denotes the class of local feedbacks $\mathcal{S}_L = \{S = (S_1,\dots,S_k)\}$. Given the drift term of the system :

$$b : \mathbb{R}^+ \times \mathbb{R}^n \times \mathcal{V} \to \mathbb{R}^n$$
$$\quad t \quad x \quad u \quad b(t,x,u)$$

with $\qquad n = \sum_{i\in I} n_i, \quad \mathcal{V} = \prod_{i\in I} \mathcal{V}_i,$

- the diffusion term :

$$\sigma : \mathbb{R}^+ \times \mathbb{R}^n \to M_n$$
$$t \quad x \qquad \sigma(t,x)$$

with M_n the set of matrices (n,n) and $a = \frac{1}{2}\sigma\sigma^*$ where $*$ denotes the transposition

- the instantaneous cost :

$$c = \mathbb{R}^+ \times \mathbb{R}^n \times \mathcal{V} \to \mathbb{R}^+$$
$$t \quad x \quad u \qquad c(t,x,u)$$

then boS [resp coS] denotes the functions $\mathbb{R}^+ \times \mathbb{R}^n \to \mathbb{R}^n$

$$[\text{resp } \mathbb{R}^+ \times \mathbb{R}^n \to \mathbb{R}^+] \quad b(t,x,S(t,x)) \; [\text{resp } c(t,x,S(t,x))]$$

Then if X^S denotes the diffusion (boS,a) (drift boS, and diffusion term σ) and P_μ^S its measure defined on $\Omega = C(\mathbb{R}^+, \mathbb{R}^n)$ with μ the law of the initial condition we want to solve

$$\underset{S\in\mathcal{S}_L}{\mathrm{Min}} \; \mathbb{E}_{P_\mu^S} \int_0^T \mathrm{CoS}(t,\omega_t)dt$$

where $\omega \in \Omega$, T denotes the time horizon. We have here a team of I players working to optimize a single criterion

A simple way to obtain the optimality conditions is to consider another formulation of this problem : the control of the Fokker Planck equation that is :

$$\underset{S\in\mathcal{S}_L}{\mathrm{Min}} \; J^S = \int_Q \mathrm{CoS}(t,x)p^S(t,x)dt\,dx$$

with p solution of

$$\mathcal{L}_S^* \, p^S = 0$$

$$p^S(o,.) = \mu$$

with $\quad Q = [0,T] \times \mathcal{O} \quad$ and $\mathcal{O} = \mathbb{R}^n$

$$\mathcal{L}_S = \frac{\partial}{\partial t} + \sum_j b_j\mathrm{o}S \frac{\partial}{\partial x_j} + \sum_{i,j} a_{ij} \frac{\partial^2}{\partial x_i \partial x_j}$$

μ the law of the initial condition.

Than we have :

Theorem 1

A N.S.C. for $J^R \geq J^S$, R $S \in \mathcal{S}_L$, is that :

(1) $$H(t,R,p^R,V^S) \geq H(t,S,p^R,V^S) \quad \text{pp in } t$$

with

(2) $$\begin{cases} H(t,R,p,V) = \int_{\mathcal{O}} [CoR(t,x) + \sum_i b_i oR(t,x) \frac{\partial V}{\partial x_i} (t,x)]\, p(t,x)dx \\ \mathcal{L}_R^* p^R = 0 \quad p^R(o,.) = \mu \ ; \ \mathcal{L}_S \, V^S + CoS = 0, \quad V^S(T,.) = 0 \end{cases}$$

Remark 1. From this theorem the Pontriagyn condition can be obtained, that is a necessary condtion of optimality of the strategy S is that: p,V,S satisfy

$$H(t,S,p^S,V^S) = \underset{R \in \mathcal{S}_L}{\text{Min}} \ H(t,R,p^S,V^S) \ ;$$

(3) $$\begin{cases} \mathcal{L}_S^* p^S = 0 \quad , \quad p(o,.) = \mu \quad ; \\ \mathcal{L}_S V^S + CoS = 0 \quad , \quad V^S(T,.) = 0. \end{cases}$$

A proof is given in J.L. Lions [13].

Remark 2. This theorem give an algorithm to improve a given strategy R that is :

Step 1 : compute p^R

Step 2 : solve backward simultaneously

(4) $$\begin{cases} \mathcal{L}_S V^S + CoS = 0 \qquad V^S(T,.) = 0 \\ S \in \text{Arg} \underset{Z}{\text{Min}} \ H(t,Z,p^R,V^S) \end{cases}$$

By this way we obtain a better strategy S.

A fixed point of the application $R \rightarrow S$ will satisfy the conditions (3).

We see that one iteration (4) of this algorithm is more expensive than the computation cost of the solution of the H.J.B. equation.

II-2. Uncoupled dynamic systems.

This is the particular case where b_i is a function of x_i and u_i, $\forall i \in I$

$$b_i : \mathbb{R}^+ \times \mathbb{R}^{n_i} \times \mathcal{V}i \to \mathbb{R}^{n_i}$$
$$t \quad x_i \quad u_i \qquad b_i(t,x_i,u_i)$$

and the noises are not coupled between the subsystems that is :

$$\sigma_i : \mathbb{R}^+ \times \mathbb{R}^{n_i} \to M_{n_i}$$
$$t \quad x_i \qquad \sigma_i(t,x_i)$$

In this situation we have

$$p^R = \prod_{i \in I} p_i^{R_i}$$

with p_i^{Ri} solution of

$$(5) \qquad \mathcal{L}^*_{i,R_i} p_i^{R_i} = 0 \qquad p_i^{R_i}(o,.) = \mu_i \quad \text{with } \mu = \prod_{i \in I} \mu_i$$

and

$$\mathcal{L}_{i,R_i} = \frac{\partial}{\partial t} + \sum_{k \in I_i} b_k o R_i(t,X) \frac{\partial}{\partial X_k} + \sum_{k,l \in I_i} a_{kl} \frac{\partial^2}{\partial X_k \partial X_l}$$

with $\quad I_i = \{ \sum_{j<i} n_j < k \leq \sum_{j<i+1} n_j \}.$

Let us denote by

$$(6) \qquad C_i^R o R_i : \mathbb{R}^+ \times \mathbb{R}^{ni} \to \mathbb{R}^+$$
$$t \quad x_i \qquad \int CoR(t,x) \prod_{j \neq i} p_j^{R_j}(t,X_j) dX_j$$

That is the conditional expectation of the instantaneous cost knowing the information only on the local subsystem i.

We have the following sufficient conditions to be optimal player by player :

Theorem 2. A sufficient condition for a strategy S to be optimal player by player is that the following conditions are satisfied :

(7) $\underline{\underset{R_i}{\text{Min}} [\mathcal{L}_{i,R_i} V_i + C_i^R o R_i] = o,\ i \in I}$;

with $C_i^R o R_i$ defined by (6) and (5)

The optimal cost is $\mu_1(V_1)..=\mu_I(V_I)$ with $\mu_i(V_i) = \int_{\mathbb{R}^{n_i}} \mu_i(dX_i)V_i(o,X_i)$

Remark 3. The theorem 3 gives an algorithm to compute a feedback optimal player by player

given $\varepsilon,\ \nu \in \mathbb{R}^+$

Step 1) Choose $i \in I$
Solve (7)
if : $\mu_i(V_i) \leq \nu - \varepsilon$ than $\nu := \mu_i(V_i)$
$R_i := \underset{R_i}{\text{Arg Min}} \{\mathcal{L}_{i,R_i} V_i + C_i^R o R_i\}$

if not choose another $i \in I$ until

$$\mu_i(V_i) \geq \nu - \varepsilon,\ \forall i \in I.$$

Step 2) When $\mu_i(V_i) \geq \nu - \varepsilon,\ \forall i \in I$, than $\varepsilon := \frac{\varepsilon}{2}$, go to step 1.

By this algorithm we obtain a decreasing sequence $\nu^{(n)}$ which converges to a cost optimal player by player.

For a proof of a discrete version of this algorithm see Quadrat-Viot [2].

Remark 4. The interpretation of $V_i(t,X_i)$ $i \in I$ in terms of the variables of theorem 1 is :

$$V_i(t,x_i) = \int V(t,x) \prod_{j \neq i} p^{R_j}(t,X_j)dX_j$$

Remark 5. In this situation we have to solve a coupled system of P.D.E. but each of them is on a space of small dimension. By this way we can optimize , in the class of local feedback, systems which are not reachable by H.J.B. equation. An application to hydropower systems is given in Delebecque-Quadrat [4].

II-3. Systems having the product form property.

The property that a system has its dynamic uncoupled is very restrictive in this paragraph, we show systems which have their invariant measure uncoupled, they are limit of network of queues of Jackson type. This property can be used to apply to them the results of II-2 for the corresponding ergodic control problem that is:

$$\underset{S}{\text{Min}} \lim_{T\to\infty} \frac{1}{T} \int_0^T C_\circ S(\omega_t)dt$$

Given B a generator of a Markov chain defined on $E = \{1,2,\dots,n\}$, a function $\begin{matrix} E \times \mathbb{R} \to \mathbb{R} \\ (i,x) \quad u_i(x) \end{matrix}$ a matrix $\sigma \in M_n$, $A = \frac{1}{2}\sigma\sigma^*$, D a diagonal matrix satisfying :

(8) $$DB^* + BD + 2A = 0$$

Theorem 3

The invariant measure of probability p of the diffusion ($b = Bu$, $a=A$) such that (8) is true has the product form property that is :

(9) $$p(x) = C \prod_{i=1}^{n} p_i(x_i) , \qquad i \in E$$

(10) $$p_i(x_i) = \exp - \frac{1}{d_{ii}} \int_0^{x_i} u_i(s)ds$$

where C is a constant of normalization.

Demonstration : The Fokker-Planck equation can be written :

(11) $$-\text{div}\,[bp] + \text{div}\,[A \text{ grad } p] = 0$$

Let us make the change of variables $p = \exp V$ in (11), we obtain

$$(\text{grad } V, b-A \text{ grad } V) + \text{div}\,(b-A \text{ grad } V) = 0$$

Using (10), we have :

(12) $$(D^{-1}u, (B + AD^{-1})u) + \text{tr}\,[(B + AD^{-1}) \text{ grad } u] = 0$$

The quadratic part in (u) of (12) is equal to 0 if and only if :

$$D^{-1}B + B^*D^{-1} + 2D^{-1}AD^{-1} = 0$$

which can be written :

$$BD + DB^* + 2A = 0$$

which is (8).

We have also $\mathrm{tr}\,[B + AD^{-1}]\,\mathrm{grad}\, u = 0$. Indeed grad u is diagonal because u_i is a function of x_i only and the coefficient of $\frac{\partial u_i}{\partial x_i}$ is $b_{ii} + a_{ii} / d_{ii}$ which is equal to zero by (8).

Remark 6. This class of diffusion processes are quite natural if we see them as the limit process when $N \to \infty$, obtained from Jackson network of queues by the scaling $x \to \frac{x}{N}$, $t \to \frac{t}{N^2}$.

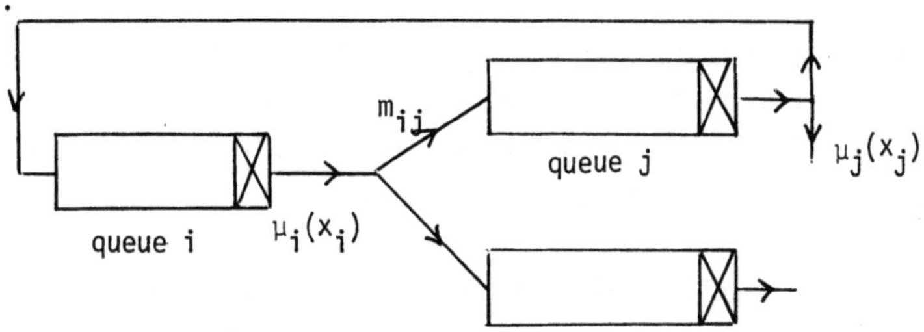

where $\mu_i(x_i)$ is the output rate of the queue i, m_{ij} is the probability of a customer leaving the queue i to go to the queue j.

The correlation of the noise given by (8) corresponds to system for which the noise satisfies a conservation law (for example the total number of customer in a closed network of queues).

Remark 7. We can now apply the result of II-2 to compute the optimal local feedback for systems having the product form property and an ergodic criterion. Indeed :

$$\mathrm{Min}\ \frac{1}{T}\int_0^T C\circ S\quad(\omega_t)dt = \int C\circ S\ (x)\ p(x)dx$$

$$p(x) = \prod_{i=1}^{n} p_i(x_i)$$

and p_i satisfies :

$$-\frac{\partial}{\partial x_i}[u_i p_i] + \frac{\partial^2}{\partial x_i^2}[d_{ii}\ p_i] = 0, \qquad i \in E$$

$$\int p_i(x_i)\ dx_i = 1$$

II-4. Remarks on decoupling.

Another way to use the results of II-2 when the dynamic is coupled is to do a change of feedback let us consider the simpler case

$$b : \mathbb{R}^n \times U \to \mathbb{R}^n \qquad \text{with } u \in \mathbb{R}^n$$
$$\quad x \quad\ \ u \quad b(x,u)$$

we use the feedback transformation $v = b(x,u)$ to decouple the drift terms. Now v is the control and we can apply the results of II-2 to compute the best local feedback $v_i = S_i(x_i)$.

Then the solution in u of

(13) $$b(x,u) = S(x)$$

gives the best feedback among the class that we can call "local decoupling feedbacks".

One difficulty with this approach is for example the constraints on the control : the image by b of an hypercube is not in general an hypercube and if we take for constraints on the new control $v \in V(x) \subset b(x,\mathcal{U})$ with $V(x)$ an hypercube of $\mathbb{R}^n$, the loss of optimality can become unacceptable.

This approach is well studied for deterministic linear and non linear systems Womham [8], Isidori [17] and in the dynamic programming litterature Larson [11] .

III - OPTIMIZATION IN A PARAMETRIZED CLASS OF FEEDBACKS BY MONTE CARLO TECHNIQUES.

We have seen in §2 that we are able to compute the optimal local feedback only in particular cases. Moreover, sometimes the local information is not good ; we can have, a priori, an idea on a better one and would like to use this a priori information to solve a simpler problem than the general one. A way to do that is to parametrize the feedback, optimize the open loop parameter by a Monte Carlo technique. More precisely given the stochastic control problem

(1) $$\begin{cases} dX_t = b(t,x_t,u_t)dt + dw_t & x_t \in \mathbb{R}^n, \quad u_t \in \mathbb{R}^m \\ \text{Min } E \int_0^T C(t,x_t,u_t)dt & \end{cases}$$

we make the feedback transformation

(2) $$u(t) = S(t,x_t,v_t) \qquad v_t \in \mathbb{R}^p$$

where $S : \mathbb{R}^+ \times \mathbb{R}^n \times \mathbb{R}^p \to \mathbb{R}^m$ is given.

We use for the approximation of the probability law of the noise P, the distribution

$$\mu = \frac{1}{r} \sum_{j=1}^{r} \delta_{\omega_j}(\omega)$$

where ω_j are trajectories of the noise obtained by random generation perhaps after a discretization time if we want to avoid the difficulties of the non existence of a solution trajectory by trajectory to (1). And now we have to solve :

(3)
$$dx_t^j = b(t,x_t^j,S(t,x_t^j,v_t))dt + dw_t^j$$
$$\min_v \frac{1}{r} \sum_{j=1}^{r} \int_0^T C(t,x_t^j,S(t,x_t^j,v_t))dt$$

w_t^j denotes here a particular trajectory of the noise. Thus, at the end, we have to solve a deterministic dynamic control problem. For that, we can use a gradient technique or the Pontryagin principle. For discrete time system the convergence properties of this approach has been studied in Quadrat-Viot [15]. Application to the French hydropower system is currently done at EDF now. Feedbacks on the demand of electricity and the level of water in the local dam are optimized with success by this technique (Lederer - Colleter [2]).

The idea of the stochastic gradient method is the same as the former one but we use a recursive way to optimize. The recursivity is on the index of the trajectory of the noise generated. The problem (1) (2) can be reduced to the problem

$$\min_{v \in V} \mathbb{E}\, J(v)$$

is a situation where we are able to compute $\frac{\partial J}{\partial v}$ by adjoint state technique here

$$J(v) = \int_0^T C(t,x_t,S(t,x_t,v_t))dt.$$

Moreover, we can consider that after discretization v is finite dimensional. Then the stochastic gradient algorithm is the following recursive way to improve the parameter v

$$v_{r+1} = P_V \{v_r - \rho_r \frac{\partial J}{\partial v}(v_r,\omega_r)\} \qquad \rho_r \in \mathbb{R}^+, \quad \forall r \in \mathbb{N} ,$$

$$\sum_{r \in N} \rho_r = \infty \quad , \quad \sum_{r \in N} \rho_r^2 < \infty$$

ω_r denotes a generated random realization of the stochastic parameter in the definition of J(v), for our problem (1), (2) that is a realization of the Wiener process w_t. P_V denotes the projection on the set V.

In a convex situation which is not in general the case for the problem (1), (2), we can give some global convergence results.

Theorem 1

On the hypotheses

1) $v \to J(\omega,v)$ convex $\forall\omega$;

2) $\omega \to J(\omega,v)$ is L' , $\forall v$

3) $\sup_{\substack{v\in V\\ \omega\in\Omega}} |\frac{\partial J}{\partial v}(v,\omega)| \le q$;

4) $\mathbf{E}\, J(v) - \bar{J}^* \ge C\ \ell^2(v)$

where $\bar{J}^*$ denotes the optimal cost and $\ell(v)$ denotes the distance of v to the optimal set.

5) V a bounded convex set, we have $\lim_n \mathbf{E}\ \ell^2(v_n) = 0$ and moreover if $\rho_r = \dfrac{1}{cr + \frac{q^2}{y_o C}}$

with $y_0 = \mathbf{E}\ \ell^2(v_0)$ we have :

$$\mathbf{E}\ \ell^2(v_r) \le \frac{1}{\frac{c^2}{q^2} r + \frac{1}{y_0}}$$

The proof of this theorem can be found in Dodu-Goursat-Hertz-Quadrat-Viot [5] a lot of similar results can be found in Polyak [14] and in the reference of this paper. In Kushner-Clark [12] local convergence are proved in the non convex case.

The following result shows that in some sense the stochastic gradient algorithm is optimal. We suppose that :

6) the noise is finite valued and we denote by $v_\mu = \text{Argmin}\ \mathbf{E}_\mu\ J(v)$ and we suppose that

7) $v \to J(\omega,v)$ is twice differentiable and uniformly convex $\forall\omega \in \Omega$, then we have D.G.H.Q.V. [5].

Theorem 2

On the hypothese 6) and 7) we have :

$$E_\mu\ (\hat{v} - v\ )^{\otimes 2} \ge \frac{1}{r} H_\mu^{-1}\ Q_\mu\ H_\mu^{-1}$$

with

$$H_\mu = \frac{\partial^2}{\partial v^2} \mathbf{E}_\mu\ J(v)$$

$$Q_\mu = \mathbf{E}_\mu (\frac{\partial J}{\partial v}(v_\mu))^{\otimes 2}$$

for all unbiased statistic v of v_μ defined on $(\Omega,\mu)^{\otimes r}$.

If we remark that $\frac{1}{C}$ is an estimation of H_μ^{-1} and q^2 an estimate of Q_μ we see that in a certain sense the speed of convergence of the stochastic gradient technique is optimal.

We have applied this algorithm to the problem of the optimization of the investment of a transmission network of electricity D.G.H.Q.V. [5]. The comparison with a sophisticated simplex approach shows that the stochastic gradient mathematic is undoubtly better.

IV - PERTURBATION METHODS.

By perturbation methods we can reduce a difficult problem to a simpler one. In this chapter we study the small intensity noise case. In this situation it is possible to build an affine control which leads to ε^4 error with respect to the optimal control, where ε denotes diffusion term.

We consider the following stochastic control problem :

(1) $$\begin{cases} dX_t = f(x_t,u_t)dt + \varepsilon dw_t \quad , \quad x_t \in \mathbb{R}^n \; , \; u_t \in \mathbb{R}^m \\ V^\varepsilon(o,y) = \underset{u}{\text{Min}}\, \mathbf{E}\, [\int_o^T C(x_t,u_t)dt \mid X(o) = y]. \end{cases}$$

where ε belongs to $\mathbb{R}^+$ and is small.

We denote by

(2) $$H(x,u,p) = p\, f(x,u) + C(x,u)$$

We suppose that :

(3) $$u \to f(x,u) \quad \text{is linear ;}$$

(4) $$(v, \frac{\partial^2 C}{\partial u^2} v) \geq k\, |v|^2 \quad \text{where k is a positive constant, } \forall x \text{ ;}$$

Let us consider the deterministic control problem

(5) $$dX_t = f(x_t,u_t)dt$$

$$V(o,y) = \underset{u}{\text{Min}}\{\int_o^T C(x_t,u_t)dt \mid X(o) = y\} \text{ ;}$$

and denote by $u_0(t)$ the optimal open loop deterministic control.

The second variation calculus around the optimal trajectory of (6) Cruz [3] gives the quadratic form "osculatrice" of the optimal cost V around the optimal trajectory. This quadratic form is defined by the (n,n) time dependent matrix P solution of the Riccati equation :

(6) $\quad \mathring{P} + PA + A'P - PSP + Q = 0 \qquad\qquad P(T) = 0$

where

(7) $\quad A = f_x - f_u H_{uu}^{-1} H'_{ux}$

(8) $\quad S = f_u H_{uu}^{-1} f'_u$

(9) $\quad Q = H_{xx} - H_{ux} H_{uu}^{-1} H'_{ux}$

are evaluated along the optimal trajectory of (5) on the hypotheses that :

(10) $\quad H_{uu} > 0, \quad H_{xx} - H_{ux} H_{uu}^{-1} H'_{ux} \geq 0$

Let us consider the following affine control :

(11) $\quad u_f(t,X(t)) = u_o(t) + K(t) [X(t) - X_o(t)]$

where $X_o(t)$ denotes the optimal trajectory of the deterministic control problem (5), X(t) the actual trajectory of the diffusion process (1_1) when the control is (11), and K(t) is defined by

(12) $\quad K(t) = H_{uu}^{-1} (H'_{ux} + f'_u P)(t)$

evaluated on the optimal trajectory $X_o(t)$.

We have :

Theorem

On the hypotheses (3), (4), (10) and (f,c) twice differentiable, the affine control build on the deterministic control problem, used in the stochastic control problem leads to a loss of optimality of order $\mathcal{O}(\varepsilon^4)$.

Ideas of the proof : Fleming [6]has shown that the optimal deterministic feedback used in the stochastic control problem leads to an error of $\mathcal{O}(\varepsilon^4)$. But in the estimation of the proof he does not need the optimal deterministic control but a control which gives exact V, $\frac{\partial V}{\partial y}$, $\frac{\partial^2 V}{\partial y^2}$ along the optimal trajectory of the deterministic control problem.

Using for example Cruz [3] we know that the affine feedback (11) has this property and thus the result is proved.

REFERENCES.

[1] BASKET - CHANDY - MUNTZ - PALACIOS : Open Closed and Mixed Network of Queues with Different Class of Customers, JACM 22, pp. 248-260

[2] COLLETER - LEDERER : Rapports internes EDF sur la gestion des réservoirs hydroélectriques français

[3] CRUZ : Feedback Systems, MacGraw Hill, 1972.

[4] DELEBECQUE - QUADRAT : Contribution of Stochastic Control Singular Perturbation Averaging and Team Theories to an Example of Large Scale System : Management of Hydropower Production, IEEE AC, April 1978.

[5] DODU - GOURSAT - HERTZ - QUADRAT - VIOT : "Méthodes de Gradient Stochastique pour l'Optimisation des Investissements dans un Réseau Electrique", EDF Bulletin Série C, n°2, 81, pp. 133-164.

[6] FLEMING : Stochastic Control for Small Noise Intensities, Brown University Report, April 1970.

[7] HOLLAND :-Small Noise Open Loop Control, SIAM J. Control, 12, August 1974.
-Gaussian Open Loop Control Problems, SIAM J. Control, 13, August 1975.

[8] ISIDORI - KRENER - GORI GIORGI - MONACO : Non Linear Decoupling via Feedback a Differential Geometric Approach, IEEE, AC 26, n°2, April 1981.

[9] JACKSON : Jobshop Like Queueing Systems, Management Science, 10, pp. 131-142, 1963.

[10] KELLY : Reversibility and Stochastic Networks, J. Wiley and Sons, 1979.

[11] KORSAK - LARSON : Dynamic Successive Approximation Technique with Convergence Proofs, Automatica, 1969.

[12] KUSHNER - CLARK : Stochastic Approximation Methods for Constrained and Unconstrained Systems, Springer Verlag, 1978.

[13] LIONS : Contrôle Optimal des Systèmes Distribués, Dunod, 1968.

[14] POLYAK : Subgradient Methods : a Survey of Soviet Research in Non Smooth Optimization, eds. C. Lemaréchal & R. Mifflin, Pergamon Press, 1978.

[15] QUADRAT - VIOT : Méthodes de Simulation en Programmation Dynamique Stochastique, RAIRO, April 1973, pp. 3-22.

[16] " " : Product Form and Optimal Local Feedback for Multiindex Markov Chains, Allerton Conference, 1980.

[17] WONHAM : Linear Multivariable Control : A Geometric Approach , New York, Springer Verlag, 1979.

UNNORMALIZED CONDITIONAL PROBABILITIES AND OPTIMALITY FOR PARTIALLY OBSERVED CONTROLLED JUMP MARKOV PROCESSES

by

Raymond Rishel
Department of Mathematics
University of Kentucky
Lexington, KY 40506

I. INTRODUCTION

Optimality conditions for controlled partially observed jump Markov processes have been given in a number of papers, for instance as in [1],[4],[5],[9]. The objective of this paper is to give a simple proof of necessary conditions for optimality for partially observed controlled jump Markov processes similar to the simple proof for completely observed controlled jump Markov processes given in [6]. By making use of the concepts of unnormalized conditional probabilities and expectations of a partially observed jump Markov process, it will be possible to formulate two intermediate deterministic control problems whose optimal control is the optimal control for the stochastic problem. The Pontryagin maximum principle applies to these deterministic problems to give necessary conditions for optimality. The two problems are dual to each other in that the adjoint variables for one problem are the negatives of the state variables for the other and vice versa.

In order to give a short discussion of the control problem, the discussion of unnormalized and normalized conditional expectations is given in an appendix. Thus the paper really consists of two almost equal parts, the body discussing the control problem and the appendix discussing unnormalized conditional expectations.

Unnormalized conditional expectations are important in their own right. For instance if y is a process obtained from x by aggregating the states of x together, and if x is conditionally Markov given y, then the conditional distributions of the jump times and locations of the aggregated process y may be expressed in terms of unnormalized conditional probabilities. This relationship is given in Corollary A1 of the appendix.

Unnormalized conditional expectations for partially observed jump Markov processes were used in [8], Theorem 2, and in [7]. However in both of these papers they are used mainly as a vehicle to obtain normalized conditional expectations rather than being studied for their own importance.

II. PROBLEM FORMULATION

A jump process x is a stochastic process whose paths are right-continuous step functions. For these processes there is a one-to-one correspondence

$$x \leftrightarrow (x_0,\tau_1,x_1,\tau_2,\dots,x_n,\tau_n,\dots) \tag{1}$$

between the process x and the sequence of its jump times and jump locations. If we use the convention that $\tau_0 \equiv 0$, this correspondence is defined by

$$x(t) = x_n \quad \text{if} \quad \tau_n \le t < \tau_{n+1} . \tag{2}$$

If g is some function on the state space of x which is not one to one and $y(t) = g(x(t))$ is observed, call the process partially observed. Call x(t) the unobserved process and y(t) the observed process. The observed process y(t) will also be a jump process. Let

$$n(t) = \{\text{no. of jumps of } x \text{ in } [0,t]\}$$

$$k(t) = \{\text{no. of jumps of } y \text{ in } [0,t]\} .$$

When $n(t)=n$, x on $[0,t]$ corresponds to

$$(x_0,\tau_1,x_1,\dots,\tau_n,x_n) . \tag{3}$$

If we also have $k(t)=k$, we must have $k \le n$ and y on $[0,t]$ corresponds to

$$(y_0,\bar\tau_0,y_1,\dots,\bar\tau_k,y_k) \tag{4}$$

where $y_0,\dots,y_k$ are the values of $(g(x_0),\dots,g(x_n))$ with repetitions suppressed and $(\bar\tau_1,\dots,\bar\tau_k)$ is that subset of $(\tau_1,\dots,\tau_n)$ at which there is a jump of g(x) at each $\bar\tau$. For brevity let

$$X_n = (x_o,\tau_1,x_1,\dots,\tau_n,x_n) \tag{5}$$

and let

$$Y_k = (y_0,\bar\tau_1,y_1,\dots,\bar\tau_k,y_k) . \tag{6}$$

Call X_n the history of x up to the n-th jump and Y_k the observed history up to the k-th jump. If $\bar\tau_k \le \tau_n < \bar\tau_{k+1}$ then Y_k is the observed history corresponding to X_n.

We shall want to formulate the concept of a controlled family of partially observed jump Markov processes. For convenience we shall take these to be jump processes with values in the finite set of integers $(1,\dots,N)$. The controlled processes of the family will be specified in terms of a controlled generator

$$A(t,u) = (a_{ij}(t,u)) \qquad i,j = 1,\dots,N \tag{7}$$

which satisfies

$$a_{ij}(t,u) \ge 0 \quad , \quad a_{ii}(t,u) = -\sum_{j\ne i} a_{ij}(t,u) \tag{8}$$

and a family of control functions

$$\{u(t,Y_k)\}$$

with values in a closed set U of E^r.

The family $\{u(t,Y_k)\}$ of functions specifies a control in the sense that if there is an observed process y and $k(t)=k$ and the observed history is Y_k then the control $u(t,Y_k)$ is applied at time t.

Given a control $\{u(t,Y_k)\}$ the probability measure of the corresponding partially observed jump process x can be constructed as follows. Construct the conditional probability of the time and location of the next jump given the history up to the n-th jump by defining

$$P[\tau_{n+1}\le t, x_{n+1}=j|X_n]$$
$$= \sum_i 1_{\{\tau_n\le t, x_n=i\}} \int_{\tau_n}^t a_{ij}(s,u(s,Y_k)) \exp\left[\int_{\tau_n}^s a_{ii}(r,u(r,Y_k))dr\right] ds . \tag{9}$$

where in (9) Y_k is the observed history corresponding to X_n.[1]

In terms of a given initial distribution

$$P[x_0=i] = P_i \tag{10}$$

and the conditional distributions (9) construct finite dimensional distributions of the sequence

$$(x_0,\tau_1,x_1,\ldots,\tau_n,x_n,\ldots) \tag{11}$$

and then extend this to a probability measure in the usual way. Then define the process x corresponding to the control $\{u(t,Y_k)\}$ by (2).

Since the generator $(a_{ij}(t,u))$ depends on the past only through the state i, and the value of the control u, we call the process constructed a controlled partially observed jump Markov process. This is a misnomer since the conditional distribution (9) of the next jump depends on the current state, $x_n=i$, and on the past measurements Y_k. Thus the process is not Markov. It would have been Markov had a control depending only on the current value $y_k=g(i)$ of the measurements been used. Of course since there are only partial observations, there is advantage to be gained by using controls which depend on the past measurements. We shall continue to use the term partially observed jump Markov process, even though the process is not Markov.

For the process constructed it follows from (9) that the control applied at time t is $u(t,Y_{k(t)})$. We shall often abbreviate by writing u(t) for $u(t,Y_{k(t)})$.

To formulate an optimization problem let c(t,i,u) denote a cost rate and T a fixed terminal time. Formulate the control problem of finding the control in a given class of controls so that the corresponding expected total cost

$$E\left[\int_0^T c(s,x(s),u(s))ds\right] \tag{13}$$

is a minimum.

III. UNNORMALIZED AND NORMALIZED CONDITIONAL PROBABILITIES

Optimality conditions will be expressed in terms of conditional expectations and conditional probabilities conditioned with respect to the past of the process x and also in terms of conditional probabilities and expectations conditioned with respect

[1]The notation 1_A denotes the characteristic function of the set A, that is the function which is one on A and zero on the complement of A.

to the past of the measurements y. Let

$$\sigma_t = \sigma[x(s)\ ;0\le s\le t]$$

$$\mu_t = \sigma[y(s)\ ;0\le s\le t]$$

$$\sigma(X_n) = \sigma[x_0,\tau_1,x_1,\ldots,\tau_n,x_n]$$

$$\sigma(Y_k) = \sigma[\bar{y}_0,\bar{\tau}_1,\bar{y}_1,\ldots,\bar{\tau}_n,\bar{y}_k]$$

denote the σ-fields generated respectively by the past of the process x, by the past of the process y, by the jump times and locations up through the n-th jump of the process x, by the jump times and jump locations up through the k-th jump of the observed process y.

A fundamental observation about any jump process, such as say y, is that on the set $\bar{\tau}_k \le t < \bar{\tau}_{k+1}$ every μ_t measurable random variable may be written as a $\sigma(Y_k)$ measurable random variable. Using this it easily follows as is shown for instance in [2], Theorems 5 and 4, that for any random variable z for which $E|z| < \infty$ that

$$E[z|\mu_t] = \sum_{k=0}^{\infty} 1_{\{\bar{\tau}_k\le t<\bar{\tau}_{k+1}\}} \frac{E[z\ 1_{\{\bar{\tau}_k\le t<\bar{\tau}_{k+1}\}}\ |\,Y_k]}{E[1_{\{\bar{\tau}_k\le t<\bar{\tau}_{k+1}\}}\ |\,Y_k]}\ . \tag{14}$$

Thus call

$$E[z\ 1_{\{\bar{\tau}_k\le t<\bar{\tau}_{k+1}\}}\ |\,Y_k] \tag{15}$$

the unnormalized conditional expectation of z on $\{\bar{\tau}_k\le t<\bar{\tau}_{k+1}\}$ and call

$$\frac{E[z\ 1_{\{\bar{\tau}_k\le t<\bar{\tau}_{k+1}\}}\ |\,Y_k]}{E[1_{\{\bar{\tau}_k\le t<\bar{\tau}_{k+1}\}}\ |\,Y_k]} \tag{16}$$

the normalized conditional expectation of z on $\{\bar{\tau}_k\le t<\bar{\tau}_{k+1}\}$.

Notice that both the normalized and unnormalized conditional expectations evaluated at $t=\bar{\tau}_k$ agree with

$$E[z\ |\ Y_k]\ . \tag{17}$$

Define normalized and unnormalized conditional expectations for the x-process conditioning with respect to σ_t and $\sigma(X_n)$ in a similar manner to that above for the y-process.

IV. ESTIMATION EQUATIONS (FILTERING FORMULAS)

In this section we will define various unnormalized and normalized conditional probabilities and expectations involved in deriving necessary conditions for optimality of the control problem. Proofs of the relations and properties stated in this section for these quantities are given in the appendix.

Let $Q_i(t,Y_k)$ denote the unnormalized conditional probability that x(t)=i given the history Y_k of the observed process y. That is

$$Q_i(t,Y_k) = E[1_{\{x(t)=i\}}\ 1_{\{\bar{\tau}_k\le t<\bar{\tau}_{k+1}\}}\ |\,Y_k]\ . \tag{18}$$

THEOREM 1: On the interval $\bar{\tau}_k \le t < T$, for each i for which $g(i)=y_k$, $Q_i(t,Y_k)$ satisfies the system of differential equations

$$\frac{d}{dt} Q_i(t,Y_k) = a_{ii}(t,u(t,Y_k))Q_i(t,Y_k) + \sum_{\{j:g(i)=g(j)\}} a_{ji}(t,u(t,Y_k))\, Q_j(t,Y_k) \tag{19}$$

and the boundary condition

$$Q_i(\bar{\tau}_k,Y_k) = P[x(\bar{\tau}_k)=i \mid Y_k] \tag{20}$$

is satisfied. If $g(i)\ne y_k$, $Q_i(t,Y_k)=0$.

Define $J_u(t,X_n)$, the normalized conditional expected cost from time t onward conditioning with respect to the entire past of x by

$$J_u(t,X_n) = \frac{E\left[1_{\{\tau_n \le t < \tau_{n+1}\}} \int_t^T c(s,x(s),u(s))ds \mid X_n\right]}{E[1_{\{\tau_n \le t<\tau_{n+1}\}} \mid X_n]} . \tag{21}$$

THEOREM 2: If i denotes the value of x_n, the quantities (Y_k,i) are sufficient statistics for expressing the conditional remaining cost (21) in the sense that when $x_n=i$ and Y_k is the observed process corresponding to X_n, then $J_u(t,X_n)$ can be written

$$J_u(t,X_n) = J_u(t,Y_k,i) \tag{22}$$

as a function of (t,Y_k,i).

THEOREM 3: On the interval $\bar{\tau}_k \le t \le T$, $J_u(t,Y_k,i)$ satisfies the system of differential equations

$$\frac{d}{dt} J_u(t,Y_k,i) \qquad i=(1,\dots,N)$$

$$= -c(t,i,u(t,Y_k)) - \sum_{\{j:g(i)=g(j)\}} J_u(t,Y_k,j)\, a_{ij}(t,u(t,Y_k)) \tag{23}$$

$$- \sum_{\{j:g(i)\ne g(j)\}} J_u(t,Y_k,t,g(j),j)\, a_{ij}(t,u(t,Y_k)) - J_u(t,Y_k,i)\, a_{ii}(t,u(t,Y_k)) .$$

At time T the boundary condition

$$J_u(T,Y_k,i) = 0 \tag{24}$$

is satisfied.

In Thoerem 3, $Y_{k+1} = (Y_k,t,g(j))$ so that $J_u(t,Y_{k+1},j) = J_u(t,Y_k,t,g(j),j)$ is the quantity defined by (22) and (21).

THEOREM 4: The conditional remaining cost from an observed jump time $\bar{\tau}_k$ onward (or from time $\bar{\tau}_0 \geq 0$ onward if k=0) conditioned with respect to the observed history Y_k satisfies

$$E\left[\int_{\bar{\tau}_k}^{T} c(s,x(s),u(s))ds \mid Y_k\right] \tag{25}$$
$$= \int_{\bar{\tau}_k}^{T} \sum_i \left[c(t,i,u(t,Y_k)) + \sum_{\{j:g(j)\neq g(i)\}} J_u(t,Y_k,t,g(j),j)a_{ij}(t,u(t,Y_k))\right] Q_i(t,Y_k)dt$$

and also satisfies

$$E\left[\int_{\bar{\tau}_k}^{T} c(s,x(s),u(s))ds \mid Y_k\right] = \sum_i J_u(\tau_k,Y_k,i) \quad P[x(\bar{\tau}_k)=i \mid Y_k] . \tag{26}$$

In Theorem 4, $Y_{k+1} = (Y_k,t,g(j))$, so that

$$J_u(t,Y_{k+1},j) = J_u(t,Y_k,t,g(j),j)$$

is the quantity defined through (22) and (21).

V. THE RELATED DETERMINISTIC CONTROL PROBLEMS

Fix an integer r. Let $\{u^*(t,Y_k)\}$ be an optimal control for the problem formulated in Section II. In this section we shall formulate two deterministic control problems whose solution agrees with $u^*(t,Y_r)$. These problems look at minimizing the conditional remaining cost from the r-th jump onward over a class of controls which differ from the optimal control only between the r-th and (r+1)-st observed jump.

To begin this discussion consider the class of controls $\{v(t,Y_k)\}$ such that $v(t,Y_k) = u^*(t,Y_k)$ if $k\neq r$. That is, these controls agree with the optimal control except perhaps between the r-th and (r+1)-st jump of the observed process. Since controls of the above class agree up to time $\bar{\tau}_r$ with the optimal control the following theorem follows from the standard argument used to obtain the principle of optimality.

THEOREM 5: If v is any control of the class above

$$E\left[\int_{\bar{\tau}_r}^{T} c(s,x(s),v(s))ds \mid Y_r\right] \geq E\left[\int_{\bar{\tau}_r}^{T} c(s,x(s),u^*(s))ds \mid Y_r\right] \tag{27}$$

almost surely.

Since any control v of the class above agrees with the optimal control u* from the (r+1)-st observed jump onward, another standard argument implies that

$$J_v(\bar{\tau}_{r+1},Y_{r+1},j) = J_{u^*}(\bar{\tau}_{r+1},Y_{r+1},j) . \tag{28}$$

If $\bar{\tau}_{r+1} = t$ and then $Y_{r+1} = (Y_r,t,g(j))$, equation (28) can be also written as

$$J_v(t,Y_k,t,g(j),j) = J_{u^*}(t,Y_k,t,g(j),j) . \tag{29}$$

Theorem 5 and formulas (25),(19),(20) suggest we fix r and Y_r and abbreviate by defining $v(t) = v(t,Y_r)$, $Q_i(t) = Q_i(t,Y_r)$ and define the deterministic optimal control problem with states Q_i and control v given by:

Problem A. Choose v(t) on $[\bar{\tau}_r,T]$ with values in U to minimize

$$\int_{\bar{\tau}_r}^{T} \sum_i \left[c(t,i,v(t)) + \sum_{\{j:g(j)\neq g(i)\}} J_{u*}(t,Y_r,t,g(j),j)\ a_{ij}(t,v(t)) \right] Q_i(t)dt. \quad (30)$$

Subject to $Q_i(t)$ being solutions of

$$\frac{d}{dt} Q_i(t) = a_{ii}(t,v(t))\ Q_i(t) + \sum_{\{j:g(j)=g(i)\}} a_{ij}(t,v(t))\ Q_j(t) \qquad \text{if } g(i) = y_r \quad (31)$$

with boundary condition

$$Q_i(\bar{\tau}_r) = P[x(\bar{\tau}_r)=i \mid Y_r] \quad (32)$$

Theorem 5 and formulas (26),(23),(24) suggest we again fix r and Y_r, abbreviate by defining $\nu(t) = \nu(t,Y_r)$, $J_i(t) = J_v(t,Y_r,i)$ and define the deterministic optimal control problem with states J_i and control v of:

Problem B. Choose v(t) on $[\bar{\tau}_r,T]$ with values in U to minimize

$$\sum_{i=0}^{N} J_i(\bar{\tau}_r)\ P[x(\bar{\tau}_r)=i \mid Y_r] \quad (33)$$

subject to $J_i(t)$ being solutions of

$$\frac{d}{dt} J_i(t) = -c(t,i,v(t)) - J_i(t)\ a_{ii}(t,v(t)) - \sum_{\{j:g(i)=g(j)\}} a_{ij}(t,v(t))J_j(t)$$

$$- \sum_{\{j:g(i)\neq g(j)\}} J_{u*}(t,Y_r,t,g(j),j)\ a_{ij}(t,v(t)) \qquad i = 0,1,\dots,N \quad (34)$$

with boundary condition

$$J_i(T) = 0 \qquad i = 0,1,\dots,N\ . \quad (35)$$

VI. OPTIMALITY CONDITIONS

If $u^*(t)$ is an optimal control for Problem A, a calculation using the Pontryagin principle shows that the adjoint equations of Problem A are

$$\frac{d}{dt} \phi_i(t) = c(t,i,u^*(t)) + J_{u*}(t,Y_r,t,g(j),j)\ a_{ij}(t,u^*(t))$$

$$- \phi_i(t)\ a_{ii}(t,u^*(t)) - \sum_{\{j:g(i)\neq g(j)\}} \phi_j(t)\ a_{ij}(t,u^*(t)) \quad (36)$$

with transversality conditions

$$\phi_j(T) = 0\ . \quad (37)$$

Thus (34), (35) and the uniqueness of systems of linear differential equations implies the adjoint variables of Problem A are the negatives of the state variables of Problem B.

A similar calculation with Pontryagin's principle shows that the adjoint equations of Problem B are the same as the state equations of Problem A, and that the transversality conditions are the negatives of the boundary conditions (32). Thus the adjoint variables of Problem B are the negatives of the state variables of Problem A. Putting these together with the maximum condition of Pontryagin's principle gives the following type of duality for Problems A and B.

THEOREM 6: A necessary condition that $u^*(t)$ be an optimal control for either Problem A or Problem B is that almost surely with respect to Lebesgue measure on $[\bar{\tau}_k,T]$

$$\min_{v\in U}\left\{\sum_i\left[c(t,i,v)+\sum_{\{j:g(i)\neq g(j)\}} J_{u^*}(t,Y_r,t,g(j),j)a_{ij}(t,v)\right]Q_i(t)\right.$$
$$\left.+\sum_i\left[J_i(t)\ a_{ii}(t,v)+\sum_{\{j:g(i)=g(j)\}} a_{ij}(t,v)J_j(t)\right]Q_i(t)\right\} \tag{38}$$

is attained by the value of the optimal control $v = u^*(t)$, where $Q_i(t)$ is the solution of (31) with $v(t) = u^*(t)$ and with boundary condition (32) and $J_i(t)$ is the solution of (34) with $v(t) = u^*(t)$ and with boundary condition (35).

Putting Theorem 6 together with Theorem 5 gives:

THEOREM 7: A necessary condition that $\{u^*(t,Y_k)\}$ be an optimal control for the stochastic control problem of Section II is that almost surely with respect to the distribution of Y_k and almost surely with respect to Lebesgue measure on $[\bar{\tau}_k,T]$ that

$$\min_{v\in U}\left\{\sum_i\left[c(t,i,v)+\sum_{\{j:g(i)\neq g(j)\}} J_{u^*}(t,Y_k,t,g(j),j)\ a_{ij}(t,v)\right]Q_i(t,Y_k)\right.$$
$$\left.+\sum_i\left[J_{u^*}(t,Y_k,i)\ a_{ii}(t,v)+\sum_{\{j:g(i)=g(j)\}} a_{ij}(t,v)J_{u^*}(t,Y_k,j)\right]Q_i(t,Y_k)\right\} \tag{39}$$

is attained by $v = u^*(t,Y_k)$ where $Q_i(t,Y_k)$ is the solution of (19) with control $u^*(t,Y_k)$ and boundary condition (20) and $J_{u^*}(t,Y_k,i)$ is the solution of (23) with control $u^*(t,Y_k)$ and boundary condition (24).

APPENDIX

VII. UNNORMALIZED CONDITIONAL PROBABILITIES

In this appendix, x is a jump process and y defined by $y(t) = g(x(t))$ is the observed process. To shorten the notation slightly we shall assume that the conditional distributions $P[\tau_{n+1}\leq t,x_{n+1}=j \mid X_n]$ have the form

$$P[\tau_{n+1}\leq t,x_{n+1}=j \mid X_n]$$
$$=\sum_i 1_{\{\tau_n\leq t,x_n=i\}}\int_{\tau_n}^{t} a_{ij}(s,Y_k)\ \exp\left[\int_{\tau_n}^{s} a_{ii}(r,Y_k)dr\right]ds\ . \tag{40}$$

Thus (9) is of this form with

$$a_{ij}(s,Y_k) = a_{ij}(s,u(s,Y_k)) \ .$$

The following four formulas which are consequences of (40) are important for later work.

$$E\,[1_{\{\tau_n \le t < \tau_{n+1}\}} \mid X_n] = \sum_i 1_{\{\tau_n \le t, x_n = i\}} \exp\left[\int_{\tau_n}^{t} a_{ii}(r,Y_k)dr\right] \tag{41}$$

$$\begin{aligned} &P[\tau_{n+1} \le t, x_{n+1} = j \mid X_n] \\ &= \sum_i 1_{\{\tau_n \le t, x_n = i\}} \int_{\tau_n}^{t} a_{ij}(s,Y_k)\ E[1_{\{\tau_n \le s < \tau_{n+1}\}} \mid X_n]ds \end{aligned} \tag{42}$$

$$\begin{aligned} &1_{\{\bar\tau_k \le \tau_n < \bar\tau_{k+1}\}}\ P[\tau_{n+1} \le t, x_{n+1} = j \mid X_n] \\ &= \sum_i 1_{\{\bar\tau_k \le \tau_n < \bar\tau_{k+1}, x_n = i\}} \int_{\bar\tau_k}^{t} a_{ij}(s,Y_k)\ E[1_{\{\tau_n \le s < \tau_{n+1}\}} \mid X_n]ds \end{aligned} \tag{43}$$

$$\begin{aligned} &E\,[1_{\{x_n = i, \bar\tau_k \le \tau_n < \bar\tau_{k+1}, \tau_n \le t < \tau_{n+1}\}} \mid X_n] \\ &= 1_{\{x_n = i, \bar\tau_k \le \tau_n \le \bar\tau_{k+1}, \tau_n \le t\}} \\ &\quad + 1_{\{x_n = i, \bar\tau_k \le \tau_n \le \bar\tau_{k+1}\}} \int_{\bar\tau_k}^{t} a_{ii}(s,Y_k)\ E[1_{\{\tau_n \le s < \tau_{n+1}\}} \mid X_n]ds \end{aligned} \tag{44}$$

To obtain (41) since $1_{\{\tau_n \le t\}}$ is $\sigma(X_n)$ measurable

$$\begin{aligned} E\,[1_{\{\tau_n \le t < \tau_{n+1}\}} \mid X_n] &= 1_{\{\tau_n \le t\}}\ E\,[1_{\{t < \tau_{n+1}\}} \mid X_n] \\ &= 1_{\{\tau_n \le t\}}[1 - P[\tau_{n+1} \le t \mid X_n]] \ . \end{aligned} \tag{45}$$

Using (40) and (8) and integrating gives

$$P\,[\tau_{n+1} \le t \mid X_n] = \sum_i 1_{\{\tau_n \le t, x_n = i\}}\left[1 - \exp\left[\int_{\tau_n}^{t} a_{ii}(r,Y_k)dr\right]\right]. \tag{46}$$

Substituting (46) in (45) now gives (41).

To obtain (42) notice from (41) that

$$
\begin{aligned}
&\sum_i 1_{\{\tau_n\le t,x_n=i\}} \int_{\tau_n}^{t} a_{ij}(s,Y_k)\ E[1_{\{\tau_n\le s<\tau_{n+1}\}}\ |\ X_n]ds \\
&= \sum_i 1_{\{\tau_n\le t,x_n=i\}} \int_{\tau_n}^{t} a_{ij}(s,Y_k) \sum_m 1_{\{\tau_n\le s,x_n=m\}} \exp\left[\int_{\tau_n}^{s} a_{mm}(r,Y_k)dr\right] ds \\
&= \sum_i 1_{\{\tau_n\le t,x_n=i\}} \int_{\tau_n}^{t} a_{ij}(s,Y_k)\ 1_{\{\tau_n\le s,x_n=i\}} \exp\left[\int_{\tau_n}^{s} a_{ii}(r,Y_k)dr\right] ds \\
&= \sum_i 1_{\{\tau_n\le t,x_n=i\}} \int_{\tau_n}^{t} a_{ij}(s,Y_k) \exp\left[\int_{\tau_n}^{s} a_{ii}(r,Y_k)dr\right] ds\ .
\end{aligned} \tag{47}
$$

The next to the last equality in (47) follows because to have $1_{\{\tau_n\le t,x_n=i\}}1_{\{\tau_n\le s,x_n=m\}}\neq 0$ we must have $i=m$. The last equality follows because on the set $\{\tau_n\le t,x_n=i\}$ for s in the range of integration $\tau_n \le s \le t$ we have $1_{\{\tau_n\le s,x_n=i\}}=1$.

To obtain (43) notice from (42), since

$$1_{\{\tau_n\le s\}}\ E[1_{\{\tau_n\le s<\tau_{n+1}\}}|X_n] = E[1_{\{\tau_n\le s<\tau_{n+1}\}}\ |\ X_n]\ , \tag{48}$$

that

$$
\begin{aligned}
&1_{\{\bar\tau_k\le\tau_n<\tau_{k+1}\}}\ P[\tau_{n+1}\le t,x_{n+1}=j\ |\ X_n] \\
&= 1_{\{\bar\tau_k\le\tau_n<\bar\tau_{k+1}\}} \sum_i 1_{\{\tau_n\le t,x_n=i\}} \int_{\tau_n}^{t} 1_{\{\tau_n\le s\}}\ a_{ij}(s,Y_k)\ E[1_{\{\tau_n\le s<\tau_{n+1}\}}\ |\ X_n]ds\ .
\end{aligned} \tag{49}
$$

The lower limit τ_n in (49) may be replacod by $\bar\tau_k$, because on the set $\{\bar\tau_k\le\tau_n\}$, $1_{\{\tau_n\le s\}}$ is zero for $\bar\tau_k \le s < \tau_n$. In addition

$$1_{\{\tau_n\le t,x_n=i\}}\ 1_{\{\tau_n\le s\}} = 1_{\{\tau_n\le s\}}\ 1_{\{x_n=i\}}\ . \tag{50}$$

Thus (43) follows from (49).

Sincc

$$P[\tau_n\le t<\tau_{n+1}\ |\ X_n] = 1_{\{\tau_n\le t\}}(1 - P[t\le\tau_{n+1}\ |\ X_n]) \tag{51}$$

(44) follows from (43) and (8).

We shall often use a simple but very important generalization of the law of iterated conditional expectations stated by Makowski in [3], Theorem A1, which applies in the case in which there may not be global inclusion between two sigma fields.

THEOREM: Let σ and σ' be two σ-fields. Let A be a set which is in both σ and σ' and for which

$$\sigma \cap A \subset \sigma' \cap A\ . \tag{52}$$

Then if z is a random variable for which

$$E|z| < \infty$$

it follows that

$$E\ [1_A\ E[z\ |\sigma']\ |\sigma] = E\ [1_A\ z\ |\sigma]\ . \tag{53}$$

In particular the set $\{\bar\tau_k\le\tau_n\}$ is both $\sigma(X_n)$ and $\sigma(Y_k)$ measurable and

$$1_{\{\bar\tau_k\le\tau_n\}}\ \sigma(Y_k) \subset 1_{\{\tau_k\le\tau_n\}}\ \sigma(X_n)\ . \tag{54}$$

Thus

$$E[E[1_{\{\tau_k \le \tau_n\}}] \, E[z \mid X_n] \mid Y_k] = E\,[1_{\{\tau_k \le \tau_n\}} \, z \mid Y_k] \, . \tag{55}$$

Call (53) or (55) the generalized law of iterated conditional expectations

THEOREM A1: Given the history Y_k of the first k jumps of the observed process, the conditional probability that the next observed jump $\bar{\tau}_{k+1}$ occurs before t and that at this time the unobserved process x takes on the value j is given by

$$P\,[\bar{\tau}_{k+1} \le t, x(\bar{\tau}_{k+1}) = j \mid Y_k]$$

$$= \begin{cases} \displaystyle\int_{\bar{\tau}_k \wedge t}^{t} \sum_i Q_i(s,Y_k)\, a_{ij}(s,Y_k)ds & \text{if} \quad g(j) \ne y_k \\ 0 & \text{if} \quad g(j) = y_k \end{cases} \tag{56}$$

The form of the conditional probability distribution of the observed process and the conditional probabilities of the unobserved process at the time of an observed jump are given in the following two corollaries which follow immediately from Theorem A1.

COROLLARY A1:

$$P[\bar{\tau}_{k+1} \le t, y_{k+1} = \ell \mid Y_k]$$

$$= \int_{\bar{\tau}_k \wedge t}^{t} \sum_{\{j:g(j)=\ell\}} \sum_i Q_i(s,Y_k)\, a_{ij}(s,Y_k)ds \, .$$

COROLLARY A2:

$$P[x(\bar{\tau}_{k+1}) = j \mid Y_{k+1}]$$

$$= \begin{cases} \dfrac{\sum_i Q_i(\tau_{k+1},Y_k)\, a_{ij}(\tau_{k+1},Y_k)}{\sum_{\{j:g(j)=y_{k+1}\}} \sum_i Q_i(\tau_{k+1},Y_k)\, a_{ij}(\tau_{k+1},Y_k)} & \text{if} \quad g(j) = y_{k+1} \\ 0 & \text{if} \quad g(j) \ne y_{k+1} \end{cases}$$

PROOF OF THEOREM A1: Since $\bar{\tau}_{k+1}$ is a jump time of the observed process, the state j that the unobserved process jumps to must satisfy $g(j) \ne y_k$. Thus it is impossible at time $\bar{\tau}_{k+1}$ for the unobserved process to be in a state j for which $g(j) = y_k$, and the conditional probability (56) is 0 for these states. In the remainder of the proof assume that $g(j) \ne y_k$.

Since $g(j)\neq y_k$ and since $\{\bar{\tau}_k\}$ are the jumps of the observed process y and $\{\tau_n\}$ the jump of the unobserved process x, we have that

$$\{\bar{\tau}_{k+1}\le t, x(\bar{\tau}_{k+1})=j\} = \bigcup_n \{\bar{\tau}_k\le\tau_n<\bar{\tau}_{k+1}, \tau_{n+1}\le t, x_{n+1}=j\} . \tag{59}$$

Thus the monotone convergence theorem for conditional expectations implies

$$P[\bar{\tau}_{k+1}\le t, x(\bar{\tau}_{k+1})=j \mid Y_k] = \sum_n P[\bar{\tau}_k\le\tau_n<\bar{\tau}_{k+1}, \tau_{n+1}\le t, x_{n+1}=j \mid Y_k] . \tag{60}$$

The generalized law of iterated conditional expectations implies

$$P[\bar{\tau}_k\le\tau_n<\bar{\tau}_{k+1}, \tau_{n+1}\le t, x_{n+1}=j \mid Y_k]$$

$$= E\,[1_{\{\bar{\tau}_k\le\tau_n\}}\; E\,[1_{\{\bar{\tau}_k\le\tau_n<\bar{\tau}_{k+1}, \tau_{n+1}\le t, x_{n+1}=j\}} \mid X_n] \mid Y_k] . \tag{61}$$

The $\sigma(X_n)$ measurability of $1_{\{\bar{\tau}_k\le\tau_n<\bar{\tau}_{k+1}\}}$ and (42) imply

$$E\,[1_{\{\bar{\tau}_k\le\tau_n<\bar{\tau}_{k+1}, \tau_{n+1}\le t, x_{n+1}=j\}} \mid X_n]$$

$$= 1_{\{\bar{\tau}_k\le\tau_n<\bar{\tau}_{k+1}\}}\; E\,[1_{\{\tau_{n+1}\le t, x_{n+1}=j\}} \mid X_n] \tag{62}$$

$$= \sum_i 1_{\{\bar{\tau}_k\le\tau_n<\bar{\tau}_{k+1}, x_n=i\}} \int_{\bar{\tau}_k}^{t} a_{ij}(s,Y_k)\; E[\tau_n\le s<\tau_{n+1} \mid X_n]\, ds .$$

Thus substituting (62) in (61) and then (61) in (60) interchanging integration, summation and conditional expectation; using the generalized law of iterated conditional expectations; and using that

$$\bigcup_n \{\bar{\tau}_k<\tau_n<\bar{\tau}_{k+1}, x_n=i, \tau_n\le s<\tau_{n+1}\} = \{\bar{\tau}_k\le s<\bar{\tau}_{k+1}, x(s)=i\} \tag{63}$$

gives

$$P[\bar{\tau}_{k+1}\le t, x(\bar{\tau}_{k+1})=j \mid Y_k]$$

$$= \int_{\bar{\tau}_k}^{t} \sum_i \left\{E\,[1_{\{\bar{\tau}_k\le s<\bar{\tau}_{k+1}, x(s)=i\}} \mid Y_k]\; a_{ij}(s,Y_k)\right\} ds \tag{64}$$

$$= \int_{\bar{\tau}_k}^{t} \sum_i Q_i(s,Y_k)\; a_{ij}(s,Y_k)\, ds ,$$

which is the conclusion of Theorem A1.

THEOREM 1: If $g(i)=y_k$, on the interval $\bar{\tau}_k \le t < T$, $Q_i(t,Y_k)$ satisfies the system of differential equations

$$\frac{d}{dt} Q_i(t,Y_k) = a_{ii}(t,u(t,Y_k))\; Q_i(t,Y_k)$$

$$+ \sum_{\{j:g(i)=g(j)\}} a_{ji}(t,u(t,Y_k))\; Q_j(t,Y_k) \tag{19}$$

and the boundary condition

$$Q_i(\bar{\tau}_k,Y_k) = P[x(\bar{\tau}_k)=i \mid Y_k] \tag{20}$$

is satisfied. If $g(i)\neq y_k$, $Q_i(t,Y_k)=0$.

PROOF OF THEOREM 1: If $g(i)\neq y_k$, the sets involved in the definition of $Q_i(t,Y_k)$ have an empty intersection, and $Q_i(t,Y_k)$ is the conditional expectation of a zero random variable and hence is zero. Assume in the remainder of the proof that $g(i)=y_k$.

The monotone convergence theorem for conditional expectations implies

$$E\,[1_{\{x(t)=i\}}\,1_{\{\bar{\tau}_k\le t<\bar{\tau}_{k+1}\}} \mid Y_k] = \sum_n E\,[1_{\{x(t)=i,\bar{\tau}_k\le t<\bar{\tau}_{k+1},\tau_n\le t<\tau_{n+1}\}} \mid Y_k]\,. \tag{65}$$

Since $\bar{\tau}_k$ are the jumps of the observed process y and τ_n the jumps of x the following set equality holds:

$$\{\bar{\tau}_k\le t<\bar{\tau}_{k+1},\tau_n\le t<\tau_{n+1}\} = \{\bar{\tau}_k\le\tau_n<\bar{\tau}_{k+1},\tau_n\le t<\tau_{n+1}\}\,. \tag{66}$$

The generalized law of iterated conditional expectations (66) and (44) imply

$$\begin{aligned}
&E\,[1_{\{x(t)=i,\bar{\tau}_k\le t<\bar{\tau}_{k+1},\tau_n\le t<\tau_{n+1}\}} \mid Y_k]\\
&= E[1_{\{x_n=i,\bar{\tau}_k\le\tau_n<\bar{\tau}_{k+1},\tau_n\le t<\tau_{n+1}\}} \mid Y_k]\\
&= E[1_{\{\bar{\tau}_k\le\tau_n\}}\,E\,[1_{\{x_n=i,\bar{\tau}_k\le\tau_n<\bar{\tau}_{k+1},\tau_n\le t<\tau_{n+1}\}} \mid X_n] \mid Y_k]\\
&= E\,[1_{\{x_n=i,\bar{\tau}_k\le\tau_n<\bar{\tau}_{k+1},\tau_n\le t\}} \mid Y_k]\\
&\quad + E\,[1_{\{x_n=i,\bar{\tau}_k\le\tau_n<\bar{\tau}_{k+1}\}} \int_{\bar{\tau}_k}^{t} a_{ii}(s,Y_k)\,E\,[1_{\{\tau_n\le s<\tau_{n+1}\}} \mid X_n]ds \mid Y_k]\,.
\end{aligned} \tag{67}$$

Writing $\{\bar{\tau}_k\le\tau_n\} = \{\bar{\tau}_k=\tau_n\} \cup \{\bar{\tau}_k\le\tau_{n-1}\}$, we have

$$\begin{aligned}
&E\,[1_{\{x_n=i,\bar{\tau}_k\le\tau_n<\bar{\tau}_{k+1},\tau_n\le t\}} \mid Y_k]\\
&= 1_{\{\bar{\tau}_k\le t\}}\,E\,[1_{\{x_n=i,\bar{\tau}_k=\tau_n\}} \mid Y_k]\\
&\quad + E\,[1_{\{x_n=i,\bar{\tau}_k\le\tau_{n-1}<\bar{\tau}_{k+1},\tau_n<\bar{\tau}_{k+1},\tau_n\le t\}} \mid Y_k]\,.
\end{aligned} \tag{68}$$

Notice since we are assuming $g(i)=y_k$, that $x_n=i$ and $\tau_{n-1}<\bar{\tau}_{k+1}$ imply that $\tau_n<\bar{\tau}_{k+1}$. Thus the condition $\tau_n<\bar{\tau}_{k+1}$ may be dropped in the last term of (68). Now by the generalized law of iterated conditional expectations and (42)

$$\begin{aligned}
&E\,[1_{\{x_n=i,\bar{\tau}_k\le\tau_{n-1}<\bar{\tau}_{k+1},\tau_n\le t\}} \mid Y_k]\\
&= E\,[1_{\{\bar{\tau}_k\le\tau_{n-1}\}}\,E\,[1_{\{\bar{\tau}_k\le\tau_{n-1}<\bar{\tau}_{k+1},\tau_n\le t,x_n=i\}} \mid X_{n-1} \mid Y_k]\\
&= E\left[\sum_j 1_{\{\bar{\tau}_k\le\tau_{n-1}<\bar{\tau}_{k+1},x_{n-1}=j\}} \int_{\bar{\tau}_k}^{t} a_{ji}(s,Y_k)\,E[1_{\{\tau_{n-1}\le s<\tau_n\}} \mid X_n]ds \mid Y_k\right]\,.
\end{aligned} \tag{69}$$

Combining (69), (68) and (67), using the generalized law of iterated conditional expectations, interchanging integration, summation and conditional expectation gives

$$E\,[1_{\{x(t)=i,\bar{\tau}_k\le t<\bar{\tau}_{k+1},\tau_n\le t<\tau_{n+1}\}}\mid Y_k]$$

$$= 1_{\{\tau_k\le t\}}\, E\,[1_{\{x_n=i,\bar{\tau}_k=\tau_n\}}\mid Y_k]$$
$$+ \int_{\bar{\tau}_k}^{t} \sum_j a_{ji}(s,Y_k)\, E\,[1_{\{\bar{\tau}_k\le\tau_{n-1}<\bar{\tau}_{k+1},\tau_{n-1}\le s<\tau_n,x_{n-1}=j\}}\mid Y_k]\, ds \tag{70}$$
$$+ \int_{\tau_k}^{t} a_{ii}(s,Y_k)\, E\,[1_{\{\bar{\tau}_k\le\tau_n<\bar{\tau}_{k+1},\tau_n\le s<\tau_{n+1},x_n=i\}}\mid Y_k]\, ds\ .$$

Using (66) and summing (70) over n gives

$$E\,[1_{\{x(t)=i,\bar{\tau}_k\le t<\bar{\tau}_{k+1}\}}\mid Y_k]$$

$$= 1_{\{\tau_k\le t\}}\, E\,[1_{\{x(\bar{\tau}_k)=i\}}\mid Y_k]$$
$$+ \int_{\bar{\tau}_k}^{t} \sum_j a_{ji}(s,Y_k)\, E\,[1_{\{\bar{\tau}_k\le s<\bar{\tau}_{k+1},x(s)=j\}}\mid Y_k]\, ds \tag{71}$$
$$+ \int_{\bar{\tau}_k}^{t} a_{ii}(s,Y_k)\, E\,[1_{\{\bar{\tau}_k\le s<\bar{\tau}_{k+1},x(s)=i\}}\mid Y_k]\, ds\ .$$

Using the definition (18) of $Q_i(t,Y_k)$ we have from (71) that

$$Q_i(t,Y_k) = 1_{\{\bar{\tau}_k\le t\}}\, P[x(\bar{\tau}_k)=i\mid Y_k]$$
$$+ \int_{\bar{\tau}_k}^{t} \left[a_{ii}(s,Y_k)\, Q_i(s,Y_k) + \sum_j a_{ji}(s,Y_k)\, Q_j(s,Y_k)\right] ds \tag{72}$$

from which Theorem 1 follows by differentiation.

VIII. CONDITIONAL REMAINING COSTS

THEOREM A2: If z is any $\sigma\{\tau_{n+1},x_{n+1},\ldots,\tau_{n+k},x_{n+k},\ldots\}$-measurable random variable for which $E|z| < \infty$, then $E[z \mid X_n]$ can be expressed as a function of x_n and Y_k where Y_k is the observed history corresponding to X_n.

PROOF: It can be shown using (40) that the theorem is true for the characteristic function of any $\sigma(\tau_{n+1},x_{n+1},\ldots,\tau_{n+k},x_{n+k})$-measurable set. The class of random variables for which Theorem A2 is true is closed under monotone limits and products. Therefore, the theorem follows from the monotone class theorem.

Theorem 2 follows by applying Theorem A2 to both the numerator and denominator in the definition (21) of $J_u(t,X_n)$.

THEOREM 3: On the interval $\bar{\tau}_k \le t \le T$, $J_u(t,Y_k,i)$ satisfies the system of differential equations

$$\begin{aligned}\frac{d}{dt} J_u(t,Y_k,i) &= -c(t,i,u(t,Y_k)) \\ &- \sum_{\{j:g(i)=g(j)\}} J_u(t,Y_k,j)\, a_{ij}(t,u(t,Y_k)) \\ &- \sum_{\{j:g(i)\neq g(j)\}} J_u(t,Y_k,t,g(j),j)\, a_{ij}(t,u(t,Y_k)) \\ &- J_u(t,Y_k,i)\, a_{ii}(t,u(t,Y_k)) \; .\end{aligned} \tag{23}$$

At time T the boundary condition

$$J_u(T,Y_k,i) = 0 \tag{24}$$

is satisfied.

PROOF OF THEOREM 3: Breaking the integral in (73) into the integral between t and the next jump τ_{n+1} and using that between the n-th and (n+1)-st jump, $x(t)=x_n$, and $u(t)=u(t,Y_k)$ and then using the formulas (41) and (56) giving the distribution of τ_{n+1} and x_{n+1}, we have

$$\begin{aligned}&E\left[1_{\{\tau_n\le t<\tau_{n+1}\}} \int_t^T c(s,x(s),u(s))ds \mid X_n\right] \\ &= E\left[1_{\{\tau_n\le t<\tau_{n+1}\}}\left(\int_t^{T\wedge\tau_{n+1}} c(s,x(s),u(s))ds + \int_{T\wedge\tau_{n+1}}^T c(s,x(s),u(s))ds\right) \mid X_n\right] \\ &= E\left[1_{\{\tau_n\le t\}} \int_t^T \sum_i 1_{\{\tau_n\le s<\tau_{n+1},x_n=i\}}\, c(s,i,u(s,Y_k))ds \mid X_n\right] \\ &\quad + E\left[1_{\{\tau_n\le t<\tau_{n+1}\}}\, E\left[\int_{T\wedge\tau_{n+1}}^T c(s,x(s),u(s))ds \mid X_{n+1}\right] \mid X_n\right] \\ &= 1_{\{\tau_n\le t\}} \int_t^T \sum_i 1_{\{x_n=i\}} \exp\left[\int_{\tau_n}^s a_{ii}(r,Y_k)dr\right] c(s,i,u(s,Y_k))ds \\ &\quad + E\left[1_{\{\tau_n\le t<\tau_{n+1}\}} J_u(\tau_{n+1},X_{n+1}) \mid X_n\right] \\ &= \sum_i 1_{\{\tau_n\le t,x_n=i\}} \int_t^T \exp\left[\int_{\tau_n}^s a_{ii}(r,Y_k)dr\right] c(s,i,u(s,Y_k))ds \\ &\quad + \sum_i 1_{\{\tau_n\le t,x_n=i\}} \int_t^T \sum_j J_u(\tau_{n+1},X_n,\tau_{n+1},j)\, a_{ij}(\tau_{n+1},Y_k) \exp\left[\int_{\tau_n}^{\tau_{n+1}} a_{ii}(r,Y_k)dr\right] ds \; .\end{aligned} \tag{73}$$

Now since

$$J_u(t,X_n) = \frac{E\left[1_{\{\tau_n \le t < \tau_{n+1}\}} \int_t^T c(s,x(s),u(s))ds \mid X_n\right]}{E\left[1_{\{\tau_n \le t < \tau_{n+1}\}} \mid X_n\right]}$$

and from (41)

$$E\left[1_{\{\tau_n \le t < \tau_{n+1}\}} \mid X_n\right] = \sum_i 1_{\{\tau_n \le t, x_n = i\}} \exp\left[\int_{\tau_n}^t a_{ii}(r,Y_k)dr\right] . \tag{41}$$

Dividing (73) by (41) gives

$$J_u(t,X_n) = \sum_i 1_{\{\tau_n \le t, x_n = i\}} \int_t^T \exp\left[\int_t^s a_{ii}(r,Y_k)dr\right] c(s,i,u(s,Y_k))ds$$

$$+ \sum_i 1_{\{\tau_n \le t, x_n = i\}} \int_t^T \sum_j J_u(s,X_n,s,j)a_{ij}(s,Y_k) \exp\left[\int_t^s a_{ii}(r,Y_k)dr\right] ds . \tag{74}$$

From Theorem 2 we have that if Y_k is the observed history corresponding to X_n and $x_n=i$ that

$$J_u(t,X_n) = J_u(t,Y_k,i)$$

and if $X_{n+1} = (X_n,t,j)$ and $g(j)=y_k$, then Y_k is the observed history corresponding to X_{n+1} and

$$J_u(t,X_{n+1}) = J_u(t,Y_k,j)$$

and if $X_{n+1} = (X_n,t,j)$ and $g(j) \ne y_k$, then $Y_{k+1} = (Y_k,t,g(j))$ is the observed history corresponding to X_{n+1} and

$$J_u(t,X_{n+1}) = J_u(t,Y_k,t,g(j),j) .$$

Making these substitutions in (74) gives for $t \ge \tau_n$ that

$$\begin{aligned} J_u(t,Y_k,i) &= \int_t^T \exp\left[\int_t^s a_{ii}(r,Y_k)dr\right] c(s,i,u(s,Y_k))ds \\ &+ \int_t^T \left[\sum_{\{j:g(j)=y_k\}} J_u(t,Y_k,j)a_{ij}(s,Y_k) + \sum_{\{j:g(j)\ne y_k\}} J_u(s,Y_k,s,g(j),j)a_{ij}(s,Y_k)\right] \exp\left[\int_t^s a_{ii}(r,Y_k)dr\right] ds . \end{aligned} \tag{75}$$

Theorem 3 follows by differentiating (75) with respect to t.

THEOREM 4: The conditional remaining cost from an observed jump time $\bar{\tau}_k$ onward (or from time $\bar{\tau}_0=0$ onward if k=0) conditioned with respect to the observed history Y_k satisfies

$$\begin{aligned} &E\left[\int_{\bar{\tau}_k}^T c(s,x(s),u(s))ds \mid Y_k\right] \\ &= \int_{\bar{\tau}_k}^T \sum_i \left[c(t,i,u(t,Y_k)) + \sum_{\{j:g(j)\ne g(i)\}} J_u(t,Y_k,t,g(j),j)a_{ij}(t,u(t,Y_k))\right] Q_i(t,Y_k)dt \end{aligned} \tag{25}$$

and also satisfies

$$E\left[\int_{\bar\tau_k}^{T} c(s,x(s),u(s))ds \mid Y_k\right] = \sum_i J_u(\tau_k,Y_k,i)\ P[x(\bar\tau_k)=i \mid Y_k] \ . \tag{26}$$

PROOF OF THEOREM 4:

$$E\left[\int_{\bar\tau_k}^{T} c(s,x(s),u(s))ds \mid Y_k\right]$$

$$= E\left[\int_{\bar\tau_k}^{T\wedge\bar\tau_{k+1}} c(s,x(s),u(s))ds + \int_{T\wedge\bar\tau_{k+1}}^{T} c(s,x(s)u(s))ds \mid Y_k\right] . \tag{76}$$

Using the generalized law of iterated conditional expectations

$$E\left[\int_{T\wedge\bar\tau_{k+1}}^{T} c(s,x(s),u(s))ds \mid Y_k\right]$$

$$= E\left[E\left[\sum_n 1_{\{\tau_{k+1}=\tau_n\}}\ E\left[\int_{T\wedge\bar\tau_{k+1}}^{T} c(s,x(s),u(s))ds \mid X_n\right] \mid Y_{k+1}\right] \mid Y_k\right] . \tag{77}$$

Now

$$E\left[\int_{T\wedge\bar\tau_{k+1}}^{T} c(s,x(s),u(s))ds \mid X_n\right] = J_u(\bar\tau_{k+1},X_n) \tag{78}$$

and on the set $\{\bar\tau_{k+1}=\tau_n\}$, Y_{k+1} is the observed history corresponding to X_n. Thus by Theorem 2

$$\sum_n 1_{\{\tau_{k+1}=\tau_n\}}\ J_u(\bar\tau_{k+1},X_n) = \sum_n\sum_j 1_{\{\tau_{k+1}=\tau_n,x_n=j\}}\ J_u(\bar\tau_{k+1},Y_{k+1},j)$$

$$= \sum_j 1_{\{x(\bar\tau_{k+1})=j\}}\ J_u(\bar\tau_{k+1},Y_{k+1},j) \ . \tag{79}$$

Hence from (76),(77),(78),(79)

$$E\left[\int_{\bar\tau_k}^{T} c(s,x(s),u(s))ds \mid Y_k\right]$$

$$= E\left[\int_{\bar\tau_k}^{T} 1_{\{\bar\tau_k\le s<\bar\tau_{k+1}\}}\ c(s,x(s),u(s))ds \mid Y_k\right] + E\left[\sum_j 1_{\{x(\tau_{k+1})=j\}}\ J_u(\bar\tau_{k+1},Y_{k+1},j) \mid Y_k\right]$$

$$= E\int_{\bar\tau_k}^{T} \sum_i E\left[1_{\{\bar\tau_k\le s<\bar\tau_{k+1},x(s)=i\}} \mid Y_k\right] c(s,i,u(s,Y_k))ds \tag{80}$$

$$+ \int_{\bar\tau_k}^{T} \sum_j J_u(s,Y_k,s,g(j),j) \sum_i Q_i(s,Y_k)\ a_{ij}(s,Y_k)\ ds$$

$$= \int_{\bar\tau_k}^{T} \sum_i \left[c(s,i,u(s,Y_k)) + \sum_{\{j:g(j)\neq y_k\}} J_u(s,Y_k,s,g(j))\ a_{ij}(s,Y_k)\right] Q_i(s,Y_k)ds$$

which gives (25).

To obtain (26), notice using the generalized law of iterated conditional expectations and Theorem 2 that

$$
\begin{aligned}
&E\left[\int_{\bar{\tau}_k}^{T} c(s,x(s),u(s))ds \mid Y_k\right] \\
&= E\left[\sum_n 1_{\{\bar{\tau}_k=\tau_n\}} E\left[\int_{\bar{\tau}_k}^{T} c(s,x(s),u(s))ds \mid X_n\right] \mid Y_k\right] \\
&= E\left[\sum_n \sum_i 1_{\{\tau_k=\tau_n, x_n=i\}} J_u(\bar{\tau}_k, Y_k, i) \mid Y_k\right] \qquad (81) \\
&= E\left[\sum_i 1_{\{x(\tau_k)=i\}} J_u(\bar{\tau}_k, Y_k, i) \mid Y_k\right] \\
&= \sum_i P[x(\bar{\tau}_k)=i \mid Y_k] \, J_u(\bar{\tau}_k, Y_k, i) \, .
\end{aligned}
$$

REFERENCES

[1] R. Boel, P. Varaiya, "Optimal Control of Jump Processes," SIAM Journal on Control and Optimization," Vol. 15 (1977).

[2] M. Kohlman, A. Makowski, R. Rishel, "Representation Results for Jump Processes with Application to Optimal Stopping," Stochastics, Vol. 4 (1980).

[3] A. Makowski, "Local Optimality Conditions for Optimal Stopping," Stochastics 1982.

[4] R. Rishel, "A Minimum Principle for Controlled Jump Processes," Springer Lecture Notes in Economics and Mathematical Systems, Vol. 107 (1975).

[5] R. Rishel, "The Minimum Principle, Separation Principle and Dynamic Programming for Partially Observed Jump Processes," IEEE Transactions on Automatic Control, Volume AC-23 (1978).

[6] R. Rishel, "Optimality for Completely Observed Jump Processes," IEEE Transactions on Automatic Control, Vol. AC-22 (1977).

[7] R. Rishel, "State Estimation for Partially Observed Jump Processes," Journal of Mathematical Analysis and Applications, Vol. 65 (1978).

[8] M. Rudemo, "State Estimation for Partially Observed Markov Chains," Journal of Mathematical Analysis and Applications, Vol. 44 (1973).

[9] A. Segal, "Optimal Control of a Finite State Markov Process," IEEE Transactions on Automatic Control, Vol AC-22 (1977).

ON NORMAL APPROXIMATION IN BANACH SPACES

V. V. Sazonov

Steclov Mathematical Institute

Moscow, U.S.S.R.

During recent years considerable progress was made in the study of the accuracy of normal approximation given by the central limit theorem in Banach spaces. In this talk I intend to survey some important new results in the area. Consider first a somewat simplier case of Hilbert space.

Let $X_1, X_2, \dots$ be a sequence of independent identically distributed (i.i.d.) random variables with values in a separable Hilbert space $\mathcal{H}$. The inner product and norm in $\mathcal{H}$ will be denoted by $(\cdot,\cdot)$ and $|\cdot|$ respectively. Suppose that

$$E|X_j|^2 < \infty, \; EX_j = 0,$$

and let V be the covariance operator of X_j, i.e.

$$E(X_j, h)^2 = (Vh, h), \quad h \in \mathcal{H};$$

recall that V is linear bounded nonnegative symmetric operator with finite trace $\operatorname{tr} V$. Denote

$$S_n = n^{-1/2} \sum_1^n X_j$$

and let Y be a Gaussian random variable with values in $\mathcal{H}$ and with $EY = 0$ and covariance operator V.

The Central Limit Theorem states that the probability measures P_n induced on $\mathcal{H}$ by S_n converge weakly to the Gaussian probability Q induced on $\mathcal{H}$ by Y.

We are interested in the estimation of

$$D_n(\mathcal{A}) = \sup\{|P_n(A) - Q(A)|;\ A \in \mathcal{A}\},$$

where $\mathcal{A}$ is some class of Borel subsets of $\mathcal{H}$. It is well known (see e.g. [15]) that contrary to the finite dimensional case $D_n(\mathcal{A})$ in general do not tend to zero as $n \to \infty$ if $\mathcal{A}$ is rich enough, e.g. if $\mathcal{A}$ contains the class of all halfspaces or the class of all balls. In what follows we will be concerned mainly with the case when $\mathcal{A} = \mathcal{S}_h$, the class of all balls with a fixed centre h.

The first estimates of $D_n(\mathcal{S}_h)$ (apart from logarithmic speed obtained in [17]) were of the type $O(n^{-\alpha})$, $\alpha < 1/2$, and were valid when $E|X_j|^m < \infty$ for some $m \geqslant 3$ (depending on the estimate) (see [11], [13], [16] and Yurinskii's theorem in [15]). Under some additional (rather restrictive) conditions on the probability measure induced by X_1 better speeds were obtained in [4] and [5].

Denote λ_k the kth eigenvalue of V (we assume that eigenvalues of V are numbered in the decreasing order). In 1979 F.Götze [7] proved that if $E|X_1|^6 < \infty$ and $\lambda_{31} > 0$ then

$$D_n(\mathcal{S}_0) = O(n^{-1/2});$$

moreover, if $E|X_1|^8 < \infty$ and $\lambda_{53} > 0$ then

$$D_n(\mathcal{S}_0) = O(n^{-1+\varepsilon}) \qquad (1)$$

for any $\varepsilon > 0$. Later F.Götze was able to show (see [10]) that (1) is true if $E|X_1|^4 < \infty$ and $\lambda_j > 0,\ j \geqslant 1,\ \sum_{j=m}^{\infty} \lambda_j = O(m^{-p})$ for some $p > 0$.

In 1981 V.V.Yurinskii [18-19] under the assumption that $X_j,\ j = 1, 2, \ldots$ are independent (but not necessary identically distributed), $EX_j = 0$, all X_j have the same covariance matrix V, proved that

$$D_n(\mathcal{S}_h) \leq c \sup_{1 \leq j \leq n} E|X_j|^3 \left(1 + (|h|(\operatorname{tr} V)^{-1/2})^3\right) n^{-1/2}, \qquad (2)$$

where c depends only on the eigenvalues of $V(\operatorname{tr} V)^{-1}$. Note that V.V. Yurinskii also generalized this result to spaces ℓ_{2k} where k is a positive integer. In this more general case the result has the same form only power 3 in square brackets in (2) is replaced by 3(2k-1).

Recently B.A.Zalesskii was able to extend and partially improve these results: assuming $E|X_j|^{3+\delta}<\infty$, $0\le\delta<1$, he proved that

$$D_n(\mathcal{S}_0)=o(n^{-(1+\delta)/2});$$

in particular if $E|X_j|^4<\infty$ then for any fixed $\varepsilon>0$

$$D_n(\mathcal{S}_0)=o(n^{-1+\varepsilon})$$

(see [20] ,[21]).

In the same paper B.A.Zalesskii pointed out that the speed in (2) is the best possible; namely for any $h\neq 0$ he constructed a sequence $\{X_j\}$ with $E|X_j|^m<\infty$ for any $m>0$ such that

$$\liminf n^{1/2}D_n(\mathcal{S}_h)>0.$$

He also proved that there exist a sequence $\{X_j\}$ with $E|X_j|^4<\infty$ such that

$$\liminf nD_n(\mathcal{S}_0)>0.$$

Two main methods were employed to obtain the abovementioned estimates. One method consists in approxomating the indicator function $\chi_{S_r(h)}$ of the ball $S_r(h)=\{x\in H:|x-h|<r\}$ by a smooth function $f_{r,h}(x)$ thus replacing the estimation of

$$(P_n-Q)(S_r(h))=\int\chi_{S_r(h)}(x)(P_n-Q)(dx)$$

by the estimation of

$$\int f_{r,h}(x)(P_n-Q)(dx). \qquad (3)$$

To estimate (3) the difference P_n-Q. is represented as

$$P_n-Q=\Big(\sum_{j=0}^{n-1}P_{(n)}^{*j}*Q_{(n)}^{*(n-j-1)}\Big)*(P_{(n)}-Q_{(n)}),$$

where $P_{(n)}$ is the probability measure induced by $n^{-1/2}X_1$, $Q_{(n)}$ is normal $(0, n^{-1/2}V)$. Thus one has to estimate terms of the type

$$\int f_{r,h}(x) P_{(n)}^{*j} * Q_{(n)}^{*(n-j-1)} * (P_{(n)} - Q_{(n)})(dx) = \int f_{r,h}(x+y) P_{(n)}^{*j} * Q_{(n)}^{*(n-j-1)}(dx)(P_{(n)} - Q_{(n)})(dy).$$

For this purpose $f_{r,h}(x+y)$ is expanded by Taylor's formula with respect to y . Since probability measures $P_{(n)}, Q_{(n)}$ have the same first and second moments, the integration of the first three terms of the expantion gives zero. These are the main lines of the first method. The best speed obtained by applying it is $n^{-1/6}$. However the method permitted to construct estimates with explicit dependence on the parameters of the probability measure induced by X_1 (see [15]).

The second method consists in application of the Fourier analysis to the measure on the real line induced by $|S_n - h|^2$. Usually the random variables $X_1, \ldots, X_n$ are at first truncated, i.e. replaced by $X_1', \ldots, X_n'$, where $X_j' = X_j \chi_{\{|X_j| \le f(n)\}}$ (χ_E is the indicator function of E). The function $f(n)$ is chosen in such a way that

$$D_n'(J_h) = \sup\{|P(|S_n - h| < r) - P(|S_n' - h| < r)|, r > 0\},$$

where $S_n' = n^{-1/2}\sum_1^n X_j'$, is small enough. The characteristic function f_n' of $|S_n' - h|^2$ is smooth and one may apply successfully the Fourier analysis. By an Esseen inequality we have

$$D_n'(J_h) \le c\left(\int_{-T}^{T} |f_n'(t) - \varphi(t)||t|^{-1} dt + cT^{-1}\right),$$

where $\varphi(t)$ is the characteristic function of $|Y - h|^2$. To estimate $|f_n'(t) - \varphi(t)|$ for small values of t an expansion of $f_n' - \varphi$ similar to the expansion of $P_n - Q$ in the first method described above is used. Then (as for larger t) Götze-Yurinskii estimates of the characteristic functions such as f_n' (see [7], [18]) are applied. E.G. for any $A > 0$, $\varepsilon_1, \varepsilon_2 > 0$ with $\varepsilon_1 + \varepsilon_2 \le 1$ if $|t| \le c_1 (\min(\varepsilon_1, \varepsilon_2) n \ln n)^{-1/2}$ then

$$|f_n'(t)| \le (c_2 n^{-A} + E\exp\{ic_3 t_0(\varepsilon_1, \varepsilon_2)^{1/2} n^2 (\eta_1, \eta_2)\})^{1/4},$$

where $t_0 = \min\{|t|, (\max(\varepsilon_1, \varepsilon_2) n \ln n)^{-1/2}\}$ and η_1, η_2 are independent centered Gaussian $\mathcal{H}$-valued random variables with the same covariance operator as X_j' and constants c_j are independent of h and n.

In special cases and under additional assumptions better estimates may be obtained. Consider, for example, $\mathcal{H} = L_2(0,1)$ and let

$$X_j = \delta(x - \xi_j) - x, \quad 0 \le x \le 1,$$

where $\xi_j, j = 1, 2, \dots$ are independent and uniformly distributed on $[0,1]$, $\delta(u) = 1$ if $u > 0$, $= 0$ if $u \le 0$. Note that in this case

$$|S_n|^2 = \int_0^1 \Big(n^{-1/2} \sum_{j=1}^{n} (\delta(x - \xi_j) - x)\Big)^2 dx$$
$$= n \int_0^1 (U_n(x) - x)^2 dx$$
$$= W_n^2,$$

where $U_n(x)$ is the empirical distribution function corresponding to $\xi_1, \dots, \xi_n$, and thus

$$D_n(\mathcal{S}_0) = \sup\{|P(W_n < r) - F(r)|, r > 0\}$$

where $F(r)$ is the limiting distribution function for $P(W_n < r)$. In other words the estimation of $D_n(\mathcal{S}_0)$ in this special case means the estimation of the speed of convergence in ω^2-criterion in statistics. Recently Yu.V.Borovskih [3] proved that

$$|P(W_n < r) - F(r)| \le c(1 + r^2)^{-1} n^{-1}.$$

There exist several generalizations of these results to the case of random variables with values in a separable Banach spase B (see e.g. [1], [2], [6], [8], [9], [12] - [14]. Here we will mention two recent results. The first one belongs to F.Götze [9] and is more general also in the sence that it deals with a larger class of sets then balls. Suppose $X_1, X_2, \dots$ are i.i.d. random variables with values in B and such that $E\|X_1\|^4 < \infty$ ($\|\cdot\|$ is the norm in B), $EX_1 = 0$. Assume also that the probability measure P_n induced by $n^{-1/2}\sum_1^n X_j$ converges weakly to some Gaussian Borel measure Q on B . Let now F be a real functional on B such that

(i) F is 4 times continuously Fréchet differentiable and

$$\|F^{(j)}(x)\| \le C_F(1+\|x\|)^r \quad , \ 1 \le j \le 4$$

$$\|F^{(4)}(x) - F^{(4)}(y)\| \le C_F(1+\|x\|+\|y\|)^r \|x-y\|^\beta$$

for any $x, y \in B$ and some $C_F > 0$, $r \geq 0$, $1 \geq \beta > 0$,

(ii) $Q\{x : E(F^{(1)}(x)(X_1))^2 \le \varepsilon\} = O(\varepsilon^{c/\beta})$

as $\varepsilon \downarrow 0$ where c is large enough,

then

$$\sup\{|(P_n - Q)(x : F(x) \le r)| ; r > 0\} = O(n^{-1/2}). \tag{4}$$

Note that in Hilbert space the condition (i) on $F(x) = |x|$ is satisfied. If B is an arbitrary Banach space this result can be applied to $F(x) = \|\cdot\|$ only if $\|\ \|$ is smooth which is not always true. E.g. the supremum norm in $B = C(S)$, the space of continuous real functions on a compact metric space S , is not Fréchet differentiable. We will mention here a recent result for $C(S)$-valued random variables obtained by V.Bentkus [1]. Let ρ be a continuous pseudometric on S , $N_\rho(S)$ be the smallest number of open ρ-balls of radius ε covering S , and let $H_\rho = \ln N_\rho(\varepsilon)$ be the metric entropy of S . For $\varphi \in C(S)$ denote

$$|\varphi|_\rho = \|\varphi\| + \sup\{|\varphi(s) - \varphi(t)| / \rho(s,t) ; s,t \in S\}$$

and let

$$\theta(\varepsilon) = \int_0^\varepsilon H_\rho^{1/2}(u)\,du, \ \varepsilon > 0, \quad \rho_\theta(s,t) = \theta(\rho(s,t)).$$

As before $X_1, X_2, \ldots$ are i.i.d. random variables with values in $C(S)$, $EX_1(s) = 0$ for any $s \in S$ and $S_n = n^{-1/2}\sum_1^n X_j$. Then if $\theta(1) < \infty$

$$A = E|X_1|_\rho^3 + E|X_1|_{\rho_\theta}^3 < \infty$$

$$N_\rho(\varepsilon) \le c_1 \varepsilon^{-\alpha},$$

$\alpha > 0$, and for some $c_2 > 0$ and all $r, \tau > 0$

$$P(r-\tau \le \|Y\| \le r+\tau) \le c_2 \tau,$$

Y being a Gaussian $C(S)$-valued random variable with $EX_1(s)X_1(t) = EY(s)Y(t)$ for any $s, t \in S$, we have

$$D(s_0) = \sup\{|P(\|S_n\| < r) - P(\|Y\| < r)|;\ r > 0\}$$

$$\le c(\rho)\, c_2^{3/4} A^{1/4} n^{-1/8} \ln n. \tag{5}$$

Inequality (4) is obtained by a method similar to the Fourier method in Hilbert space outlined above. Estimate (5) is obtained by applying a finite dimentional approximation and using appropriate finite dimentional estimates.

References

1. Bentkus, V., Estimates of the speed of convergence in the central limit theorem in $C(S)$, to appear in Dokl. Akad. Nauk SSSR.

2. Bernotas, V. and Paulauskas, V., Non-uniform estimates in the central limit theorem in some Banach spaces, Litovsk. Mat. Sb., 19(1979), 23-43.

3. Borovskih, Yu., Estimates of characteristic functions with applications to ω^2-ststistics, to appear in Teor. Verojatnost. i Primenen.

4. Borovskih, Yu. and Račkauskas, A., Asymptotic behaviour of distributions in Banach spaces, Litovsk. Mat. Sb., 19(1979), 39-54.

5. Čebotarev, V.I., Estimates of the speed of convergence in the central limit theorem in ℓ_p , Sibirsk. Mat. Z., 20(1974), 1099-1116.

6. Gene, E.M., Bounds of the speed of convergence in the central limit theorem in $C(S)$, Z. Wahrsch. verw. Geb. 36(1976), 317-331.

7. Götze, F., Asymptotic expansions for bivariate von Mises functional Z. Wahrsch. verw. Geb. 50(1979), 333-355.

8. ———, On Edgeworth expansions in Banach spaces, Ann. Probab., 9(1981), 852-859.

9. ———, On the rate of convergence in the central limit theorem in Banach spaces, Preprints in Statistics, University of Cologne, 68(1981), pp. 1-34.

10. ———, Convergence rate in the central limit theorem in Hilbert space, 14th European Meeting of Statisticeans, Wroclaw 1981, Abstracts, p. 35.

11. Kuelbs, J., Kurtz, T., Berry-Esseen estimates in Hilbert space and an application to the law of the iterated logarithm, Ann. Probab., 2(1974), 387-407.

12. Paulauskas, V., The estimate of the rate of convergence in the central limit theorem in $C(S)$, Litovsk. Mat. Sb., 16(1976), 167-201.

13. ———, On the rate of convergence in the central limit theorem in some Banach spaces, Teor. Verojatnost. i Primenen., 21 (1976), 775-791.

14. ———, Paulauskas, V., On convergence of some functionals of sums of independent random variables in a real Banach space, Litovsk. Mat. Sb., 16(1979), 103-121.

15. Sazonov, V.V., Normal approximation - some recent advances. Lecture Notes in Math. vol.879, Springer, Berlin (1981).

16. Ulyanov, V.V., On the estimation of the rate of convergence in the central limit theorem in a real separable Hilbert space, Mat. Zametki, 28(1980), 465-473.

17. Vakhania, N.N. and Kandelaki, N.P., On the estimation of the speed of convergence in the central limit theorem in Hilbert space, Trans. Comput. Centre Acad. Sci. Georgian SSR., 10:1 (1969), 150-160.

18. Yurinskii, V.V., On the accuracy of Gaussian approximation for the probability of hitting a ball, Teor. Verojatnost. i Primenen., 2(1982), 270-278.

19. ———, Estimate of the accuracy of normal approximation of the probability of hitting a ball, Dokl. Akad. Nauk SSSR, 258 (1981), 557-558.

20. Zalesskii, B.A., Estimates of the accuracy of normal approximation in a Hilbert space, Teor. Verojatnost. i Primenen., 2(1982), 279-285.

21. ———, On the rate of convergence in the central limit theorem in Hilbert space, to appear in Dokl. Acad. Nauk SSSR.

A CLASS OF PROBLEMS IN THE OPTIMAL CONTROL OF DIFFUSIONS WITH FINITELY MANY CONTROLS

Diane D. Sheng
Bell Laboratories
Holmdel, New Jersey 07733
USA

1. INTRODUCTION

In this paper we consider a class of problems in which the process to be controlled is fundamentally a Brownian motion on the nonnegative real line with either absorption or reflection at the origin. At each point in time one of a finite number of controls must be employed, and the controls chosen influence the evolution of the Brownian motion process. Operational costs and linear holding costs are continuously incurred at a rate depending on the control in use and the state of the Brownian motion process. Lump-sum switching costs are incurred instantaneously each time there is a switch in controls, and these switching costs are dependent on the two controls involved in the switch. Further, if the barrier at the origin is absorbing there is a cost for hitting the boundary. We allow a general set of control strategies where the current control may depend in an arbitrary, measurable way on past states and past controls. The objective is to determine a strategy for selecting a control at each point in time, so as to minimize the total expected cost discounted over an infinite planning horizon. These problems are related to those studied by Doshi [4,5], Mandl [10], and Pliska [13], differences being in the form of the holding costs or switching costs; and also are related to work by Arkin, Kolemaev, and Shiryaev [1], Beneš [2], Davis and Varaiya [3], and Fleming [6], differences being in the controllable nature of the drift coefficient and/or diffusion coefficient of the diffusion processes. Additionally, our focus emphasizes explicit production of optimal control strategies.

Diffusion models in general, and controlled Brownian motion, in particular, have been used by Foschini [7], Iglehart [8], and others in modeling water reservoirs, cash management inventories and other input-output systems. As an application of the work presented here, consider a service facility with infinite capacity that can be operated at different service rates. (For example, the rate at which a packet switch transports data might be varied by changing the number or speed of outgoing transmission channels.) Suppose the costs associated with this service facility are a linear customer-holding cost, an operational cost charged at a rate dependent on the prevailing service rate, and switching costs charged with each change in service rate. The problem is to choose the service rate at each point in time so as to minimize the service facility's total expected costs. The process to be controlled, then is the number of customers in the service facility and indeed, it is often reasonable to represent the number of customers in a queueing system by a Brownian motion process. (This is because, with sufficient congestion, the discreteness of individual customers becomes relatively unimportant and is justified by limit theorems; see, for example, Rath [14].)

It is conjectured here that there always exists an optimal Markov control strategy that is of a particular band form, but this has been shown only for two avaiable controls. Necessary and sufficient conditions are given for such a "band strategy" to be optimal, and optimal band strategies are explicitly produced for a few special cases.

2. FORMULATION

The data of our control problems are the set $A = \{1,2,\ldots,N\}$ of available actions or control modes, and real constants r_i, μ_i, and σ_i for each $i \in A$. These constants are the operational cost rate, drift parameter, and variance parameter, respectively, for control mode i. We assume that $\sigma_i > 0$ for all $i \in A$. Also given are nonnegative switching costs K_{ij} for $i,j \in A$ which satisfy

(2.1) $K_{ii} = o$ for each $i \in A$, and

(2.2) $K_{ij} \leq K_{ia} + K_{aj}$ for each $i,j,a \in A$.

Finally, there is a holding cost rate h, a boundary cost R, an interest rate $\alpha > 0$, and a boundary parameter $\lambda \in \{0,1\}$.

The state of the system at time t is denoted by X(t) for $t \geq 0$, and the state space by $S = [0,\infty)$. Whenever control mode i is employed, the state of the system changes locally like a Brownian motion with infinitesimal drift μ_i and infinitesimal variance σ_i^2. At the origin the process is absorbed if $\lambda = 0$, or instantaneously reflected if $\lambda = 1$.

There are continuous costs incurred at each point in time which are a function of the current state of the system and current action. Denote these cost by $g\colon S \times A \to \mathbb{R}$ where

$$g(x,i) = \begin{cases} hx + r_i & \text{if } x > 0 \\ (1-\lambda)\alpha R & \text{if } x = 0. \end{cases}$$

Additionally, the lump sum switching costs K_{ij} are incurred instantaneously whenever there is a change from control mode i to mode j. Condition (2.2) guarantees that is cheaper to switch directly from action i to j, than to switch from i to any other intermediary action and then to j. Both the continuous and switching costs are discounted so that a total cost of C incurred at time t is equivalent to a cost of $Ce^{-\alpha t}$ incurred at time zero.

Based on the switching costs, we can divide up the action space into disjoint equivalence classes. For each $i \in A$, let $\theta(i) = \{j \in A: K_{ij} = K_{ji} = 0\}$ denote the set of all control modes that can be reached from i and from which i can be reached with zero switching costs. Denote the M disjoint equivalence classes by $A_1, A_2, \ldots, A_M$. It costs nothing to switch in either direction among actions in any one equivalence class, and there is a positive switching cost for switching in some direction between actions in different equivalence classes. Let $C_{k\ell}$ for each $k, \ell \in \{1,2,\ldots,M\}$ represent the switching costs between equivalence classes. That is,

(2.3) $C_{k\ell} = K_{ij}$ for each $i \in A_k$, $j \in A_\ell, k, \ell \in \{1,2,\ldots,M\}$,

(2.4) $C_{kk} = 0$ for each $k \in \{1,2,\ldots,M\}$, and

(2.5) $C_{k\ell} \leq C_{km} + C_{m\ell}$ for each $k, \ell, m \in \{1,2,\ldots,M\}$.

Start with a probability space (Ω, F, P) on which is defined a standard Brownian motion $B = \{B(t); t \geq 0\}$, and let $\{F_t; t \geq 0\}$ be the increasing family of sub-σ-fields generated by B. We allow the following class of control strategies.

Definition. An admissible strategy is a function

$\pi: \Omega \times (0,\infty) \to A$ s.t.

(2.6) $\pi(\omega,t)$ is jointly measurable in ω and t,

(2.7) $\pi(;t)$ if F_t measurable for each $t > 0$, and

(2.8) the function $\theta^*(\omega,t) = \theta(\pi(\theta,t))$ has only finitely many discontinuities in each finite interval of time.

Hereafter, we suppress the dependence of π on ω, and augment π by including at time zero an initial action $\pi(0) = i$.

We now associate with each admissible strategy π, initial state x, and initial action i, a controlled Itô process. For each admissible strategy π, $x \in S$, and $i \in A$, let X and Y be the unique pair of nonanticipating process s.t.

(2.9) $$X(t) = x + \int_0^t \mu_{\pi(\mu)}\, du + \int_0^t \sigma_{\pi(u)}\, dB(u) + Y(t) \quad \text{for each } t \geq 0, \text{ and}$$

(2.10) $Y(\cdot)$ is continuous and nondecreasing with $Y(0) = 0$ and grows only when $X(t) = 0$.

That is, $X(t) = Z(t) + Y(t)$ where

$$Z(t) = x = \int_0^t \mu_\pi(u)\, du + \int_0^t \sigma_\pi(u)\, dB(u), \text{ and}$$

$$Y(t) = \left[-\inf_{0 \leq u \leq t} Z(u) \right]^+ \text{ for each } t \geq 0.$$

Denote the <u>controlled process</u> generated by π, x, and i by $\{X(t|x,i,\pi); t \geq 0\}$. These controlled processes are precisely as follows: If $\lambda = 1$ then $X(t|x,i,\pi) = X(t)$, and if $\lambda = 0$ then $X(t|x,i,\pi) = X(t \wedge T)$ where $X(t)$ is defined in (2.9) and $T = \inf\{t \geq 0: X(t) = 0\}$.

<u>Remarks</u>. (1) Even though an admissible strategy, is a measurable function of the Brownian path, it can be viewed as a function of the history of the action process and controlled process. However, if early actions were in some sense "randomized" later actions would not be contained in the Brownian σ-fields. This issue of action randomization is not considered here.

(2) We need not require a fixed initial state x. Instead we could fix an inital distribution for X(0) such that X(0) has finite quadractic mean.

We now associate with each admissible strategy π, a value function V_π, which given initial state $x \in S$ and initial control mode $i \in A$ represents the expected total discounted cost generated by π, x, and i. For each pair of equivalence classes k and ℓ in $\{1,2,\ldots,M\}$ define $\{Q_{k\ell}(t|x,i,\pi); t \geq 0\}$ to be the counting process corresponding to the number of switches made under strategy π from within class k to class ℓ during [0,t]. The <u>value function</u> for strategy π is then defined on $S \times A$ by

$$V_\pi(x,i) = E\left[\int_0^\infty e^{-\alpha t}[g(X(t|x,i,\pi), \pi(t))]dt + \sum_{k=1}^M \sum_{\ell=1}^M \int_0^\infty C_{k\ell} e^{-\alpha t}\, dQ^*_{k\ell}(t|x,i,\pi)\right]$$

where $Q^*_{k\ell}(\cdot|x,i,\pi)$ is the counting measure associated with counting process $Q_{k\ell}(\cdot|x,i,\pi)$, and where by convention

$$\int_0^t C_{k\ell} dQ^*_{k\ell}(u|x,i,\pi) = 0 \quad \text{if} \quad k = \ell.$$

<u>Lemma 1</u>. For an admissible strategy π and for each $i \in A$, $|V_\pi(x,i) - hx/\alpha|$ is bounded for each $x \in S$ if and only if $E\left[\int_0^\infty e^{-\alpha t} dQ^*_{k\ell}(t|x,i,\pi)\right] < \infty$ for each k,ℓ in $\{1,2,\ldots,M\}$ where $C_{k\ell} > 0$.

Proof. In the case of absorption

$$E\left[\left|\int_0^\infty e^{-\alpha t} hX(t|x,i,\pi)dt - hx/\alpha\right|\right] \le \frac{|h|\mu^*}{\alpha^2}$$

where $\mu^* = \max_i\{|\mu_i|\}$, and in the case of reflection

$$E\left[\left|\int_0^\infty e^{-\alpha t} hX(t|x,i,\pi)dt - hx/\alpha\right|\right] \le \frac{2|h|\mu^*}{\alpha^2} + \frac{|h|\sqrt{\sigma^*}}{\alpha\sqrt{2\alpha}}$$

where $\sigma^* = \max\{\sigma_i^2\}$; see Sheng [16]. Therefore

$E\left[\left|\int_0^\infty e^{-\alpha t} g(X(t|x,i,\pi), \pi(t))dt - hx/\alpha\right|\right]$ is bounded for each x and i,

and $|V_\pi(x,i) - {}^{hx}/\alpha|$ is likewise bounded if and only if

$E\left[\sum_{k=1}^{M}\sum_{\ell=1}^{M}\int_0^\infty C_{k\ell}e^{-\alpha t}dQ_{k\ell}^*(t|x,i,\pi)\right]$ is. □

The optimal value function $V_*: S \times A \to R$ is defined by $V_*(x,i) = \inf_\pi V_\pi(x,i)$ where the infinum is taken over all admissible strategies. The admissible strategy π is called (x,i)-optimal if $V_\pi(x,i) = V_*(x,i)$. Given initial state $x \in S$ and initial action $i \in A$, our control problem is to construct an admissible strategy that is (x,i)-optimal.

3. OPTIMAL BAND STRATEGIES

Of special interest is a subset of admissible strategies which we call band strategies. Each band stratety is based on a band function which dictates changes in control mode as a function of the current state of the system and current control.

Definition. A band function is a function $f: S \times A \to A$ s.t.

(2.11) for each $i \in A$, $f(x,i)$ has finitely many discontinuities in $x \in S$ and $f(o,i) = \lim_{x\downarrow o} f(x,i)$,

(2.12) for each $(x,i) \in S\times A$, $f(x,f(x,i)) = f(x,i)$

(2.13) for each $i \in A$, the class continuation set $I_i = \{x \in S: f(x,i) \in \theta(i)\}$ is an open subset of S, and

(2.14) if y is a closed boundary point of the action continuation set $I^i = \{x \in S: f(x,i) = i\}$ for some $i \in A$, then there exists $j \in \theta(i)$ where $j = \lim_{\substack{x\to y \\ x\notin I^i}} f(x,i)$ and $f(y,j) = i$.

The state space can be divided into a finite number of intervals such that for $i \in A$, $f(x,i)$ is constant across any particular interval. Condition (2.12) guarantees that it will not be possible to make more than one change in control mode at any one point in time. For each $i \in A$, the action continuation set defined in (2.14) denotes those states for which f will continue with control mode i, and the class continuation set defined in (2.13) denotes those states for which f will continue with an action from equivalence class $\theta(i)$. Set I_i is the finite union of intervals open in S and set I^i is the finite union of intervals open or closed in I_i.

For example, suppose that we are given three contol modes with $K_{12} = K_{21} = 0$, $K_{13} = K_{23} > 0$, and $K_{31} = K_{32} > 0$. Then $A_1 = \{1,2\}$, $A_2 = \{3\}$, $C_{12} = K_{13} = K_{23}$, and $C_{21} = K_{31} = K_{32}$. In Figure 1 we illustrate the following band function:

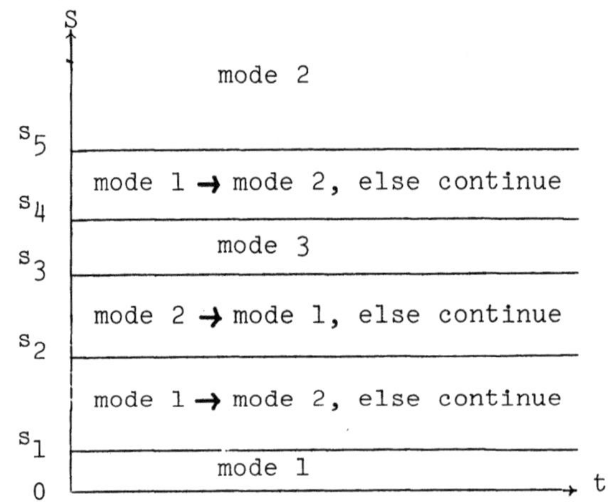

FIGURE 1. AN ILLUSTRATIVE BAND FUNCTION

$$f(x,1) = \begin{cases} 1 & \text{if} \quad x \in [0,s_1] \\ 2 & \text{if} \quad x \in (s_1,s_2] \\ 1 & \text{if} \quad x \in (s_2,s_3) \\ 3 & \text{if} \quad x \in [s_3,s_4] \\ 2 & \text{if} \quad x \in (s_4,\infty), \end{cases}$$

$$f(x,2) = \begin{cases} 1 & \text{if} \quad x \in [0,s_1] \\ 2 & \text{if} \quad x \in (s_1,s_2] \\ 1 & \text{if} \quad x \in (s_2,s_3) \\ 3 & \text{if} \quad x \in [s_3,s_4] \\ 2 & \text{if} \quad x \in (s_4,\infty), \quad \text{and} \end{cases}$$

$$f(x,3) = \begin{cases} 1 & \text{if} \quad x \in [0,s_1] \\ 3 & \text{if} \quad x \in (s_1,s_5) \\ 2 & \text{if} \quad x \in [s_5,\infty) \end{cases} .$$

Note that $I^1 = [0,s_1]\cup(s_2,s_3)$, $I^2 = (s_1,s_2]\cup(s_4,\infty)$, $I^3 = (s_1,s_5)$, $I_1 = I_2 = [0,s_3)\cup(s_4,\infty)$, $I_3 = (s_1,s_5)$, and that (2.14) holds at s_1 and s_2.

We now introduce some more notation concerning band functions. For each $i \in A$, let $0 = s_0^i < s_1^i <\ldots< s_{n(i)}^i < s_{n(i)+1}^i = +\infty$ be the discontinuity points in S of $f(x,i)$. Further, denote by b_p^i for $p = 0,1,2,\ldots,n(i)$, the constant value of $f(x,i)$ on (s_p^i,s_{p+1}^i).

The following theorem shows that given an initial state and control mode, each band function corresponds to a unique admissible strategy.

Theorem 1. If f is a band function and $(x,i) \in S\times A$, then there exists a unique admissible strategy π s.t.

(3.1) $\pi(t) = f(X(t|x,i,\pi), \pi(t))$ a.s. for each $t > 0$.

Proof. Starting from initial state x and action i, we construct admissible strategy π and its associated controlled process $X(\cdot|x,i,\pi)$, that together satisfy (3.1), by sequentially doing so on intervals of the sort $[\tau_n,\tau_{n+1})$ where $\{\tau_n;n=0,1\ldots\}$ is a sequence of increasing Brownian stopping times.

Let $a_1 = f(x,i)$ and $p \in \{0,1,\ldots,n(a_1)\}$ be s.t. $x \in \left[s_p^{a_1},s_{p+1}^{a_1}\right]$ and $b_p^{a_1} = a_1$. Define $\{\xi_1(t);\ t \geq 0\}$ to be the unique reflected Brownian motion process beginning in state x with infinitesimal drift μ_{a_1} and infinitesimal variance $\sigma_{a_1}^2$. Let ξ_1 be the first hitting time of $\left\{s_p^{a_1}, s_{p+1}^{a_1}\right\}$ by ξ_1 and begin construction of action process π and state process X by specifying that $\pi(t) = a_1$ and $X(t) = \xi_1(t)$ on $(0,\tau_1)$.

Suppose that $\xi_1(\tau_1) = s_p^{a_1}$. Now either i) $s_p^{a_1}$ is a closed boundary point of I^{a_1}, ii) $s_p^{a_1}$ is an open boundary point of I^{a_1} but not a boundary point of I_{a_1}, of iii) $s_p^{a_1}$ is a boundary point of I_{a_1}.

Cases i) and ii) differ very slightly and will now be handled together. Let $a_2 \in \theta(a_1)$ and $\bar{p} \in \{0,1,2,\ldots,n(a_2)\}$ be s.t. $b_{\bar{p}}^{a_2} = a_2$, $s_{\bar{p}+1}^{a_2}$, =

$s_p^{a_1}$, $f(s_{\bar{p}+1}^{a_2}, a_2) = a_1$ for i), $f(s_{\bar{p}+1}^{a_1}, a_1) = a_2$ for ii). A result of Nakao [12] guarantees the pathwise unique existence of the process $\{\psi_2(t);\ t \geq \tau_1\}$ s.t.

$$\psi_2(t) = s_p^{a_1} + \int_{\tau_1}^{t} \mu\left(\psi_2(u)\right)du + \int_{\tau_1}^{t} \sigma\left(\psi_2(u)\right)dB(u) \qquad \text{for } t \geq \tau_1$$

where

$$\mu(y) = \begin{cases} \mu_{a_2} & \text{if} \quad y < s_p^{a_1} \\ \mu_{a_1} & \text{if} \quad y > s_p^{a_1} \end{cases},$$

$$\sigma(y) = \begin{cases} \sigma_{a_2} & \text{if} \quad y < s_p^{a_1} \\ \sigma_{a_1} & \text{if} \quad y > \sigma_p^{a_1} \end{cases},$$

$\mu(s_p^{a_1}) = \mu_{a_1}$ and $\sigma(s_p^{a_1}) = \sigma_{a_1}$ for i), and $\mu(s_p^{a_1}) = \mu_{a_2}$ and $\sigma(s_p^{a_1}) = \sigma_{a_2}$ for ii). This is to be distinguished from uniqueness in distribution, and the importance lies in the fact that the solution process ψ_2 can be constructed on our Brownian motion process, so that $\psi_2(t)$ is a unique measureable function of $\{B(u);\ \tau_1 \leq u \leq t\}$; see Watanabe and Yamada [17]. Let $\{\xi_2(t);\ t \geq \tau_1\}$ be the uniquely reflected ψ_2 process and define τ_2 to be the first hitting time of $\left\{s_{p+1}^{a_1}, s_{\bar{p}}^{a_2}\right\}$ by the process ξ_2. We further extend our definitions of π and X by

$$\pi(t) = \begin{cases} a_2 & \text{if} \quad \xi_2(t) < s_p^{a_1} \\ a_1 & \text{if} \quad \xi_2(t) > s_p^{a_1} \end{cases},$$

$\pi(t) = a_1$ if $\xi_2(t) = s_p^{a_1}$ for i), and $\pi(t) = a_2$ if $\xi_2(t) = s_p^{a_1}$ for ii), and $X(t) = \xi_2(t)$ on $[\tau_1, \tau_2)$.

For iii), let $a_2 = f(s_p^{a_1}, a_1)$ and $\bar{p} \in \{0,1,2,\ldots,n(a_2)\}$ be s.t. $b_{\bar{p}}^{a_2} = a_2$ and $s_p^{a_1} \in \left[s_{\bar{p}}^{a_2}, s_{\bar{p}+1}^{a_2}\right]$. If $s_{\bar{p}}^{a_2} < s_p^{a_1} < s_{\bar{p}+1}^{a_2}$, we proceed as in the second paragraph of this proof, thereby further defining π and X on $[\tau_1, \tau_2)$ where $\tau_2 > \tau_1$ a.s. If $s_p^{a_1} = s_{\bar{p}}^{a_2}$ or $s_{\bar{p}+1}^{a_2}$, we proceed as in the previous paragraph.

Similarly, we handle all of the possibilities where $\xi_1(\tau_1) = s_{p+1}^{a_1}$. Process π and X will then have been defined from time zero until stopping time τ_2 with (3.1) satisfied on $(0,\tau_2)$. Following the same procedures we next define the process ξ_3, stopping time $\tau_3 > \tau_2$ a.s., and $\pi(t) = f(X(t), \pi(t))$ on $[\tau_2,\tau_3)$ where $X(t) = \xi_3(t)$.

In this way we continue to build the action process π and state process X, and if the sequence of stopping times $\{\tau_n;\ n\geq 1\}$ approaches infinity almost surely, then $\pi(t)$ and $X(t)$ are well-defined for all $t \geq 0$. Observing that there exists a minimal bandwidth

$$\min_{i\in A}\ \min_{p\in\{0,1,\ldots,n(i)\}} \left(s^i_{p+1} - s^i_p\right)$$

of positive width and using the arguments of Appendix 4 in [15], we conclude that $E[\tau_{n+1}-\tau_n] \geq b \geq 0$ for all n and some constant b. This implies that $\tau_n \to \infty$ a.s., and furthermore, that the function $\theta^*: (0,\infty) \to \{1,2,\ldots,M\}$ defined by $\theta^*(t) = \theta(\pi(t))$ has (a.s.) finitely many discontinuities in each finite interval of time. Therefore, π is an admissible strategy and the controlled process uniquely generated by π, x, and i, $\{X(t|x,i,\pi);\ t \geq 0\}$, is precisely $X(t|x,i,\pi) = X(t)$ if $\lambda = 1$ and $X(t|x,i,\pi) = X(t\wedge T)$ if $\lambda = 0$ where $t = \inf\{t \geq 0: X(t) = 0\}$.

We define the band strategies generated by a band function f (one strategy for each initial state and action pair) as the unique solutions constructed in Theorem 1. In order to characterize the value function corresponding to a band strategy we will need the following extension Itô's lemma which is proved in [16].

Lemma 2. Let X and Y be a pair processes s.t.

(3.2) $$X(t) = X(0) + \int_0 \sigma(u)dB(u) + Y(t) \quad \text{for each } t \geq 0$$

where $X(0)$ is F_0-measurable, σ is a nonzero, bounded and non-anticipating, and Y is a continuous, non-anticipating process of bounded variation. Let $f: [0,\infty) \times \mathbb{R} \to \mathbb{R}$ satisfy

(3.3) $f(t,\cdot) \in C^2_*(\mathbb{R})$ for each $t \geq 0$, and

(3.4) $f(\cdot,x)$ is continuously differentiable on $[0,\infty)$ for each $x \in \mathbb{R}$.

Then the process $\{f(t,X(t));\ t \geq 0\}$ satisfies

(3.5) $$f(t,X(t)) = f(0,X(0)) + \int_0^t \left[f_u(u,X(u) + \tfrac{1}{2}f_{xx}(u,X(u))\sigma^2(u)\right]du + \int_0^t f_x(u,X(u))\sigma(u)dB(u) + \int_0^t f_x(u,X(u))dY(u) \quad \text{for each } t \geq 0.$$

Remarks. (1) We adopt the notation $C^2_*(\mathbb{R})$ to denote the set of continuous functions $f: \mathbb{R} \to \mathbb{R}$ such that in each finite interval of $\mathbb{R}$, f' and f'' exist at all but a finite number of points; and such that $f''(x-)$ and $f''(x+)$ exist for each $x \in \mathbb{R}$.

(2) The proof of Lemma 3 relies heavily on a similar result, Theorem

2.2 in Kunita and Watanabe [9], the difference being that our class of functions include those that misbehave at a finite number of points in each finite spatial interval.

We now show the value function of a band strategy to be the solution to a standard differential equation.

Theorem 2. For each band function f and state-action pair $(x,i) \in S\times A$, let $\pi_{f,x,i}$ denote the admissible strategy uniquely corresponding to f,x, and i. The function V_f: $S\times A \to \mathbb{R}$ defined as $V_f(x,i) = V_{\pi_{f,x,i}}(x,i)$ is the unique function V: $S\times A \to \mathbb{R}$ satisfying the following for each $i \in A$:

(3.6) $$V(\cdot,i) \in C^2_*(S),$$

(3.7) $\left|V(x,i) - \frac{hx}{\alpha}\right|$ is bounded for all $x \in S$,

(3.8) $D_iV(x,i) - \alpha V(x,i) + g(x,i) = 0$ for all x in the interior of I^i, where D_i denotes the differential operator $D_i = \mu_i \frac{\partial}{\partial x} + \frac{1}{2}\sigma_i^2 \frac{\partial^2}{\partial x^2}$,

(3.9) $V(x,i) = K_{i,f(x,i)} + V(x,f(x,i))$ for all $x \notin I^i$, and

(3.10) $\lambda V'(0,i) - (1-\lambda)V(0,i) + (1-\lambda)R = 0.$

Proof. We begin by verifying that V_f satisfies (3.6)-(3.10). By Lemma 1, condition (3.7) holds if

(3.7') $$E\left[\int_0^\infty e^{-\alpha t} dQ_{k\ell}(t|x,i,\pi_{f,x,i})\right] < \infty$$

for each $(x,i) \in S\times A$ and distinct $k,\ell \in \{1,2,\ldots,M\}$. Fix $(x,i) \in S\times A$ and define $\{Y_n: n \geq 0\}$ by $Y_0 = 0$ and $Y_n = \tau_n - \tau_{n-1}$ for $n \geq 1$. where the sequence of stopping times $\{\tau_n; n \geq 0\}$ associated with f, x, i is defined as in the proof of Theorem 1. We have already established that the Y_n's are independent and that $E[Y_n] \geq b \geq 0$ for each n and some constant b. Define $\{S_n; n \geq 0\}$ by $S_n = \sum_{j=0}^{n} Y_j$ for $n \geq 0$ and $\{N(t); t \geq 0\}$ by $N(t) = \sup\{n: S_n \leq t\}$ for $t > 0$. Since

$$S_{n(t)+1} = \sum_{k=0}^{\infty}\sum_{n=0}^{k} Y_n I_{\{N(t)+1=k\}} = \sum_{n=0}^{\infty}\sum_{k=n}^{\infty} Y_n I_{\{N(t)+1=k\}},$$

we have

$$E\left[S_{n(t)+1}\right] = \sum_{n=0}^{\infty} E\left[Y_n I_{\{N(t)+1\geq n\}}\right] \geq bE[N(t) + 1].$$

Since $S_{n(t)+1} \leq t + Y_{n(t)+1}$, we get $bE[N(t)] \leq t - b + E[Y_{n(t)+1}]$. Hence $\limsup_{t\to\infty}[E[N(t)]/t] < 1/b$. Now fix distinct $k,\ell \in \{1,2,\ldots,M\}$.

Since every switch from an action A_k to an action A_ℓ occurs only at some time point τ_n, we have $Q_{k\ell}(t|x,i,\pi_{f,x,i}) \leq N(t)$ and therefore (3.7') holds.

Fix $i \in A$ and consider $x \in I^i$. Let p be s.t. $x \in [s_p^i, s_{p+1}^i]$ and let $\{Z_i(t);\ t \geq 0\}$ be the Brownian motion starting in state x with drift parameter μ_i, variance parameter σ_i, and absorption at s_p^i and s_{p+1}^i. Assign to process Z_i linear holding costs at rate h, operational costs at rate r_i, and absorption costs $K_{i,f(s_p^i,i)} + V_f(s_p^i, f(s_p^i,i))$ and $K_{i,f(s_{p+1}^i,i)} + V_f(s_{p+1}^i, f(s_{p+1}^i,i))$ at points s_p^i and s_{p+1}^i, respectively. Let $V_i(x)$ denote the conditional expectation of total discounted cost generated by Z_i and define F_i: $[0,\infty) \times [s_p^i, s_{p+1}^i] \to \mathbb{R}$ by $F_i(t,x) = e^{-\alpha t}V_i(x)$. Since

$$V_i(x) = E\left[\int_0^T e^{-\alpha t}\left[hZ_i(t) + r_i\right] dt + e^{-\alpha T}[K_{i,f(Z_i(T),i)} + V_f(Z_i(T), f(Z_i(T),i)]\right]$$

where $T = \inf\{t \geq 0:\ Z_i(t) \notin (s_p^i, s_{p+1}^i)\}$, we apply Lemma 2 and get

$$e^{-\alpha T}V_i(Z_i(T)) = V_i(x) + \int_0^T \left[-\alpha e^{-\alpha t}V_i(Z_i(t)) + \frac{1}{2} e^{-\alpha t}V_i''(Z_i(t))\sigma_i^2 + e^{-\alpha t}V_i'(Z_i(t))\mu_i\right] dt + \int_0^T e^{\alpha t}V_i'(Z_i(t))\sigma_i dB(t).$$

Taking expectations,

$$E\left[e^{-\alpha T}V_i(Z_i(T))\right] = E\left[e^{-\alpha T}[K_{i,f(z_i(T),i)} + V_i(Z_i(T), f(Z_i(T),i))]\right]$$

$$= V_i(x) + E\left[\int_0^T e^{-\alpha t}[D_iV_i(Z_i(t)) - \alpha V_i(Z_i(t))]dt\right].$$

This implies that

$$E\left[\int_0^T e^{-\alpha t}[\alpha V_i(Z_i(t)) - D_iV_i(Z_i(t))]dt\right] = E\left[\int_0^T e^{-\alpha t}[hZ_i(t) + r_i]dt\right],$$

and since $V_i(x) = V_f(x,i)$ on $[s_p^i, s_{p+1}^i]$, function V_f satisfies (3.8).

Clearly, if $x \notin I^i$, $V_f(x,i) = K_{i,f(x,i)} + V_f(x,f(x,i))$; hence (3.9) also holds. For state zero and the case of absorption, $V_f(0,i) = R$ for each $i \in A$; thereby validating (3.10) when $\lambda = 0$. In the case of reflection, if $0 \in I^i$ then $V_f'(0,i) = 0$ by (3.8), and if $0 \notin I^i$ then $0 \in I^{f(0,i)}$ and $V_f'(0,i) = V_f'(0,f(0,i)) = 0$; thereby validating (3.10) when $\lambda = 1$.

For every $(x,i) \in S\ A$, we have $x \in I^{f(x,i)}$. If x is an interior point of $I^{f(x,i)}$, then as seen above $V_f'(x,f(x,i))$ and $V_f''(x,f(x,i))$ exist, where V_f' and V_f'' denote the first and second partial spatial derivatives. Suppose now that x is a closed boundary point of $I^{f(x,i)}$; let $f(x,i) = a$ and p s.t. $x \in [s_p^a, s_{p+1}^a]$. Assume that $x = s_p^a$. (The case $x = s_{p+1}^a$ is similar.) Then there exists $j \in \theta(a)$ and $\bar{p}$ s.t. $s_{\bar{p}+1}^j = s_p^a$, $b_{\bar{p}}^j = j$, and $f(s_{\bar{p}+1}^j, j) = a$. As in the proof of Theorem 1, let $\{\psi(t); t \geq 0\}$ be the unique process defined for each $y \in (s_{\bar{p}}^j, s_{p+1}^a)$ by

$$\psi(t) = y + \int_0^t \mu(\psi(u))du + \int_0^t \sigma(\psi(u))dB(u)$$

where

$$\mu(y) = \begin{cases} \mu_j & \text{if} \quad y < s_{\bar{p}+1}^j \\ \mu_a & \text{if} \quad y \geq s_{\bar{p}+1}^j, \end{cases}$$

and

$$\sigma(y) = \begin{cases} \sigma_j & \text{if} \quad y < s_{\bar{p}+1}^j \\ \sigma_a & \text{if} \quad y \geq s_{\bar{p}+1}^j . \end{cases}$$

Impose on process ψ the linear holding costs, operational costs, and switching costs associated with process $X(\cdot|x,i,\pi_{f,x,i})$ and absorption upon hitting $s_{\bar{p}}^j$ or s_{p+1}^a at costs $K_{j,f(s_{\bar{p}}^j,j)} + V_f\left(s_{\bar{p}}^j, f(s_{\bar{p}}^j,j)\right)$ and $K_{a,f(s_{p+1}^a,a)} + V_f\left(s_{p+1}^a, f(s_{p+1}^a,a)\right)$, respectively. Let $V_{aj}(y)$ denote the conditional expectation of the total discounted cost generated by ψ in this setting and observe that $V_{aj}(y)$ is continuously differentiable on $(s_{\bar{p}}^j, s_{p+1}^a)$ with a second derivative except at the points $s_{\bar{p}}^j$, s_p^a and s_{p+1}^a.

Thus, if x is not a boundary point of I^i then $V_f'(x,i)$ and $V_f''(x,i)$ exist. If x is a closed boundary point of I^i, then $V_f'(x,i)$ exists; and if x is an open boundary point of I^i but not a boundary point if $I_{\theta(i)}$, then again $V_f'(x,i)$ exists. Finally, if x is a boundary point of $I_{\theta(i)}$, then $V_f'(x,i)$ does not necessarily exist. So, (3.6) holds, and V_f is indeed a solution to (3.6)-(3.10).

Suppose now that function V: $S \times A \to \mathbb{R}$ also satisfies (3.6)-(3.10). Then letting $\Delta(x,i) = V(x,i) - V_f(x,i)$ for each $(x,i) \in S \times A$, we would find that the function Δ satifies these conditions for each $i \in A$:

(3.11) $\left|\Delta(x,i)\right|$ is bounded in S,

(3.12) $D_i\Delta(x,i) - \alpha\Delta(x,i) = 0$ for all $x \in I^i$,

(3.13) $\Delta(x,i) = \Delta(x, f(x,i))$ for all $x \notin I^i$, and

(3.14) $\lambda\Delta'(0,i) - (1-\lambda)\Delta(0,i) = 0.$

The second-order differential equation (3.12) implies that

$$\Delta(x,i) = \gamma_1 e^{\beta_1 x} + \gamma_2 e^{\beta_2 x} \quad \text{for each } x \in I^i,$$

where β_1 is the positive root and β_2 and the negative root of the quadratic equation $\mu_i\beta + \frac{1}{2}\sigma_i^2\beta^2 - \alpha = 0$. But by (3.11), (3.13), and (3.14), it must be that $\gamma_1 = \gamma_2 = 0$. Hence, $(x,i) = 0$ for each $(x,i) \in S\times A$, and V_f is the unique solution to (3.6)-(3.10). □

Band function f, then, generates an (x,i)-optimal strategy if $V_f(x,i) = V_*(x,i)$. We call band function f (<u>everywhere</u>) <u>optimal</u>, if for each $(x,i) \in S\times A$ its corresponding band strategy is (x,i)-optimal. After proving a verification lemma, we will derive necessary and sufficient conditions for a band function to be everywhere optimal.

<u>Lemma 3</u>. Suppose that for each $i \in A$, $V: S\times A \to \mathbb{R}$ satifies (3.6), (3.7), (3.10),

(3.15) $V(x,i) \leq K_{ij} + V(x,j) \geq$ for each $(x,j) \in S\times A$, and

(3.16) $D_j V(x,i) - V(x,i) + g(x,j) \geq 0$ for each $(x,j) \in S\times\theta(i)$ where we further define D_j for those x without a second partial spatial derivative by $D_j V(x,i) = \mu_j V'(x,i) + \frac{1}{2}\sigma_j^2\left[(V''(x-,i) + V''(x+,i))/2\right]$.

Then $V(x,i) \leq V_*(x,i)$ for each $(x,i) \in S\times A$.

<u>Proof</u>. First, since $V(x,j) = V(x,i)$ for each $(x,i) \in S\times A$ and $j \in \theta(i)$, let $\bar{V}(x,k)$ be the common value of $V(x,j)$ for each $j \in A_k$. Conditions (3.6), (3.7), (3.10), (3.15), and (3.16), can be restated in terms of $\bar{V}(x,k)$ for $(x,k) \in S \times \{1,2,...,M\}$ and class switching costs $C_{k\ell}$ for $k,\ell \in \{1,2,...,M\}$.

Fix $(x,i) \in S\times A$ and let π be an arbitrary admissible strategy. We want to show that $V_\pi(x,i) \geq V(x,i)$. Let $0 = T_0 < T_1 < T_2 < \ldots$ denote the discontinuity points of function $\bar{\theta}: [0,\infty) \to \{1,2,...,M\}$ where $\bar{\theta}(t) = \theta(\pi(t))$ for each $t \geq 0$. Then using (3.16),

$$\text{(3.17)} \qquad V_\pi(x,i) = E\left[\sum_{n=0}^{\infty}\left[\int_{T_n}^{T_{n+1}} e^{-\alpha t} g\big(X(t|x,i,\pi),\pi(t)\big)dt \right.\right.$$

$$\left.\left. + e^{-\alpha T_{n+1}} C_{\bar{\theta}(T_{n+1}-),\bar{\theta}(T_{n+1}+)}\right]\right]$$

$$\geq E\left[\sum_{n=0}^{\infty}\left[\int_{T_n}^{T_{n+1}} e^{-\alpha t}[\alpha\bar{V}(X(t|x,i,\pi),\bar{\theta}(t)) - D_{\pi(t)}\bar{V}(X(t|x,i,\pi),\bar{\theta}(t))]dt + e^{-\alpha T_{n+1}}C_{\bar{\theta}(T_{n+1}-),\bar{\theta}(T_{n+1}+)}\right]\right].$$

Fix n and define function F_n on $(T_n, T_{n+1}) \times S$ by $F_n(t,x) = e^{-\alpha t}\bar{V}(x,\bar{\theta}(t))$. Applying Lemma 2, we get

$$(3.18)\qquad e^{-\alpha t}\bar{V}(X(t|x,i,\pi),\bar{\theta}(t)) = e^{-\alpha T_n}\bar{V}(X(T_n|x,i,\pi)\bar{\theta}(T_n+))$$
$$+ \int_{T_n}^{t} e^{-\alpha u}\left[D_{\pi(u)}\bar{V}(X(u|x,i,\pi),\bar{\theta}(u)) - \alpha\bar{V}(X(u|x,i,\pi),\bar{\theta}(u))\right]du$$
$$\int_{T_n}^{t} e^{-\alpha u}\bar{V}'(X(u|x,i,\pi),\bar{\theta}(u))\sigma_{\pi(u)}dB(u) + \int_{T_n}^{t} e^{-\alpha u}\bar{V}'(X(u|x,i,\pi),\bar{\theta}(u)dY(u)$$

for each $(t,x) \in (T_n,T_{n+1}) \times S$, where X and Y uniquely satify

$$X(t|x,i,\pi) = X(T_n|x,i,\pi) + \int_{T_n}^{t} \mu_{\pi(u)}du + \int_{T_n}^{t} \sigma_{\pi(u)}dB(u) + Y(t)$$

and (2.10). The last integral in (3.18) has value zero since if $\lambda = 0$ then $Y = 0$, and if $\lambda = 1$ then Y grows only when $X(\cdot|x,i,\pi) = 0$ and $\bar{V}'(0,\bar{\theta}(u)) = 0$ for each $u \in (T_n,T_{n+1})$. Taking expectations in (3.18) and substituting into (3.17), we have

$$V_\pi(x,i) \geq E\left[\bar{V}(x,\theta(i)) + \sum_{n=1}^{\infty}\left[e^{-\alpha T_n}\bar{V}(X(T_n|x,i,\pi),\bar{\theta}(T_n+)) - e^{-\alpha T_n}\bar{V}(X(T_n|x,i,\pi),\bar{\theta}(T_n-)) + e^{-\alpha T_n}C_{\bar{\theta}(T_n-),\bar{\theta}(T_n+)}\right]\right]$$

$$\geq E\left[\bar{V}(x,\theta(i)) + \sum_{n=1}^{\infty} e^{-\alpha T_n}\bar{V}(X(T_n|x,i,\pi),\bar{\theta}(T_n+))\right.$$

$$- \left[C_{\bar{\theta}(T_n-),\bar{\theta}(T_n+)} + \bar{V}\big(X(T_n|x,i,\pi),\bar{\theta}(T_n+)\big) \right] + C_{\bar{\theta}(T_n-),\bar{\theta}(T_n+)} \Big] \Big]$$

$$= \bar{V}\big(x,\theta(i)\big).$$

Since $\bar{V}\big(x,\theta(i)\big) = V(x,i)$, we have as desired $V_\pi(x,i) \geq V(x,i)$ and therefore, $V_*(x,i) \geq V(x,i)$. □

We now present our necessary and sufficient conditions for a given band function to be optimal.

Theorem 3. A band function f is optimal if and only if its value function V_f satisfies the following for each $i \in A$:

(3.19) $V_f(x,i) = \min_{j \in A}\{K_{ij} + V_f(x,j)\}$ for each $x \in S$, and

(3.20) $\min_{j \in L(x,i)} \{D_j V_f(x,j) - \alpha V_f(x,j) + g(x,j)\} = 0$ for each $x \in S$,

where $L(x,i) = \{j \in A: V_f(x,i) = K_{ij} + V_f(x,j)\}$ and D_j is as in (3.16).

Proof. Suppose that V_f satisfies (3.19) and (3.20). By Theorem 2 V_f also satisfies (3.6), (3.7), and (3.10). Since $\theta(i) \subseteq L(x,i)$ for each $(x,i) \in S \times A$, (3.20) implies (3.16), and (3.19) implies (3.15). Therefore by Lemma 3, $V_f(x,i) = V_*(x,i)$ everywhere.

Now suppose that band function f is optimal. If (3.19) fails, then there exists $x \in S$ and $i,j \in A$ s.t. $V_f(x,i) > K_{ij} + V_f(x,j)$. By the continuity of $V_f(\cdot,i)$ and the optimality of f, there also exists $\epsilon > 0$ s.t. $V_f(y,i) > K_{ij} + V_f(y,j)$ for each $y \in [x-\epsilon,x+\epsilon]$, and

$$\left[K_{ij} + \frac{hx}{\alpha} + \frac{r_j}{\alpha} + \frac{\mu_j}{\alpha^2}\right]\left[1-E[e^{-\alpha T}]\right] - \frac{\mu_j}{\alpha} E[Te^{-\alpha T}] + E\left[e^{-\alpha T} V_f(\psi(T),i)\right] \leq V_f(x,i),$$

where $\{\psi(t);\ t \geq 0\}$ is the process $\psi(t) = x + \mu_j t + \sigma_j B(t)$ and $T = \inf\{t \geq 0: \psi(t) \in [x-\epsilon,x+\epsilon]\}$.

Now define admissible strategy π by $\pi(t) = j$ for $j \in (0,T]$, and $\pi(t) = f(X(t|x,i,\pi), \pi(t))$ for $t > T$ where $X(T|x,i,\pi)$ is the contolled process generated by π, x, and i. Hence,

$$V_\pi(x,i) = K_{ij} + E\left[\int_0^T e^{-\alpha t}[h\psi(t) + r_j]dt\right] + E\left[e^{-\alpha T} V_f X\big(T|x,i,\pi),j\big)\right]$$

$$< K_{ij} + E\left[\int_0^T e^{-\alpha t}[h\psi(t) + r_j]dt\right] + E\left[e^{-\alpha T}[V_f\big(\psi(T),i\big) - K_{ij}]\right] \leq V_f(x,i).$$

This contraducts the optimality of f; so f optimal implies (3.19).

Now suppose that f is optimal and (3.20) fails. Then

there exists $x \in S\setminus\{0\}$ $i,j \in A$ and $\epsilon > 0$ s.t. $V_f(x,i) = K_{ij} + V_f(x,i)$ and $D_j V_f(y,j) - V_f(y,j) + g(y,j) < 0$ for each y $[x-\epsilon, x+\epsilon]$. Defining the process ψ, the stopping time T, and the admissible strategy π exactly as above, we have

$$V_\pi(x,i) = K_{ij} + E\left[\int_0^T e^{-\alpha t} g(\psi(t),j)dt\right| + E\left[e^{-\alpha T} V_f(\psi(T),j)\right]$$

$$< K_{ij} + E\left[\int_0^T e^{-\alpha t}\left[\alpha V_f(\psi(t),j) \quad - D_j V_f(\psi(t),j)\right]dt\right]$$

$$+ E\left[e^{-\alpha T} V_f(\psi(T),j)\right].$$

Lemma 2 further implies that

$$E\left[e^{-\alpha T} V_f(\psi(T),j)\right] = V_f(x,j) + E\left[\int_0^T e^{-\alpha t}[D_j V_f(\psi(t),j) - \alpha V_f(\psi(t),j)]dt\right],$$

and hence $V_\pi(x,i) < K_{ij} + V_f(x,j) = V_f(x,i)$. Again we have contradicted the optimality of f, thereby providing that (3.20) is also necessary for f to be optimal. □

We can summarize the optimality conditions (3.19) and (3.20) by demanding that the optimal value function V_f satify the satisfy the following single condition for each $(x,i) \in S\times A$:

(3.21) $$\min_{j\in A}\left[K_{ij} + V_f(x,j) - V_f(x,i)] + t[D_j V_f(x,j) - \alpha V_f(x,j) + g(x,j)]\right\} = 0$$

for all small enough t.

Condition (3.21) is a lexicographic minimum condition since it requires first that (3.19) hold and second that (3.20) hold. Note that (3.21) is the appropriate Bellman-Hamilton-Jacobi equation for our control problem.

We conjecture that there always exists an optimal band function and, moreover, that is has a special "finite-critical-number" form. Hereafter, we label the control modes so that $\mu_1 \geq \mu_2 \ldots \geq \mu_N$.

Conjecture. There exists an optimal band function f_* s.t.

(3.21) for each $i \in A$, the class continuation set I_i is an open interval of S,

(3.22) if $\lambda = 1$ and $h > 0$ then for each $i \in A$, $f_*(x,i)$ is increasing in x on I_i,

(3.23) if $\lambda = 1$ and $h < 0$ then for each $i \in A$, $f_*(x,i)$ is decreasing in x on I_i,

(3.24) if $\lambda = 0$ or, $\lambda = 1$ and $h = 0$, then for each $i \in A$, $r_{f_*}(x,i)$ is decreasing in x on I_i, and

(3.25) for each $i \in A$, $V'_{f_*}(\cdot,i)$ is continuous of S.

Remarks. (1) Our cost structure can be simplified to one where each of the operational costs r_i is nonnegative. This is accomplished by redefining cost function $g\colon S\times A \to \mathbb{R}$ as

$$g(x,i) = \begin{cases} hx + \bar{r}_i & \text{if} \quad x > 0 \\ (1-\lambda)\alpha\bar{R} & \text{if} \quad x = 0 \end{cases}$$

where $\bar{r}_i = r_i - r_*$ for $i=1,2,\ldots,N$, $\bar{R} = R - \frac{r_*}{\alpha}$, and $r_* = \min\{r_1\}$.

(2) In the case of absorption the cost structure can be simplified further so that there are zero holding costs as well. To see this observe that for any admissible strategy π and $(x,i) \in S\times A$,

$$V_\pi(x,i) = E\left[\int_0^T e^{-\alpha t}[hX(t|x,i,\pi) + r_{\pi(t)}]dt + Re^{-\alpha T} + \sum_{k=1}^{M}\sum_{\ell=1}^{M}\int_0^\infty C_{k\ell}e^{-\alpha t}dQ^*_{k\ell}(t|x,i,\pi)\right]$$

where T is the time of absorption; and if we change the order of integration,

$$V_\pi(x,i) = \frac{hx}{\alpha} + E\left[\int_0^\infty e^{-\alpha t}\left[\frac{h\mu(t)}{\alpha} + r(t)\right]dt + Re^{-\alpha T} + \sum_{k=1}\sum_{\ell=1}\int_0^\infty C_{k\ell}e^{-\alpha t}dQ^*_{k\ell}(t|x,i,\pi)\right.$$

where

$$\{\mu(t),r(t)\} = \begin{cases} \{\mu_\pi(t),r_{\pi(t)}\} & \text{if} \quad t \le T \\ \{0,0\} & \text{if} \quad t > T \end{cases}$$

Therefore we have an equivalent control problem if we define $g\colon S\times A \to \mathbb{R}$ as

$$g(x,i) = \begin{cases} \bar{r}_i & \text{if} \quad x > 0 \\ \alpha\bar{R} & \text{if} \quad x = 0\,, \end{cases}$$

where $\bar{r}_i = \frac{h\mu_i}{\alpha} + r_i - \bar{r}_*$ for $i=1,2,\ldots,N$ $\bar{R} = R - \frac{\bar{r}_*}{\alpha}$, and $\bar{r}_* = \min \left\{\frac{h\mu_i}{\alpha} + r_i\right\}$.

(3) Together, conditions (3.21), (3.22), and (3.23) imply that for the optimal band function, the class continuation set I_i is at most

one open interval in S and the action continuation set I^i is at most one interval contained in I_i, for each $i \in A$. Moreover, within a class continuation set as the state of the controlled process increases, it is optimal to use a control mode with faster drift downwards in the case of positive holding costs ($h > 0$) and a control mode with faster drift upwards in the case of negative holding costs.

(4) As a consequence of remark (3) and the properties of band strategies, the total number of discontinuities in $x \in S$ for the optimal band function is at most $M(M-1) + \sum_{k=1}^{M} (N_k-1) = M^2 - 2M + N$, where N_k is the number of actions in equivalence class A_k. The $M(M-1)$ term accounts for possible switching between action equivalence classes. (There are $M(M-1)/2$ pairs of classes and for each pair A_k and A_ℓ, two switching numbers are involved. At one, switching out of A_k into A_ℓ occurs and at the other, switching out of A_ℓ into A_k occurs.) The (N_k-1) term accounts for possible switching between actions within class A_k. (To partition the A_k-class continuation interval into its possible action continuation intervals, (N_k-1) switching numbers are involved.)

For example, Figure 1 depicts an admissible band function for data $N = 3$, $K_{12} = K_{21} = 0$, $K_{13} = K_{23} > 0$, and $K_{31} = K_{32} > 0$. Suppose further that $\lambda = 1$ and $h > 0$. The optimal band function might be as follows:

$$f_*(x,1) = \begin{cases} 1 & \text{if} \quad x \in [0,s_1] \\ 2 & \text{if} \quad x \in (s_1,s_3) \\ 3 & \text{if} \quad x \in [s_3,\infty), \end{cases}$$

$$f_*(x,2) = \begin{cases} 1 & \text{if} \quad x \in [0,s_1] \\ 2 & \text{if} \quad x \in (s_1,s_3) \\ 3 & \text{if} \quad x \in [s_3,\infty), \text{ and} \end{cases}$$

$$f_*(x,3) = \begin{cases} 1 & \text{if} \quad x \in [0,s_1] \\ 2 & \text{if} \quad x \in (s_1,s_2] \\ 3 & \text{if} \quad x \in (s_2,\infty) \ . \end{cases}$$

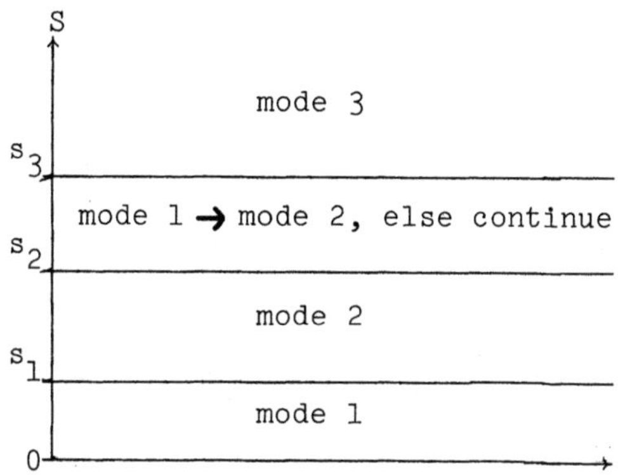

FIGURE 2. AN ILLUSTRATIVE OPTIMAL BAND FUNCTION

In Figure 2 we illustrate f_*. Note that f_* is a "three-critical-number" strategy.

4. EXPLICIT SOLUTIONS

The above conjecture has previously been proven when there are two available control modes; see Sheng [15] for the solutions to absorbing barrier problems, and Sheng [16] for the solutions to reflecting barrier problems. It has not yet been proven in all its generality for large N. As an illustration, we explicity produce here optimal band strategies for some specific cases.

(1) Consider our control problem for two avaiable control modes, reflection at the boundary, and zero switching costs. We use the following notation:

$$\beta = \frac{\mu_1 + \sqrt{\mu_1^2+2\alpha\sigma_1^2}}{\sigma_1^2}, \quad \rho = \frac{\mu_2 + \sqrt{\mu_2^2+2\alpha\sigma_2^2}}{\sigma_2^2}, \quad \nu = \frac{\mu_1 - \sqrt{\mu_1^2+2\alpha\sigma_1^2}}{\sigma_1^2},$$

$$\Delta_1 = \frac{\sigma_2^2}{2}\beta^2 - \mu_2\beta - \alpha, \quad \text{and} \quad \Delta_2 = \frac{\sigma_1^2}{2}\rho^2 - \mu_1\rho - \alpha \ .$$

Let f_1 denote the single band function of always using control mode 1; f_2, the single band function of always using control mode 2; and f_z, two-band function of always using mode 2 whenever the state of the system is above z and always using mode 1 whenever the state is below z. The solutions are indicated in the two tables below.

$\mu_1 \geq \mu_2$	$\sigma_1^2 \geq \sigma_2^2$		use strategy f_2
	$\sigma_1^2 < \sigma_2^2$	$\Delta_2 \geq (\mu_2-\mu_1)\rho - \alpha\rho r_1$	use strategy f_2
		$\Delta_2 < (\mu_2-\mu_1)\rho - \alpha\rho r_1$	use strategy f_2

TABLE 1. OPTIMAL BAND STRATEGIES WHEN $h = 1$, $r_1 > 0$, AND $r_2 = 0$

$\mu_1 \geq \mu_2$	$\sigma_1^2 \geq \sigma_2^2$	$r_2 < \frac{\mu_1-\mu_2}{\alpha}$		use strategy f_2
		$r_2 \geq \frac{\mu_1-\mu_2}{\alpha}$	$\Delta_1 \geq (\mu_1-\mu_2)\beta - \alpha\beta r_2$	use strategy f_1
			$\Delta_1 < (\mu_1-\mu_2)\beta - \alpha\beta r_2$	use strategy f_z
	$\sigma_1^2 < \sigma_2^2$	$r_2 < \frac{\mu_1-\mu_2}{\alpha}$	$\Delta_2 < (\mu_2-\mu_1)\rho - \alpha\rho r_2$	use strategy f_z
			$\Delta_2 \geq (\mu_2-\mu_1)\rho - \alpha\rho r_2$	use strategy f_2
		$r^2 \geq \frac{\mu_1-\mu_2}{\alpha}$		use strategy f_1

TABLE 2. OPTIMAL BAND STRATEGIES WHEN $h = 1$, $r_1 = 0$ AND $r_2 > 0$

In each of the two tables, the single critical number z characterizing optimal band function f_z is the unique positive solution to a transcendental equation corresponding to the condition that the optimal reward function has a continuous second derivative everywhere.

(2) Now consider the special case of N available control nodes, reflection at the boundary, positive holding costs at rate h = 1, general switching costs, and such that $\mu_1 \geq \mu_2 \ldots \geq \mu_N$, $\sigma_1^2 > \sigma_2^2 \geq \ldots \geq \sigma_N^2$, and $r_1 \geq r_2 \geq \ldots \geq r_N$. Let f_N denote the single band function of always using control mode N. The corresponding value function V_{f_N} on S×A is

$$V_{f_N}(x,i) = K_{iN} + \frac{x}{\alpha} + \frac{\mu_N}{\alpha 2} + \frac{r_N}{\alpha} + \frac{1}{\alpha\theta} e^{-\theta x}$$

where

$$= \frac{\mu_N + \sqrt{\mu_N^2 + 2\alpha\sigma_N^2}}{\sigma_N^2} .$$

Checking the optimality conditions (3.19) and (3.20), we find that f_N is optimal if and only if $K_{iN} \leq \frac{\mu_i - \mu_N}{2} + \frac{r_i - r_N}{\alpha}$ for each $i \neq N$.

(3) Finally, consider the case of two available control modes, reflection at the boundary, positive holding costs at rate h = 1, and symmetric switching costs $K_{12} = K_{21} = K > 0$. For single band function f_1 we have

$$V_{f_1}(x,1) = \frac{x}{\alpha} + \frac{\mu_1}{\alpha^2} + \frac{r_1}{\alpha} + \frac{1}{\alpha\beta} e^{-\beta x} \quad \text{for all } x \geq 0, \text{ and}$$

$$V_{f_1}(x,2) = K + \frac{x}{\alpha} + \frac{\mu_1}{\alpha^2} + \frac{r_1}{\alpha} + \frac{1}{\alpha\beta} e^{-\beta x} \quad \text{for all } x \geq 0.$$

By Theorem 3, then, f_1 is optimal if and only if

(4.1) $\frac{\mu_2 - \mu_1}{\alpha} + (r_2 - r_1) + \frac{1}{\alpha\beta}\Delta_1 e^{-\beta x} \geq \alpha K$ for all $x \geq 0$.

Similarly for single band function f_2:

$$V_{f_2}(x,1) = K + \frac{\mu_2}{\alpha^2} + \frac{r_2}{\alpha} + \frac{1}{\alpha\rho} e^{-\rho x} \quad \text{for all } x \geq 0,$$

$$V_{f_2}(x,2) = \frac{\mu_2}{\alpha^2} + \frac{r_2}{\alpha} + \frac{1}{\alpha\rho} e^{-\rho x} \quad \text{for all } x \geq 0,$$

and f_2 is optimal if and only if

(4.2) $\frac{\mu_1 - \mu_2}{\alpha} + (r_1 - r_2) + \frac{1}{\alpha\rho}\Delta_2 e^{-\rho x} \geq \alpha K$ for all $x \geq 0$.

Suppose instead we use the strategy of never switching control modes. Let f_c denote the corresponding band function; that is, $f_c(x,i) = i$ for each (x,i). Then

$$V_{f_c}(x,1) = \frac{x}{\alpha} + \frac{\mu_1}{\alpha^2} + \frac{r_1}{\alpha} + \frac{1}{\alpha\beta} e^{-\beta x} \quad \text{for all } x \geq 0,$$

$$V_{f_c}(x,2) = \frac{x}{\alpha} + \frac{\mu_2}{\alpha^2} + \frac{r_2}{\alpha} + \frac{1}{\alpha\rho}e^{-\rho x} \qquad \text{for all } x \geq 0,$$

and optimality conditions (3.19) and (3.20) reduce to

$$\begin{cases} -\alpha K \leq \dfrac{\mu_1-\mu_2}{\alpha} + r_1 - r_2 \leq \alpha K \\ \\ -\alpha K \leq \dfrac{\mu_1-\mu_2}{\alpha} + r_1 - r_2 + \dfrac{1}{\beta} - \dfrac{1}{\rho} \leq \alpha K. \end{cases}$$

The parameter combinations corresponding to (4.1), (4.2) and (4.3), however, do not exhaust the possible range of diffusion and cost parameters. For the remaining parameter combinations, we conjecture that the optimal band function has the following two-critical number form:

$$f_*(x,1) = \begin{cases} 1 & \text{if } x \in [0,Z) \\ 2 & \text{if } x \in [Z,\infty) \end{cases}$$

and

$$f_*(x,2) = \begin{cases} 1 & \text{if } x \in [0,z) \\ 2 & \text{if } x \in (z,\infty) \end{cases}$$

where $0 \leq z \leq Z \leq +\infty$. The optimal pair of switching numbers, z_* and Z_*, is chosen so as to satisfy the optimality conditions of Theorem 3. The optimality conditions for f_* are

$$-K \leq V_{f_*}(x,2) - \alpha V_{f_*}(x,1) \leq K \qquad \text{for } x \in [z,Z].$$

$$D_1 V_{f_*}(x,2) - \alpha V_{f_*}(x,2) + x + r_1 \geq \alpha K \qquad \text{for } x \in [Z,\infty), \text{ and}$$

$$D_2 V_{f_*}(x,1) - \alpha V_{f_*}(x,1) + x + r_2 \geq \alpha K \quad \text{for } x \in [0,z].$$

Therefore our two-mode symmetric-switching-cost problem is completely solved if there exist positive z_* and Z_* ($z_* < Z_*$) such that for the appropriate parameter restrictions (i.e., none of (4.1) through (4.3) holding) the following are true:

$$V_{f_*}(x,1) - V_{f_*}(x,2) \quad \text{increasing in } x \in [z_*,Z_*],$$

$$V_{f_*}(z_*,1) - V_{f_*}(z_*,2) = V_{f_*}(Z_*,2) - V_{f_*}(Z_*,1) = -K,$$

$$\frac{1}{2}\left(\sigma_1^2 - \sigma_2^2\right)V''_{f_*}(x,1) + (\mu_1-\mu_2)V'_{f_*}(x,1) \quad \text{increasing in } x \in [0,z_*],$$

$$\frac{1}{2}\left(\sigma_1^2 - \sigma_2^2\right)V''_{f_*}(z_*,1) + (\mu_1-\mu_2)V'_{f_*}(z_*,1) = r_2 - r_1 - \alpha K,$$

$$\frac{1}{2}\left(\sigma_1^2 - \sigma_2^2\right)V''_{f_*}(x,2) + (\mu_1-\mu_2)V'_{f_*}(x,2) \quad \text{increasing in } x \in [Z_*,\infty) \text{ and}$$

$$\frac{1}{2}\left(\sigma_1^2 - \sigma_2^2\right)V''_{f_*}(Z_*,2) + (\mu_1-\mu_2)V'_{f_*}(Z_*,2) = r_2 - r_1 + \alpha K.$$

REFERENCES

[1] Arkin, V. I. Kolemaev, V. A. and Shiryaev, A. N. (1964) On Finding Optimal Controls. Trudy Steklov Math. Institute LXXI, 21-25.

[2] Beneš, V. (1971) Existence of Optimal Stochastic Control Laws. SIAM J. Control 9, No. 3, 446-472.

[3] Davis, M. H. A. and Varaiya, P. (1973) Dynamic Programming Conditions for Partially Observable Stochastic Systems. SIAM J. Control 11, No. 2, 226-261.

[4] Doshi, B. T. (1978) Two Mode Control of Brownian Motion with Quadratic Loss and Switching Costs. Stoch. Proc. Appl. 6, 277-289.

[5] Doshi, B. T. (1979) Optimal Control of a Diffusion Process with Reflecting Boundaries and Both Continuous and Lump Costs. Dynamic Programming and Their Applications, ed. M. Putterman, Academic Press, New York, 269-288.

[6] Fleming, W. (1969) Optimal Continuous-Parameter Stochastic Control. SIAM Rev. 11, No. 4, 470-509.

[7] Foschini, G.J. (1982) Equilibria for Diffusion Models of Pairs of Communicating Computers-Symmetric Case. IEEE Trans. Infor. Th. 2,273-284.

[8] Iglehart, D.L.(1969) Diffusion Approximations in Collective Risk Theory. J. Appl. Prob. 6, 285-292.

[9] Kunita, H. and Watanabe, S. (1967) On Square Integrable Martingales. Nagoya Math. J. 31, 209-245.

[10] Mandl, P. (1968) Analytical Treatment of One-Dimensional Markov Processes. Springer-Verlag, New York.

[11] McKean, H. P., Jr. (1969) Stochastic Integrals. Academic Press, New York.

[12] Nakao, S.(1972) On the Pathwise Uniqueness of Solutions of One-Dimensional Stochastic Differential Equations. Osaka J. Math 19, 513-518.

[13] Pliska, S. R. (1973) Single-Person Controlled Diffusions with Discounted Costs. J. Opt. Th. Appl. 12, No. 3, 248-255.

[14] Rath, J. H. (1975) Controlled Queues in Heavy Traffic. Adv. Appl. Prob. 7, 656-671.

[15] Sheng, D. (1980) Two-Mode Control of Absorbing Brownian Motion, submitted for publication.

[16] Sheng, D. (1980) Two-Mode Control of Reflecting Brownian Motion, submitted for publication.

[17] Watanabe, S. and Yamada, T. (1971) On the Uniqueness of Solutions of Stochastic Differential Equations. J. Math. Kyoto 11, 155-167.

A Resumé of Some of the Applications of Malliavin's Calculus

Daniel W. Stroock

This research was supported in part by N.S.F. Grant MCS 80-07300.

0. Introduction:

This brief note is intended to introduce the reader to the Malliavin calculus. However, rather than attempting to explain the intricacies of Malliavin's calculus, I have decided to only sketch the ideas underlying his calculus and to concentrate on describing several of the problems to which the calculus has been successfully applied. Of course, I hope that, having seen its applications, the reader's appetite will be whetted and that to satisfy his appetite the reader will seek more information about this subject.

1. The Basic Setting:

Denote by Θ the space of, $\theta \in C([0,\infty),R^d)$ such that $\theta(0) = 0$ and let $\mathfrak{w}$ be Wiener measure on Θ. Given a mapping $\Phi : \Theta \to R^D$, the purpose of Malliavin's calculus is to provide a mechanism for studying the regularity properties of the induced measure $\mu_\Phi = \mathfrak{w} \circ \Phi^{-1}$ on R^D. In particular, Malliavin's calculus gives one a way of testing for the absolute continuity of μ_Φ and examining the smoothness of $f_\Phi = \frac{d\mu_\Phi}{dx}$ when it exists. At the same time, it is often possible to obtain regularity results about the behavior of μ_Φ as a function of Φ. Probabilists who are familiar with diffusion theory and related subjects are all too well aware of that, heretofore, the only way to attack such problems has been to identify μ_Φ as the solution to some functional equation and then invoke the regularity theory associated with that equation (cf. the discussion in the second paragraph of section 2) below).

Malliavin's idea is to work right in Wiener space $(\Theta,\mathfrak{w})$. In brief, his technique is to introduce a certain self-adjoint diffusion operator $\mathcal{L}$ on $L^2(\mathfrak{w})$. The importance of taking $\mathcal{L}$ to be a diffusion generator is that the associated bilinear map:

(1.1) $$\langle\Phi,\Psi\rangle_{\mathcal{L}} \equiv \mathcal{L}(\Phi\cdot\Psi) - \Phi\mathcal{L}\Psi - \Psi\mathcal{L}\Phi$$

will then satisfy:

(1.2) $$\langle\phi\circ\Phi,\Psi\rangle_{\mathcal{L}} = \phi'\circ\Phi\langle\Phi,\Psi\rangle_{\mathcal{L}}$$

for smooth ϕ on $R \to R$. (Equation (1.2) follows from the Itô calculus for continuous martingales; and it is in order to be dealing with continuous martingales that one needs $\mathcal{L}$ to generate a diffusion.) Given such an $\mathcal{L}$, one can integrate by parts. Namely, given $\Phi,\Psi \in \mathrm{Dom}(\mathcal{L})$ such that $\Psi/\langle\Phi,\Phi\rangle_{\mathcal{L}} \in \mathrm{Dom}(\mathcal{L})$, one has from (1.1), (1.2) and the symmetry of $\mathcal{L}$:

$$E^{W}[(\phi'\circ\Phi)\Psi] = E^{W}[\langle\phi\circ\Phi,\Phi\rangle_{\mathcal{L}}(\Psi/\langle\Phi,\Phi\rangle_{\mathcal{L}})]$$
$$= E^{W}[\phi\circ\Phi\{\Phi\mathcal{L}(\Psi/\langle\Phi,\Phi\rangle_{\mathcal{L}}) - \Psi\mathcal{L}\Phi/\langle\Phi,\Phi\rangle_{\mathcal{L}} - \mathcal{L}(\Phi\Psi/\langle\Phi,\Phi\rangle_{\mathcal{L}})\}]$$
$$= -E^{W}[\phi\circ\Phi\{2\Psi\mathcal{L}\Phi/\langle\Phi,\Phi\rangle_{\mathcal{L}} + \langle\Phi,\Psi/\langle\Phi,\Phi\rangle_{\mathcal{L}}\rangle_{\mathcal{L}}\}]$$

In particular, with $\Psi = 1$, one concludes from this that there is a $g \in L^1(\mu_\Phi)$ such that

$$E^{\mu_\Phi}[\phi'] = E^{\mu_\Phi}[\phi\cdot g] \quad , \quad \phi \in C_0^\infty(R^1) \quad . \tag{1.3}$$

It is an easy step to go from (1.3) to regularity results about μ_Φ.

There are two ingredients required before this technique can be used. First one must know that $\Phi \in \mathrm{Dom}(\mathcal{L})$. Second, one must show that $1/\langle\Phi,\Phi\rangle_{\mathcal{L}} \in \mathrm{Dom}(\mathcal{L})$. As one would expect, it is the existence of the second ingredient which is difficult to check, because it is this ingredient which contains the "non-degeneracy" of the map Φ.

Although it may appear that there is considerable lattitude in one's choice of $\mathcal{L}$, it turns out that ease of computation forces one to choose the simplest $\mathcal{L}$. The $\mathcal{L}$ chosen by Malliavin is the one known to quantum field theorists as the "number operator". For probabilists it is more illuminating to describe Malliavin's $\mathcal{L}$ as the "Ornstein-Uhlenbeck operator" associated with Wiener measure. For details, the reader is referred to the original paper by Malliavin [6], my paper [10] in which I expand Malliavin's ideas, or my recent articles [11] and [12] in which I introduce a quite different approach to understanding $\mathcal{L}$ and its associated calculus. References to other articles on this subject can be found in [11] and [12].

2. Some Applications to Diffusions:

Let

$$L = 1/2 \sum_{i,j=1}^{D} a^{ij}(x)\frac{\partial^2}{\partial x_i \partial x_j} + \sum_{i=1} b^i(x)\frac{\partial}{\partial x_j} \tag{2.1}$$

where $a : R^D \to R^D \otimes R^D$ and $b : R^D \to R^D$ are smooth functions having bounded derivatives and $a(x)$ is non-negative and symmetric for each $x \in R^D$. Suppose that $\sigma : R^D \to R^D \otimes R^D$ is a smooth function satisfying $a = \sigma\sigma^*$ and consider the Itô stochastic integral equation

$$X(T,x) = x + \int_0^T \sigma(X(t,x))d\theta(t) + \int_0^T b(X(t,x))dt \quad , \quad T > 0 \quad . \tag{2.2}$$

The cornerstone on which much of modern diffusion theory rests is the observation that the measure $P(T,x,\cdot)$ given by:

$$P(T,x,\cdot) = W \circ X(T,x)^{-1} \tag{2.3}$$

is the fundamental solution to the Cauchy initial value problem

$$(2.4) \qquad \frac{\partial u}{\partial t} = Lu \quad , \quad t > 0$$
$$u(0,\cdot) = f \quad .$$

To be more precise (by an elementary application of Itô's formula) if u is a reasonably smooth solution to (2.4) and u does not grow too fast as $|x| \to \infty$, then

$$(2.5) \qquad u(T,x) = \int f(y)P(T,x,dy)$$

Conversely, if f is smooth and has moderate growth at infinity, then one can show that the function u given by (2.5) is a solution to the Cauchy problem in (2.4). (A probabilistic proof of this latter statement can be based on the observation that $X(T,x)$ is a smooth function of x. See [7] or [13]).

Having identified the measure $P(T,x,\cdot)$ in (2.3) as the fundamental solution to (2.4), it has been customary for probabilists to read (2.3) "from right to left". That is, the theory of partial differential equations enables one to say a great deal about the fundamental solution to (2.4). (For example, in the case when $a(\cdot) \geq \varepsilon I$ for some $\varepsilon > 0$, it is a well-known consequence of the classical parametrix method that $P(T,x,dy) = p(T,x,y)dy$ where $p(T,x,y)$ is a smooth function of (T,x,y) so long as $T + |y-x|^2 > 0$ (cf. [2]).) Thus, because $P(T,x,\cdot)$ is the fundamental solution to (2.4), (2.3) allow us to conclude that, for $T > 0$, the distribution $\mathbb{W} \circ X(T,x)^{-1}$ of $X(T,x)$ under $\mathbb{W}$ will have regularity properties which, before the introduction of Malliavin's calculus, were not evident from the description of $X(T,x)$ given by (2.2).

Using Malliavin's calculus and reading (2.3) "from left to right", it is possible to not only recover some of the familiar regularity results about $P(T,x,\cdot)$ but also obtain some new results which do not seem to follow easily from the theory of partial differential equations. As we mentioned in section 1), successful application of Malliavin's calculus depends on one's proving two properties of the map $X(T,x) : \Theta \to R^D$. First one must show that $X(T,x)$ is in the domain of the operator $\mathcal{L}$. It turns out that this step is quite simple and depends only on smoothness of the functions $\sigma(\cdot)$ and $b(\cdot)$. The second, and difficult step, is to prove that $X(T,x)$ has the necessary non-degeneracy properties. Obviously, the origin of the non-degeneracy of $X(T,x)$ must be the non-degeneracy of $a(\cdot)$. The problem is to figure out how to relate the non-degeneracy of $a(\cdot)$ to that of $X(T,x)$. If $a(\cdot) \geq \varepsilon I$ for some $\varepsilon > 0$, the required non-degeneracy of $X(T,x)$ is relatively easy to prove. Thus the classical elliptic theory can be recovered from Malliavin's calculus. Not so easy, but nonetheless possible, is the proof that $X(T,x)$ is non-degenerate when L satisfies Hörmander's conditions for the hypoellipticity of $\frac{\partial}{\partial t} + L$. To be precise, rewrite L in Hörmander's form:

$$(2.6) \qquad L = 1/2 \sum_{k=1}^{d} (V^{(k)})^2 + V^{(0)} \quad ,$$

where $V^{(k)} = \sum_{j=1}^{D} V_j^{(k)}(\cdot)\frac{\partial}{\partial x_j}$ is a smooth vector field on R^D . Define $\mathcal{P}(V^{(0)},V^{(1)},\ldots,V^{(d)})$ to be the Lie algebra generated by $(ad^n V^{(0)})V^{(k)}$, $n \geq 0$ and $1 \leq k \leq d$ (here $(ad^{n+1}V^{(0)})V = [V^{(0)},(ad^n V^{(0)})V]$ and $ad^0 V^{(0)}(V) = V$). Combining Hörmander's theorem (cf. [8]) with the Schwartz kernel theorem, one can show that if $\dim(\mathcal{P}(V^{(0)};V^{(1)},\ldots,V^{(d)})(y)) = D$ for all $y \in R^D$, then, $P(T,x,dy) = p(T,x,y)dy$ with $p(T,x,\cdot) \in C^\infty(R^D)$. Using Malliavin's calculus, one can prove the same result under the assumption that $\dim(\mathcal{P}(V^{(0)},V^{(1)},\ldots, V^{(d)})(x)) = D$ (i.e. one only needs Hörmander's condition at the initial point x .) Furthermore, one can get some (admittedly crude) estimates on $p(T,x,y)$ as $T \downarrow 0$ or $y \to x$. The proof of this result can be found in section (8) of [12] . (The proof there is cased on joint work with S. Kusuoka. The bibliogrphy of [12] contains references to earlier versions of this and related results.)

Gratifying as the preceding successes of Malliavin's calculus may be, they are too close to known results to be considered real victories. To get a feeling for the sort of application in which Malliavin's calculus really comes into its own, consider the following situation. Assume that, for some $1 \leq N \leq D$, the principal N×N-minor $a^{(N)}(\cdot)$ of $a(\cdot)$ satisfies $a^{(N)}(\cdot) \geq \varepsilon I^{(N)}$ and set $X_{(N)}(T,x) = (X_1(T,x),\ldots,X_N(T,x))$. Denote by $P^{(N)}(T,x,\cdot)$ the marginal distribution of $X_{(N)}(T,\cdot)$ under $P(T,x,\cdot)$. Using Malliavin's calculus, one can show that for $T > 0$, $P^{(N)}(T,x,dy_{(N)}) = p^{(N)}(T,x,y_{(N)})dy_{(N)}$ where $p^{(N)}(T,x,y_{(N)})$ is a C^∞ function so long as $T + |x_{(N)}-y_{(N)}|^2 > 0$. (A proof of this result can be found in [11] . In a forthcoming article, S. Kusuoka and I will discuss various extensions and refinements of the result.) Observe that it is highly unlikely that the regularity of $P^{(N)}(T,x,dy_{(N)})$ could be easily derived from the theory of partial differential equations. Indeed, there is no obvious equation for $p^{(N)}(T,x,y_{(N)})$ to satisfy as a function of $y_{(N)}$. Considerations of this sort have made Malliavin's calculus a powerful tool in the study of infinite dimensional diffusion of the sort which arise in statistical mechanics. The interested reader is referred to [3] where Malliavin's calculus is applied to a continuous state Ising model.

A related application was made by D. Michel in [8] . Her idea was to use Malliavin's calculus to derive regularity properties of conditional transition functions arising in non-linear filtering theory. More recently, she and J.M. Bismut [1] have generalized her work.

3. Applications to Some Non-Markovian Situations:

At the end of section 2) we saw some examples of situations to which Malliavin's calculus applies but the theory of partial differential equations apparently does not. In this section we mention a source of examples about which the theory of partial differential equations have even less to say.

Let $\sigma : [0,\infty) \times \Theta \to R^D \otimes R^d$ and $b : [0,\infty) \times \Theta \to R^D$ be bounded progressively measurable functions which are "smooth" (in the sense of Frechet differentiability). Consider the Itô stochastic integral equation

$$X(T) = \int_0^T \sigma(t,X(\cdot))d\theta(t) + \int_0^T b(t,X(\cdot))dt \quad , \quad T \geq 0 \quad . \tag{3.1}$$

Obviously $X(\cdot)$ is not necessarily a Markov process; and, in general, it cannot be embedded in any finite dimensional Markov process. Thus it is difficult to imagine what sort of functional equation $\mu_T \equiv \mathbb{W} \circ X(T)^{-1}$ might satisfy. In particular, it seems very unlikely that one could invoke theorems from partial differential equations theory to prove regularity properties for μ_T . Nonetheless, if one assumes that $\sigma\sigma^*(\cdot) \geq \varepsilon I$, there is no probabilistic reason to suppose that μ_T is not just as smooth as in the diffusion case. Furthermore, a proof based on the Malliavin calculus should run along very much the same lines as it does in the diffusion case. In [12] , I showed that, at least on a very special case, one can indeed carry out this program. More recently, joint work with S. Kusuoka indicates that we can do the same for a much wider range of examples.

In this connection, the work of Shigekawa [9] must be mentioned. Shigekawa's interest is in regularity results for the distribution of Wiener functionals arising in Wiener's theory of homogeneous chaos. It turns out that this line of research leads quite quickly to problems best handled by algebraic geometry. For the latest progress in this direction, see the paper by S. Kusuoka [5] .

References

[1] Bismut, J.M., and Michel, D., "Diffusions conditionnelles, I., hypoellipticité partielle," J. Fnal. Anal. vol. 44 #2, pp. 174-211 (1981).

[2] Friedman, A., Partial Differential Equations of Parabolic Type, Englewood Cliffs, N.J., Prentice Hall (1964).

[3] Holley, R., and Stroock, D., "Diffusions on an infinite dimensional torus," J. Fnal. Anal., vol. 42 #1, pp. 29-63 (1981).

[4] Hörmander, L., "Hypoelliptic second order differential equations," Acta Math. 119, pp. 147-171 (1967).

[5] Kusuoka, S., "On absolute continuity of the law of a system of multiple Wiener Integrals," to appear in J. Fac. Sci. Univ of Tokyo.

[6] Malliavin, P., "Stochastic calculus of variation and hypoelliptic operators," Proc. Intern. Symp. on S.D.E.'s, Kyoto, ed. by K. Itô, Kinokuniya, Tokyo (1978).

[7] Mckean, H.P., Stochastic Integrals, Academic Press (1969).

[8] Michel, D., "Régularité des lois conditionelles en théorie du filtrage non linéaire et calcul variations stochastique," J. Fnal. Anal. 41 #1, pp. 8-36 (1981).

[9] Shigekawa, I., "Derivatives of Wiener functionals and absolute continuity of induced measures," J. Math. Kyoto Univ., 20, pp. 263-289 (1980).

[10] Stroock, D., "The Malliavin calculus and its application to second order parabolic differential equations, Part I," Math. Systems Th., 14, pp. 25-65 (1981).

[11] Stroock, D., "The Malliavin calculus, a functional analytic approach," J. Fnal. Anal., vol. 44 #2, pp. 212-257 (1981).

[12] Stroock, D., "Some applications of stochastic calculus to partial differential equations," to appear in lecture notes from the 1981 École d'Été at Saint Flour, Springer Lecture notes in Math.

[13] Stroock, D., "Topics in stochastic differential equations," to appear in Tata Inst. Lec. Notes Series, Springer-Verlag.

LARGE DEVIATIONS

S.R.S. Varadhan
Courant Institute of Mathematical Sciences
New York University
New York, NY 10012/USA

1. What are Large Deviations?

Let $\{P_n\}$ be a sequence of probability measures on Polish space X such that $P_n \Rightarrow \delta_{x_0}$ for some $x_0 \in x$ as $n \to \infty$. If A is a closed set with $A \cap \{x_0\} = \emptyset$ then $P_n(A) \to 0$. The question of large deviations is concerned with the rate of convergence of $P_n(A)$ to zero as $n \to \infty$. We will be concerned here only with the case when $P_n(A) \to 0$ at an exponential rate and the problem will be to identify the precise exponential rate in several concrete situations.

Definition. We will say that the large deviation results hold for a sequence $\{P_n\}$ on X with a functional $I(\cdot): X \to [0,\infty]$ if there is a functional $I(\cdot)$ mapping X into $[0,\infty]$ such that

(i) $0 \le I(\cdot) \le \infty$ and $I(\cdot)$ is a lower semicontinuous function of x.

(ii) $[x: I(x) \le \ell]$ is a compact subset of X for each $\ell < \infty$.

(iii) For each set A that is closed in X

$$\limsup_{n \to \infty} \frac{1}{n} \log P_n(A) \le - \inf_{x\in A} I(x) .$$

(iv) For each set G that is open in X

$$\limsup_{n \to \infty} \frac{1}{n} \log P_n(G) \ge - \inf_{x\in G} I(x) .$$

Whenever large deviation results hold for $\{P_n\}$ with a functional $I(\cdot)$ for every set A such that

$$\inf_{x\in A^0} I(x) = \inf_{x\in \bar{A}} I(x)$$

we have

$$\lim_{n\to\infty} \frac{1}{n} \log P_n(A) = - \inf_{x\in A} I(x) .$$

Moreover for every function F(x) on X which is bounded and continuous we have

Theorem.

$$\lim_{n\to\infty} \frac{1}{n} \log \left[\int \exp\{n\, F(x)\}\, P_n(dx) \right] = \sup_x \, [F(x) - I(x)]$$

This theorem is the motivation for studying large deviations because it then provides us a method for evaluating certain integrals asymptotically for large n.

2. Finite Dimensional Examples.

We will provide some examples where the large deviation results hold. We start with some fairly elementary examples in finite dimensional spaces.

Example 1. Let $X_1, X_2, \ldots, X_n, \ldots$ be a sequence of independent identically distributed random variables on the line with $M(\theta) = E\{e^{\theta X}\} < \infty$ for all $\theta \in R$. If we define

$$P_n(A) = \text{Prob}\left[\frac{X_1 + \ldots + X_n}{n} \in A\right]$$

then the large deviation results hold with

$$I(x) = \sup_\theta \, [\theta x - \log M(\theta)]$$

This result can be found in [1].

Example 2. Let us take the same situation as in example 1, but with the change that $X_1, \ldots, X_n, \ldots$ take values in R^d. Then

$$M(\theta) = E\{\exp\langle\theta, x\rangle\} \quad \text{for} \quad \theta \in R^d.$$

Again the large deviation results hold with

$$I(x) = \sup_\theta \, [\langle\theta, x\rangle - \log M(\theta)] \; .$$

This for instance can be found in [10].

3. Infinite Dimensinonal Examples.

Example 3. Suppose in example 2 we replace R^d by a Banach space X and add the hypothesis that

$$E\left\{\exp[\sigma \|x\|]\right\} < \infty \quad \text{for all} \quad \sigma > 0$$

then we have again the same results with

$$I(x) = \sup_{\theta \in X^*} [\langle \theta, x \rangle - \log M(\theta)]$$

where X^* is the dual of the Banach space X. See for instance [2].

Example 4. We now specialize the Banach space X to c[0,1], the space of continuous functions on [0,1] and the common distribution of our X valued random variables to a Gaussian process $x(t)$ with mean zero and covariance $\rho(s,t)$ which has almost surely continuous trajectories. We then obtain the following corollary

$$\lim_{\ell \to \infty} \frac{1}{\ell^2} \text{Prob} \left[\sup_{0 \le t \le 1} |x(t)| \ge \ell \right] = -(2\rho)^{-1}$$

where

$$\rho = \sup_{0 \le t \le 1} \rho(t,t) .$$

See in this context [3], [7] and [8].

Example 5. If we take X to be $C_0[0,1]$, i.e. the space of continuous function on [0,1] which vanish at the origin and P_n to be the Wiener measure with covariance $\frac{1}{n}\min(s,t)$, i.e. the measure corresponding to $\frac{1}{\sqrt{n}} \beta(\cdot)$ where $\beta(\cdot)$ is the standard Brownian motion, then again the large deviation results hold with

$$I(x) = \frac{1}{2} \int_0^1 [\dot{x}(t)]^2 \, dt$$

This is proved in [9].

Example 6. If we take $X = C_x[[0,1], R^d]$, the space of R^d valued continuous functions $x(t)$ on [0,1] with $x(0) = x$ and take for P_n the measure corresponding to the solution $x(t)$ of the stochastic differential equation

$$dx(t) = \frac{1}{\sqrt{n}} \sigma(x(t)) \, d\beta(t) + b(x(t)) \, dt$$
$$x(0) = x$$

then again the large deviation results hold with

$$I(x) = \frac{1}{2} \int_0^1 \langle \dot{x}(t) - b(x(t)), a^{-1}(x(t)) (\dot{x}(t) - b(x(t))) \rangle dt$$

This was shown in [11] when $b \equiv 0$ and in [6] and [12] for the general case.

Example 7. Let $x_1, x_2, \ldots, x_n, \ldots$ be a sequence of independent identically distributed random variables on a Polish space Y.

We take for X the space M_X of all probability distributions on Y. We take for P_n the distribution of the sample distribution based on n observation i.e.

$$P_n(A) = \text{Prob}\left[\frac{\delta_{x_1}+\ldots+\delta_{x_n}}{n} \in A\right]$$

Then again the large deviation results hold with $I(\cdot)$ defined for $\mu \in M_Y$ by

$$I(\mu) = \int \left[\log \left(\frac{d\mu}{d\alpha}\right)(y)\right] d\mu(y)$$

See [2] for details.

Remark 1. If $\mu << \alpha$ and $\log \frac{d\mu}{d\alpha} \in L_1(\mu)$ then $I(\mu)$ is given as above. Otherwise $I(\mu) = \infty$.

Remark 2. Example 1 is a special case of example 7. If $V: Y \to R$ is a real function then

$$\frac{V(x_1)+\ldots+V(x_n)}{n} = \int V(y) \left[\frac{\delta_{x_1}+\ldots+\delta_{x_n}}{n}\right] (dy)$$

Therefore

$$I_1(a) = \inf_{[\mu: \int V(y)\mu(dy)=a]} I_7(\mu)$$

When we want to refer to the $I(\cdot)$ function in a specific example we will use the subscript denoting the number of that example.

Example 8. Let $x_1, x_2, \ldots, x_n \ldots$ be a "good" Markov chain on a state space Y with transition operator $(\Pi f)(x) = \int f(y)\, \Pi(x,dy)$. We take $X = M_Y$ and consider again

$$P_n(A) = \text{Prob}\left[\frac{\delta_{x_1}+\ldots+\delta_{x_n}}{n} \in A \mid x_0 = x\right]$$

Then again we have the large deviation results with

$$I(\mu) = -\inf_{u>0} \int \log \left(\frac{\pi u}{u}\right)(y)\, \mu(dy) .$$

Application. If e^V denotes the operator of multiplication by the function $\exp[V(y)]$ then the spectral radius of the operator πe^V is given by

$$\log \sigma(\pi e^V) = \sup_{\mu} \left[\int V(y)\, \mu(dy) - I(\mu)\right] .$$

Conditions. The following conditions are sufficient to yield the result

(i) Either Y is compact or the process is "strongly" positive recurrent

(ii) π has the Feller property.

(iii) There is a reference measure with respect to which $\pi(x,dy)$ has an almost durecly positive density $\pi(x,y)$ for each $x \in Y$.

See [2] for details.

If we take $\pi(x,dy) = \alpha(dy)$ then we are back in example 7 and therefore

$$I_7(\mu) = I_8(\mu) \quad \text{with} \quad \pi(x,dy) = \alpha(dy) .$$

Example 9. We change over to continuous time Markov processes. Let T_t be a semigroup of transition operators on a state space Y and let L be the infinitesimal generator of T_t acting on bounded continuous functions on Y. Let $M_Y = X$ be as before. Let x_0 be any starting point. Define

$$P_n(A) = \text{Prob}\,[L_t \in A \mid x(0) = x_0]$$

where

$$L_t(E) = \frac{1}{t}\int_0^t \chi_E(x(s))\,ds \quad \text{for} \quad E \in \mathcal{B}(Y) .$$

Here again the large deviation results hold with

$$I(\mu) = - \inf_{\substack{u>0 \\ u\in\mathcal{D}(L)}} \int \left(\frac{Lu}{u}\right)(x)\;\mu(dx)$$

where $\mathcal{D}(L)$ is the domain of L.

Application. If we consider the operator L + V where V is multiplication by the function $V(\cdot)$ then by the maximum principle the point in the spectrum of L + V with the largest real part is real and denoting this by $\sigma(L + V)$ we get the variational formula

$$\sigma(L+V) = \sup_{\mu}\left[\int V(x)\;\mu(dx) - I(\mu)\right]$$

Remark. This example also needs conditions similar to those in example 8. See [2] for details.

Remark. If T_t happens to be self-adjoint with respect to a reference measure $\lambda(dy)$ under mild regularity assumptions

$$I(\mu) = \left\| \sqrt{-L}\;\left(\frac{d\mu}{d\lambda}\right)^{1/2}\right\|_\lambda^2 = - \left\langle L\;\left(\frac{d\mu}{d\lambda}\right)^{1/2},\;\left(\frac{d\mu}{d\lambda}\right)^{1/2}\right\rangle_\lambda$$

In particular, if

$$L = \frac{1}{2}\, \nabla\cdot a\nabla \quad \text{on} \quad R^d$$

then

$$I(\mu) = \frac{1}{8}\int \frac{\langle a\nabla f,\ \nabla f\rangle}{f}\, dx = \frac{1}{2}\int \langle a\ \nabla\sqrt{f}, \nabla\sqrt{f}\rangle\, dx$$

where $d\mu = f(x)\, dx$.

Example 10. If Y is the unit circle and $L = \frac{1}{2}\frac{d^2}{dx^2} + b(x)\frac{d}{dx}$ with $\int_Y b(y)\, dy \neq 0$ then the Markov process is not reversible with respect to any measure and if $d\mu = f(x)\, dx$, then

$$I(\mu) = \frac{1}{8}\int_Y \frac{(f')^2}{f}\, dx + \frac{1}{2}\int [b'(x)+b^2(x)]\, f(x)\, dx - \frac{1}{2}\frac{\left(\int b(x)\, dx\right)^2}{\left(\int \frac{dx}{f(x)}\right)}.$$

Example 11. Let $x_1, x_2, \ldots, x_n, \ldots$ be a sequence of independent identically distributed random variables taking values ± 1 with probability 1/2 each. Consider the point ω in the space of doubly infinite sequences defined by

$$\omega = (\ldots, x_1, \ldots, x_n, x_1, \ldots, x_n, x_1, \ldots, x_n, \ldots)$$

If T is the shift operator then $\omega, T\omega, \ldots, T^{n-1}\omega$ is a periodic orbit of period n. Then the measure

$$\frac{1}{n}\,[\delta_\omega + \delta_{T\omega} + \ldots + \delta_{T^{n-1}\omega}] = R_n(x_1, \ldots, x_n)$$

is a random stationary stochastic process. If we take X to be the space of all stationary stochastic processes on the space of sequences of ± 1, and denote by P_0 the product Bernoulli measure then the distribution of $R_n(x_1, \ldots, x_n)$, which is a measure Q_n on X, satisfies

$$Q_n \Rightarrow \delta_{P_0}\ .$$

Note that $P_0 \in X$ is a point of X. This is just the ergodic theorem. Again the large deviation results hold in this context and

$$I(P) = \log 2 - h(P)$$

where h(P) is the Kolmogorov-Sinaii entropy of the stationary process P. This result is essentially a restatement of the Shannon-Breiman McMillan theorem in information theory.

Example 12. Let us take $\Omega = \Pi\ Y$ i.e. the space of sequences with values in Y. Let us take a Markov chain in Y with transition probabilities π and replace the independent random variables of example 11 by the Markov chain. Then analogous to example 11 we have large deviation results with an $I(\cdot)$ functional defined by

$$I_{12}(P) = E^P\{I_7(P_\omega,\Pi_\omega)\}$$

where

$$P_\omega = P[x_1 \in \cdot \mid \dots x_{-1},x_0]$$

is the conditional distribution of x_1 given the past under P. Π_ω is the transition distribution $\pi(x_0,\cdot)$ where x_0 is viewed as a function of ω, i.e. the position at time 0 dictated by ω. $I_7(\mu,\alpha)$ is the I-functional of example 7

$$I_7(\mu,\alpha) = \int \log\left(\frac{d\mu}{d\alpha}(y)\right) d\mu(y) .$$

$I_7(P_\omega,\Pi_\omega)$ then depends on and is integrated with respect to P. The details will appear in [4].

Remark. Comparing this to example 11, if we take that $\pi(x,dy) = \alpha(dy)$ then Π_ω is independent of ω and then

$$I_{12}(P) = I_{11}(P) .$$

Remark. Since the one dimensional marginals of $R_n(x_1,\dots,X_n)$ are given by $\frac{1}{n}[\delta_{x_1}+\dots+\delta_{x_n}]$, example 8 is a special case of example 12 and one has the contraction principle

$$I_8(\mu) = \inf_{P:\text{Marg } P=\mu} I_{12}(P)$$

The infimum is taken over all stationary processes P with marginal distribution μ.

Remark. Let $F(x_1,\dots,x_N)$ be a function depending on a finite number of coordinates. Then if $x_1,\dots,x_n,\dots$ is a Markov chain with transition probability π we have

$$\lim_{n\to\infty} \frac{1}{n} \log E\left\{\exp[F(x_1,\dots,x_N)+F(x_2,\dots,x_{N+1})+\dots+F(x_n,\dots,x_{n+N-1})]\right\}$$
$$= \sup_P \left\{E^P[F(x_1,\dots,x_N)] - I_{12}(P)\right\} .$$

Example 13. We can carry out the analogue of example 12 in the continuous time case. Ω will be the space of cadlag functions with values in Y and X is the space of stationary measures on Ω .

We have the family R_x of Markov measures on Ω^+ corresponding to the semigroup T_t with generator L. For any $T > 0$ for any stationary process P we have the r.c.p.d. P_ω of P given the past and the Markov process $R_{\omega(0)}$ starting where ω ended at time 0. Considering both of them on the time interval $[0,T]$ we define

$$h(T,P) = E^P[I_7(P_\omega^T, R_{\omega(0)}^T)]$$

where P_ω^T, $R_{\omega(0)}^T$ are restrictions of P_ω and $R_{\omega(0)}$ to the time interval $[0,T]$. It turns out that $h(T,P)$ is linear in T and

$$h(T,P) = T\, I_{13}(P) \quad ,$$

and the large deviation results hold with this $I(\cdot)$ function. There is again a contraction principle connecting I_{13} and I_9 identical to one connecting I_8 and I_{12}.

Remark. It turns out that in examples 8 and 9 $I(\cdot)$ is a convex functional, but the ergodicity implies that in examples 12 and 13, $I(\cdot)$ is linear in its argument.

Analogous to the third remark following example 12 we have a similar formula

$$\lim_{T\to\infty} \frac{1}{T} \log E\left\{\exp \int_0^T F(\omega_s)\, ds\right\} = \sup_P [E^P\{F(\omega)\} - I_{13}(P)]$$

F is a tame function on Ω and ω_s is the path ω shifted in time by an amount s.

4. Applications.

We will look at some illustrations of these ideas through specific examples. We will not try to be precise, but just try to give the flavor.

1. Study the behavior of

$$E\left\{e^{-\int [\ell(t,x)]^\alpha\, dx}\right\} = G(t)$$

for $0 < \alpha < 1$. Here $\ell(t,x)$ is the local time at x for the standard Brownian motion.

By Brownian scaling

$$G(t) = E\left\{e^{-\tau \int [\bar{\ell}(\tau,x)]^\alpha\, dx}\right\}$$

where $\bar{\ell}(\tau,x) = \frac{1}{\tau}\,\ell(\tau,x)$ and $\tau = t^{(1+\alpha)/(3-\alpha)}$. Then

$$\lim_{\tau \pm \infty} \frac{1}{t^{\frac{1+\alpha}{3-\alpha}}} \log G(t) = - \inf_{\substack{f \geq 0 \\ \int f\,dx = 1}} \left[\int f^{\alpha}(x)\,dx + \frac{1}{8} \int \frac{(f')^2}{f}\,dx \right]$$

2. Consider

$$G(\alpha) = \lim_{t\to\infty} \frac{1}{t} \log E\left\{ \exp \int_0^t \int_0^t \frac{e^{-|\sigma - s|}}{|\beta(s) - \beta\ \sigma\ |}\,d\sigma\,ds \right\}$$

where $\beta(\cdot)$ is 3-dimensional Brownian motion.

We will show that

$$\lim_{\alpha\to\infty} \frac{G(\alpha)}{\alpha^2} = \sup_{\substack{\phi \in L_2(R^3) \\ \|\phi\|_2 = 1}} \left[2 \int\int \frac{\phi^2(x)\ \phi^2(y)}{|x-y|}\,dx\,dy - \frac{1}{2} \int |\nabla\phi|^2\,dx \right]$$

To see this, rewrite

$$\begin{aligned}
\int_0^t \int_0^t \frac{e^{-|\sigma-s|}}{|\beta(\sigma)-\beta(s)|}\,d\sigma\,ds &= 2\int_0^t ds \int_s^t \frac{e^{-(\sigma-s)}}{|\beta(\sigma)-\beta(s)|}\,d\sigma \\
&\sim 2\int_0^t ds \int_s^\infty \frac{e^{-(\sigma-s)}}{|\beta(\sigma)-\beta(s)|} \quad \text{as } t \to \infty \\
&= \int_0^t F(\omega_s)\,ds
\end{aligned}$$

where

$$F(\omega) = 2 \int_0^\infty \frac{e^{-s}\,ds}{|\omega(s) - \omega(0)|}$$

Then clearly we can expect

$$G(\alpha) = \sup_P \left[E^P\, 2 \int_0^\infty \frac{e^{-s}\,ds}{|\omega(s)-\omega(0)|} - I(P) \right]$$

By Brownian scaling

$$\frac{G(\alpha)}{\alpha^2} = \sup_P \left[E^{\ P}\, \frac{2}{\alpha^2} \int_0^\infty \frac{e^{-s/\alpha^2}\,dx}{|\omega(s)-\omega(0)|} - I(P) \right]$$

By ergodicity one sees that

$$\lim_{\alpha \pm \infty} E^P \left[\frac{1}{\alpha^2} \int_0^\infty \frac{e^{-s/\alpha^2}\,ds}{|\omega(s)-\omega(0)|} \right] = \int\int \frac{1}{|x-y|}\, f(x)\, f(y)\,dx\,dy$$

where $f(\cdot)$ is the marginal density of P.

Finally it is not unreasonable to expect

$$\lim_{\alpha\to\infty} \frac{G(\alpha)}{\alpha^2} = \sup_f \left[2 \int\int \frac{1}{|x-y|} f(x)\, f(y))\, dx\, dy - \frac{1}{8}\int \frac{|\nabla f|^2}{f}\, dx \right]$$

using the contraction principle. We can set $f = \phi^2$ for $\phi \in L_2(R^3)$ and we get the final formula. See [5] for details.

5. Counterexample.

If we take $x(t)$ to Brownian motion on R^1 with a drift, i.e.

$$L = \frac{1}{2}\frac{d^2}{dx^2} + \frac{d}{dx}$$

then direct elementary calculation shows that

$$I_9(\mu) = \frac{1}{8}\int \frac{(f')^2}{f}\, dx + \frac{1}{2}$$

where $\mu(dx) = f(x)\, dx$.

In any case $I_9(\mu) \geq \frac{1}{2}$ for all μ: The large deviation results cannot hold because if we take $A = M_Y$, the whole space, the probability is 1 and

$$\overline{\lim_{n\to\infty}} \frac{1}{n} \log 1 = 0 \leq - \inf I_9(\mu) \leq -\frac{1}{2}$$

is clearly false. That is why we need the strong positive recurrence condition.

Acknowledgments. This work was supported by NSF Grant MCS-8109183.

References

[1] Cramér, H., On a new limit theorem in the theory of probability; Colloquium on the Theory of Probability, Hermann, Paris, 1937.

[2] Donsker, M. D. and Varadhan, S.R.S., Asymptotic evolution of certain Markov process expectations for large time, III, Comm. Pure Appl. Math. 29 (1977), 389-461.

[3] Donsker, M. D. and Varadhan, S.R.S., Some problems of large deviations, Instituto Nazionale di Alta Matematica, Symposia Mathematica, Vol. XXI (1977) pp. 313-318.

[4] Donsker, M. D. and Varadhan, S.R.S., Asymptotic evaluation of certain Markov process expectations for large time, IV (to appear).

[5] Donsker, M. D. and Varadhan, S.R.S., Asymptotics of the Polaron problem,

[6] Glass, M., Perturbation of a first order equation by a small diffusion, Thesis, New York Univ. (1970).

[7] Marcus, M. B. and Shepp, L. A., Sample behavior of Gaussian processes, Sixth Berkeley Symposium on Mathematical Statistics, and Probability, Vol. II (1972) pp. 423-439.

[8] Pincus, M., Gaussian processes and Hammerstein integral equations, Trans. Amer. Math. Soc., Vol. 134 (1968) pp. 193-216.

[9] Schilder, M., Some asymptotic formulae for Wiener integrals, Trans. Amer. Math. Soc., Vol. 125 (1966) pp. 63-85.

[10] Varadhan, S.R.S., Asymptotic probabilities and differential equations, Comm. Pure Appl. Math., Vol. 19 (1966), pp. 261-286.

[11] Varadhan, S.R.S., Diffusion processes in a small time interval, Comm. Pure Appl. Math. 20 (1967) pp. 659-685.

[12] A. D. Ventcel and M. I. Freidlin, On small random perturbations of dynamical systems, Russian Math. Surveys 25 (1970) 1-55.

Lecture Notes in Control and Information Sciences

Edited by A. V. Balakrishnan and M. Thoma

MIX
Papier aus verantwortungsvollen Quellen
Paper from responsible sources
FSC® C105338

If you have any concerns about our products,
you can contact us on
ProductSafety@springernature.com

In case Publisher is established outside the EU,
the EU authorized representative is:
Springer Nature Customer Service Center GmbH
Europaplatz 3, 69115 Heidelberg, Germany

Printed by Libri Plureos GmbH
in Hamburg, Germany